Springer Series in Optical Sciences

Volume 203

Springer Series in Optical Sciences is led by Editor-in-Chief William T. Rhodes, Florida Atlantic University, USA, and provides an expanding selection of research monographs in all major areas of optics:

- lasers and quantum optics
- ultrafast phenomena
- optical spectroscopy techniques
- optoelectronics
- information optics
- applied laser technology
- industrial applications and
- other topics of contemporary interest.

With this broad coverage of topics the series is useful to research scientists and engineers who need up-to-date reference books.

More information about this series at http://www.springer.com/series/624

Vladimir Belyakov

Diffraction Optics of Complex-Structured Periodic Media

Localized Optical Modes of Spiral Media

Second Edition

Springer

Vladimir Belyakov
L. D. Landau Institute for Theoretical
Physics
Russian Academy of Sciences
Moscow, Russia

ISSN 0342-4111 ISSN 1556-1534 (electronic)
Springer Series in Optical Sciences
ISBN 978-3-030-18353-0 ISBN 978-3-319-43482-7 (eBook)
https://doi.org/10.1007/978-3-319-43482-7

This Springer imprint is published by the registered company Springer Nature Switzerland AG
The registered company address is: Gewerbestrasse 11, 6330 Cham, Switzerland

Preface to the Second Edition

A quarter of a century has passed since the publication of the *Diffraction Optics of Complex Structured Periodic Media*, Springer-Verlag. During the past time, it has been achieved a great progress practically in all fields of research related to different types of radiation interaction with the matter studied in this book. Without referring to the optical wavelength range of the electromagnetic radiation, the main advances can be summarized as follows. In the X-ray range, the mentioned advances are related to the emergence and improvement of synchrotron radiation (SR) sources and X-ray free-electron lasers (XFELs), which make it possible to study the phenomena of polarization and magnetic scattering in the X-ray wavelength range (see, for example, V. E. Dmitrienko, E. N. Ovchinnokova, S. P. Collins et al., Nature Physics 10, 202–206 (2014) doi:10.1038/nphys2859). A special mentioning deserves the progress in Mossbaur optics. The implementation of experiments with an extremely high energy resolution has become accessible in practice by the invention and the constructing the SR Mössbauer source (see G. V. Smirnov, U. van Burk, A. I. Chumakov, A. Q. R. Baron, and R. Ruffer, Phys. Rev. B **55**, 5811 (1997)). Major advances in neutron optics are obtained in research on localized neutron states (in Earth's gravity, see V. V. Nesvizhevskii, Phys. Usp, **47**, 515 (2004), condensed media whispering gallery modes, see V. V. Nesvizhevskii, A. Yu. Voronin et al. Nature Physics, v. 6, p. 114 (2010), and localized modes in periodic condensed structures, see V. A. Belyakov, JETP, **124**, 873 (2017)). In order to take into account the achievements in the above-mentioned fields, the corresponding chapters of the first edition of the book should be practically transformed into separate monographs. Therefore, in order to maintain a reasonable size of the book, only the recent achievements in photonics research, i.e., in the optical wavelength range of electromagnetic radiation, are presented in the second edition.

A new field of optical research has emerged over the last three recent decades which relates to the optics of periodic structures with the value of periodicity comparable to the optical wavelength and which has been named photonics. Photonics has revealed many important phenomena that are well applied in the optical wavelength range. However, formally all phenomena studied by photonics were known and present in X-ray diffraction. It is well known that the corresponding X-ray researches began much earlier than the photonics emerged. Nevertheless, a new term, photonics, proves to be quite adequate to the field of optical research named by this term. The reason for this is related to the fact that virtually all photonics-related effects are within the capacity of experimental observation at the contemporary level of optical techniques while most of the corresponding effects in the X-ray wavelength range are out of the capacity of experimental observation at the present level of X-ray techniques. Now, due to the appearance of synchrotron and free-electron laser X-ray sources, the already observed photonic effects in the optics seem to be accessible for observation in the X-ray wavelength range in the near future.

Photonics researches show that many photonics-related optical processes are more efficient than the corresponding processes in homogeneous media, as a known example can be called the lasing. The photonics terminology proposes the term "photonic crystal" related to a periodic medium whose period value is in the range of the optical wavelengths to distinguish the photonic crystals from the conventional crystals. To the date, a number of reviews and books have been published on linear and nonlinear optics of photonic crystals (see, for example, J. D. Joannopoulus, R. D. Meade, and J. N. Winn, *Photonic Crystals: Moulding the Flow of Light*, Princeton University Press, second edition, 2011; K. Sakoda *Optical Properties of Photonic Crystal*, Springer-Verlag, second edition, 2005; I. A. Sukhoivanov and I.V.Guryev, *Photonic Crystals*, Springer, 2014). A common feature of these publications is, practically, an absence of analytical theory of the photonic crystal optics which reflects the complexity of photonic crystals in a general case.

In this second edition of the book, a special type of photonic crystal is studied, namely a spiral medium, allowing the development of an analytical theory, as new main objects of studies are chosen localized optical modes in spiral media, the so-called edge mode (EM) and defect mode (DM). The dispersion equations determining frequencies of localized modes for EM and DM are presented and analyzed. It is demonstrated that many optical phenomena occurring at the frequencies of localized modes reveal unusual properties which can be used for efficient applications of the corresponding phenomena, for example, efficient frequency conversion, low-threshold lasing. The influence of phenomenological parameters on the localized mode characteristics of the structures under consideration is, for certainty, analyzed for chiral liquid crystals (CLCs). The cases of non-absorbing, absorbing (amplifying) CLC layers are considered in detail for EM and DM. The cases of active defect layers for defect modes are also considered. Options for suppressing absorption at localized mode frequency for CLC with local anisotropic absorption (Borrmann effect) are also investigated. The theory presented in the book allows to describe the known experimental results on the efficient low-threshold distributed feedback (DFB) lasing at the localized mode frequencies in chiral liquid crystals in the framework of a single approach, to find localized mode frequencies in their connection with the parameters of the object under consideration and, some time, to predict an unknown effect. Main experimental results related to the developed theory are also discussed. The presented analytic results related, for example, to the low-threshold DFB lasing and to the abnormally strong absorption at the localized mode frequencies, have as a direct application of the corresponding results to real physical objects such as chiral liquid crystals, so can also be applied to the conventional photonic crystals (not spiral) as a first approximation. No doubt that the analytical results presented in the book can be used as a useful guide for the understanding of unusual optical properties of photonic crystals in general.

What is concerned with the structure of the book it contains, with the exception of new chapters devoted to the properties of the localized mode, also introductory chapters devoted to the "prephotonic era" in the optics of spiral media. The inclusion in the second edition of the book the known "prephotonic era" material is justified as by the intention to present an autonomous from many scattered over various journals original papers on the subject book as well by the fact that in English there is no easy accessible book on the optics of spiral media (a great review on the subject by Belyakov, V. A.; Dmitrienko, V. E. *Optics of Chiral Liquid Crystals*; In Soviet Scientific Reviews/Section A: Physics Reviews. (Ed.) Khalatnikov, I. M. Harwood Academic Publisher: London, UK, 1989 is not available in an electronic version).

The author expresses his sincere gratitude to S. V. Semenov, the author's co-author in numerous articles on the subject of the book, and is grateful to Sofiya Tarshina, Ivan Belyakov, and Dariya Chufarova for the valuable technical assistance in the preparation of the manuscript.

Moscow, Russia
December 2017

Vladimir Belyakov

Preface to the First Edition

Little time has elapsed since 1988 when a Russian edition of this book was published, but most of the problems touched on here were under intensive investigation then. This indicates that the subject of the book was and is quite urgent, and as a consequence, a number of new publications on the subject have appeared since the first edition. It is why in the course of preparing a new edition, I decided that many new results ought to be reflected in it.

Because I wanted to retain the general style of the original book and its overall features, the problem of integrating new material was solved in the following way. Three chapters, devoted to the fields of the most significant achievements, were added: on polarization phenomena in X-ray optics, on X-ray magnetic scattering, and on Mössbauer filtration of synchrotron radiation. Only minor alterations were made to the other chapters; their end-of-chapter reference sections were also updated.

I hope that the English edition of this book will give easier access to my Western colleagues to the original Soviet results. Soviet authors contributed quite significantly to the subject of this book and look forward to developing joint investigations of the coherent interaction of different kind of radiations with matter. I would also like to express my gratitude to the Springer-Verlag editing staff for collaboration in this project.

Moscow, Russia

Vladimir Belyakov

Preface to the Russian Edition

Rapid progress in experimental techniques in the past 15–20 years, coupled with the ever-increasing variety of the objects of investigation, have advanced the study of a series of phenomena which were beyond the scope of researchers and which are very important for various applications.

Among such phenomena are the numerous unusual effects observed in coherent interaction of various electromagnetic and corpuscular radiations with regular (spatially periodic) media having complicated structures. These are polarization characteristics of light scattering, the unusually high gyrotropy of chiralliquid crystals, which changes its sign depending on frequency, nuclear scattering of gamma and synchrotron radiation sensitive to magnetic order in crystals, various anomalies in angular and frequency distributions of the coherent radiation of fast charged particles, and similar effects in nonlinear generation of high-order optical harmonics in periodic media, etc.

It should be emphasized that since specific objects, types of radiation, and conditions under which the above effects manifest themselves are very diverse and, at first glance, have quite a different nature, they are often considered unrelated phenomena. In many studies, such separate considerations of these phenomena are detrimental, and instead of reinventing the wheel and ignoring the experience and results of other distant fields of research, investigators have only to examine, in fact, closely related ones. It is worthy to remind the reader here of the words written of L. Brillouin and M. Parodi some 30 years ago in the foreword to their now famous *Wave Propagation in Periodic Structures.* "No matter how diverse the problems are, their mathematical treatment results in equations having a series of common properties and similar solutions. This deep relation between quite different problems was very well known already to Kelvin and Rayleigh. But then it has been forgotten and only in recent works this very important statement has attracted attention anew." Unfortunately, the above affirmation sounds as urgent today, in application to modern problems, as it did then.

This book presents from one standpoint and, in fact, within the context of one approach a theoretical description of various phenomena–diffraction of Mössbauer gamma radiation by magnetically ordered crystals, optics of chiralliquid crystals, coherent radiation of fast charged particles in periodic media with a complicated structure, some problems of nonlinear optics for viii Preface to the Russian Edition periodic media, and dynamical diffraction of thermal neutrons by perfect magnetically ordered crystals. The main experimental results, if any, are explored to the extent necessary for the illustration of general qualitative effects and their comparison with theoretical calculations.

This book also includes introductory material on wave propagation in one-dimensional periodic media and the approximations describing the interaction of radiation with periodic media. The main focus of the book is the application of the kinematical and dynamical approximations of the theory of scattering to the above-mentioned phenomena, which permits one to obtain an analytical solution of the problems under consideration. The numerical methods for solving the corresponding problems are beyond the scope of this book, although the results obtained by such methods (with corresponding references) are given in some instances to illustrate the experiment or theory under consideration. This is explained not only by the fact that for most problems, the above-mentioned methods yield results with sufficient accuracy, but also by the fact that these methods most clearly illustrate the common physical nature of the phenomena under consideration. We do not concentrate on the scattering of low-energy electrons in crystals, which, in fact, is closely to the problems considered in our book. Space limitations decided this, but more important, the above approximation fails to describe with sufficient accuracy the interaction between low-energy electrons and crystals. Also, extensive literature, reviews and monographs already exist on this problem.

I believe that the consideration of all the above problems in one book will be useful for both specialists in the field and beginners, since at present corresponding data are scattered in numerous original publications or highly specialized monographs. The bulk of this book presents a first attempt to generalize the problems of diffraction optics for various media with periodicities of a nonscalar nature, but at the same time, some problems considered here may also be found in recent monographs. Thus, some problems of the dynamical theory of diffraction of Mössbauer gamma radiation are stated by V. G. Baryshevskii in his monograph, *Nuclear Optics of Polarized Media* (Minsk, Izd-vo BGU, 1976) and by M. A. Andreeva and R. N. Kuz'min in *Mössbauer Gamma Optics* (Moscow, Izd-vo MFU, 1982). Problems concerning the optics of chiral cholesteric liquid crystals in an experimental and theoretical context may be found in *Optics of Cholesteric Liquid Crystals,* written by the author together with A. S. Sonin (Moscow, Nauka, 1982). The radiation of uniformly moving charged particles in media with the periodicity of scalar characteristics not associated with channeling is considered by G. M. Garibyan and C. Yang in *X-Ray Transition Radiation* (Erevan, Izd-vo AN Arm. SSR, 1983) and V. L. Ginzburg and V. N. Tsytovich in *Transition Radiation and Transition Scattering* (Moscow, Nauka, 1984).

Quite a large part of this book is based on the results obtained at the Theoretical Laboratory of the All-Union Research Institute for Physical Preface to the Russian Edition ix Technical and Radiotechnical Measurements and in the Laboratory of

Solid State Theory of the All-Union Surface and Vacuum Research Centre and published at different times in coauthorship with Yu. M. Aivazian, R. Ch. Bokun, A. G. Grigorian, V. E. Dmitrienko, A. A. Grigor'ev, L. V. Maslov, S. M. Osadchii, V. P. Orlov, E. V. Smirnov, and N. V. Shipov. The author is very grateful to all of these individuals.

The author also wishes to thank V. E. Dmitrienko and M. V. Fedorov, who have read the manuscript, for their valuable remarks.

Contents

About the Author

Vladimir Belyakov graduated from Moscow Engineering Institute in 1961. Afterwards he was a postgraduate student at the I. V. Kurchatov Atomic Energy Institute in the years 1961–64. He was the head of the Laboratory of the All-Union Physics-Technical and Radio-Technical Institute 1964–82 and of Surface and Vacuum Research Centre, Moscow 1982–95. Since 1995, he is senior Researcher in L. D. Landau Institute for Theoretical Physics. Since 1982 he is part-time Professor at the Moscow Institute for Physics and Technology. He had short stays as visiting professor of the Universities Leuven (Belgium), Tokyo (Japan), Paris Sud (France), Glasgow Thrathclyde (Scotland) and others.

He is author of the following monographs: *Optics of Cholesteric Liquid Crystals*, 1982, *Optics of Chiral Liquid Crystals*, 1989; *Diffraction Optics of Complex Structured Periodic Media*, 1988, 1992; *Optics of Photonic Crystals, Publishing House of Moscow Institute of Physics and Technology*, 2013 (a textbook, in Russian).

In 1965 he got his Ph.D. degree and in 1974 he got the Doctor of Science degree. He became professor in 1980.

He got the following honors and grants of the Russian Foundation for Basic Research (RFBR): Soros Grants, INTAS Grants, Member of Russian Academy of Natural Sciences, Member of Russian Academy of Metrology, Member of International Liquid and Liquid Crystal Society of CIS in 1982.

Chapter 1
Waves in Media with One-Dimensional Periodicity (Exact Solution)

1.1 Layered Medium

Our focus is the interaction between radiation and periodic media of complicated structures. Nevertheless, we begin our consideration with simple periodic systems, since their properties are typical of wave propagation in all media; their results are general and may become the zeroth approximation for more complicated systems.

1.1.1 Harmonic Modulation of Media

Consider the propagation of an electromagnetic wave in an infinite medium with one-dimensional periodicity. Let the dielectric permittivity of the medium be modulated along the periodicity direction (z-axis). Then, the wave equation reduces to the differential equation of the form [1, 2]

$$d^2\psi/dz^2 + f(z)\psi = 0 \tag{1.1}$$

where $f(z)$ is a periodic function.

In the general case of an arbitrary periodic function $f(z)$, infinite number of harmonics in the expansion of $f(z)$ into a Fourier series (1.1) is Hill's equation. The solutions of Hill's equation are analyzed in the general form [3] but for simplicity we analyze its solutions for a specific form of $f(z)$.

In the specific case where $f(z)$ is a harmonic function, one harmonic in the Fourier expansion of $f(z)$ (1.1) reduces to the Mathieu equation:

$$d^d u/d\xi^2 + (\eta + \gamma\cos^2\xi)u = 0. \tag{1.2}$$

V. Belyakov, *Diffraction Optics of Complex-Structured Periodic Media*, Springer Series in Optical Sciences 203,
https://doi.org/10.1007/978-3-319-43482-7_1

According to Floquet's theorem, the general solution of (1.2) has the form

$$u(\xi) = D_1 A(\xi)e^{\mu\xi} + D_2 B(\xi)e^{-\mu\xi} \tag{1.3}$$

where D_1 and D_2 are arbitrary constants and $A(\xi)$ and $B(\xi)$ are periodic functions, with period equal to the period of property modulation in the medium. Expressions of form (1.3) describe the spatial variation of an electromagnetic wave—in the derivation of (1.1) it is assumed that time dependence is described by $ei\omega t$. The solution is regarded as the superposition of two waves propagating in opposite directions.

Omitting a detailed analysis [3], we concentrate on the parametric dependence of solutions to (1.2) on η and γ. We distinguish two cases: (i) μ entering (1.3) as an imaginary quantity which corresponds to non-decaying monochromatic waves and (ii) μ as a complex or real quantity which corresponds to decaying waves. The analysis of (1.2) is illustrated in Fig. 1.1 where η and γ are plotted on coordinate axes. Open regions correspond to values of η and γ for which μ is either a complex or real quantity where the wave is decaying, whereas the hatched regions correspond to values of η and γ for which μ is an imaginary quantity. Thus, the hatched regions correspond to transmission bands and open regions stop bands. If γ infinitely increases and $\eta > -\gamma$, the hatched regions become narrower tending, in the limit, to straight lines parallel to the line $\eta = -\gamma$. For $\eta < -\gamma$, i.e., below the line $\eta = -\gamma$, there are no propagating modes. At the boundaries between open and hatched regions, the real part of μ goes to zero. At the η-axis, the value of γ is zero. This corresponds to the case of an unmodulated homogeneous medium for which there are no forbidden bands. In the vicinity of axis η, $\gamma \ll \eta$, which corresponds to a small modulation of medium properties (dielectric constant in our case), wave propagation is often described rather accurately by the approximate solutions of the Maxwell equation for a periodic medium. Two neighboring curves limiting adjacent hatched regions have a common datum point and point of tangency of $(n-1)$th order on the η-axis, with abscissa $\eta = n^2$(n is an integer). Therefore, for $n = 1$ the neighboring

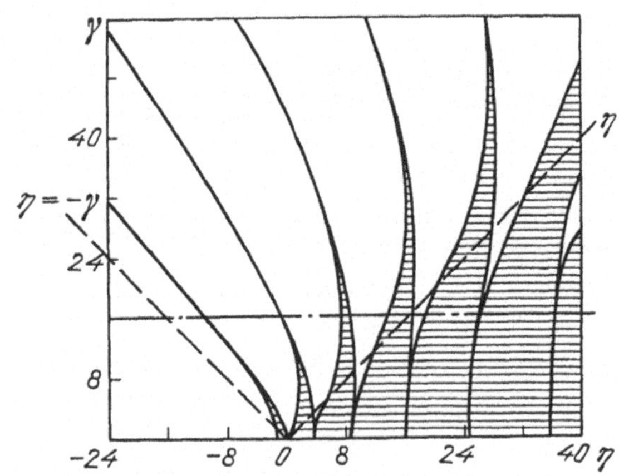

Fig. 1.1 Stability diagram for the Mathieu equation

curves originate from one point; for $n = 2$, they have a common vertical tangent; for $n = 3$, there is a common vertical tangent and curvature, etc. Note that parameter η, determined by the mean value of dielectric permittivity, ϵ_0, and parameter γ, determined by the amplitude of the spatial modulation of dielectric permittivity, ϵ_1, are proportional to wave frequency squared ω^2, but their ratio (for fixed modulation amplitude ϵ_1 and ϵ_0) is independent of frequency. Therefore, for fixed values of ϵ_0 and γ_1, the values of μ for different frequencies ω are determined by the points of the stability diagram on the line $\eta/\gamma = \epsilon_0/\epsilon_1 = \text{const}$; the particular case of such a line, $\eta = \gamma$, is shown in Fig. 1.1. Some portions of the corresponding lines are within stop bands. This means that not all the waves (with arbitrary frequencies) propagate in the periodic medium—there are some frequency ranges, the so-called forbidden bands, which have no propagating modes. In addition, the stability diagram shows that the number of "forbidden bands" is infinite and that their frequency width increases with modulation amplitude.

The dispersion curve (relation between ω and μ) which corresponds to the above band structure has the shape shown in the Brillouin diagram (Fig. 1.2). Some portions of the dispersion curve, in the first Brillouin zone ($|\mu| \leq \pi/d$ where d is the modulation period), demonstrate another important result known in optics of periodic media: the position of the break in the dispersion curve, i.e., the position of the forbidden band in momentum space, does not change with amplitude ϵ_1; however, the value of such a break, i.e., the forbidden band frequency width, depends on ϵ_1. We note that the Brillouin diagram shows only the real part of k. As follows from the stability diagram, k is real in the transmission band, whereas it is a complex quantity, ($k = i\mu\pi$), in the forbidden band. Therefore, in the range of the break on the Brillouin diagram, the imaginary component of k, the real component being fixed, depends on frequency and goes to zero at the forbidden band boundary. From the general form of the solution of the Mathieu equation (1.3), it follows that since functions A and B are periodic in an electromagnetic wave in a periodic medium, each component of (1.3) may be written as

Fig. 1.2 Brillouin diagram

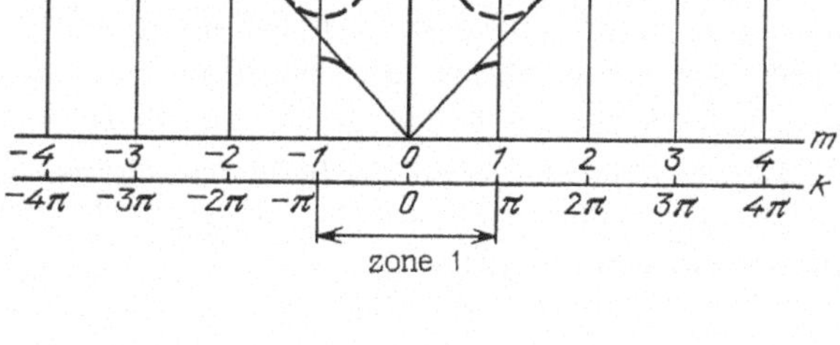

$$\psi = \sum_{s=-\infty}^{s=+\infty} A_s \exp[i(k + s\tau)z], \tag{1.4}$$

where A_s are coefficients, equal, within a constant factor, to the coefficients for the Fourier expansion of function $A(\xi)$ in (1.3). That is, generally, a monochromatic plane wave in a periodic medium is not a plane wave but an infinite superposition of plane waves determined by (1.3).

1.1.2 Properties of Eigenwaves

In many practical situations, for small modulation amplitude of dielectric permittivity, a small number of terms in expansion (1.3) are important, i.e., only some A_s coefficients are large whereas other coefficients are smaller, at least by a factor of δ, where $\delta = \epsilon_1/\epsilon_0$ is the relative modulation amplitude of dielectric permittivity or dielectric anisotropy. In the simplest case, where k is far from the forbidden band, expansion (1.4) is reduced to one term. This means that it is a good approximation to represent an electromagnetic field as a plane wave with a certain wave vector k in those regions. If k is inside the forbidden band, or in close proximity to it, an electromagnetic field cannot be represented by a plane wave, irrespective of small modulation amplitude. Thus, expansion (1.4) must be extended to two terms, with $s = 0$ and $s = n$ where n is the number of the forbidden band to which k belongs. Such a situation corresponds to electromagnetic wave diffraction by a periodic structure and approximation (1.4) with two terms is the two-wave approximation. The equation relating wave vectors of two plane waves is the Bragg condition,

$$k_n = k + n\tau. \tag{1.5}$$

For small modulation amplitude ($\delta \leq 0.1 - 0.01$), the two-wave approximation is often satisfactory from the practical standpoint and is used to quantitatively describe the interaction of radiation with periodic media. And, in the case of harmonic modulation of dielectric permittivity (and small amplitude δ), the frequency width of forbidden bands rapidly decreases with band number, n, as δ^n.

For the general form of spatial modulation of the dielectric properties of a medium i.e., if $f(z)$ in (1.1) is an arbitrary periodic function, the problem reduces to the solution of Hill's equation and yields the same qualitative results as the previously analyzed solution of Mathieu's equation [2, 3]. Hence, without a detailed analysis for the general form of the periodic function $f(z)$ in (1.1), we conclude that forbidden band frequency width may not depend on band number n in a regular way. It is essentially determined by the value of the nth harmonic of the Fourier expansion of $f(z)$ and generally increases with harmonic amplitude. In particular cases, for certain ratios between different Fourier harmonics, the frequency width of individual bands may be equal to zero [2].

Summarizing the results of the brief analysis of electromagnetic wave propagation in media with the simplest form of dielectric, property periodicity, modulation of scalar dielectric permittivity, we emphasize that, even in this simplest situation, there is no exact analytical solution to the problem. The basic principles are:

1. A monochromatic wave in a periodic medium is a superposition of an infinite number of plane waves.
2. There is an infinite number of "forbidden" frequency bands in wave propagation.
3. If a wave propagates along the direction of a medium periodicity, its polarization characteristics coincide with those of an isotropic homogeneous medium. That is, wave characteristics are independent of polarization.

The important but specific practical case of small modulation amplitude of dielectric properties permits simplification of the general statement above: the one plane wave approximation may be adequate in regions outside the forbidden band, whereas the two-wave approximation applies inside and in close proximity to it. If the modulation of dielectric properties is harmonic, the forbidden band frequency width decreases with zone number n as δ^n.

The general features of electromagnetic wave propagation in arbitrary media are the same as in periodic modulations of scalar dielectric permittivity of media. However, for arbitrary periodic media and for waves propagating at an angle to the periodicity direction, some specific properties apply which concern the polarization characteristics of the waves and fine details of the forbidden band structure. The mathematical construct used to describe wave propagation in periodic media with complicated structure is more sophisticated than that in the simplest case considered above. Nonetheless, the simple qualitative results are useful to understand the physics of more complicated situations.

1.2 Optics of Cholesterics (Exact Solution)

An example of unusual optical properties caused by structure of unidimensional periodicity is shown by cholesteric liquid crystals, the so-called cholesterics [4]. Neglecting the details in our consideration, we should like to underline here another unique aspect of optics of cholesterics and similar media: The Maxwell equations have been solved analytically and exactly for light propagation along the axes of cholesterics and chiral smectics [5, 6]. The obtained exact solution is simple and is the only example of a simple exact analytical solution of the Maxwell equations for periodic media. No other simple exact solution is known—not even for periodic structures more simple than cholesterics (see, e.g., Sect. 1.1).

1.2.1 Dielectric Properties of Cholesterics

A cholesteric is a locally uniaxial medium for electromagnetic radiation in the optical range. The anisotropy direction of the medium varies (rotates) in space along some direction normal to the local anisotropy axis in accordance with harmonic law (see Fig. 1.3). This direction, the optical axis of a cholesteric, is also called the cholesteric axis. The dielectric permittivity tensor $\hat{\epsilon}(r)$ of a cholesteric depends on the coordinates as follows [4]:

$$\hat{\epsilon}(r) = \begin{pmatrix} \overline{\epsilon} + \overline{\epsilon}\delta\cos\varphi(z) & \pm\overline{\epsilon}\delta\sin\varphi(z) & 0 \\ \pm\overline{\epsilon}\sin\varphi(z) & \overline{\epsilon} - \overline{\epsilon}\delta\cos\varphi(z) & 0 \\ 0 & 0 & \epsilon_3 \end{pmatrix} \tag{1.6}$$

where $\overline{\epsilon} = (\epsilon_1 + \epsilon_2)/2$, $\delta = (\epsilon_1 - \epsilon_2)/(\epsilon_1 + \epsilon_2)$ and $\epsilon_1, \epsilon_2 = \epsilon_3$ are the principal values of the dielectric permittivity tensor. The Z-axis is directed along the cholesteric axis; the rotation angle $\varphi(Z)$ of the local anisotropy direction around the cholesteric axis is linearly related to coordinate z and $\varphi(z) = \tau z/2$ where $\tau = 4\pi/p$. Quantity p is the period or pitch of the cholesteric helix and is equal to the distance along the cholesteric axis within which the direction of the local anisotropy makes a complete revolution about that axis. The period of dielectric properties of a cholesteric coincides with a half-pitch $p/2$ of the helix since the anisotropy directions rotated by a half-turn and the former ones are physically equivalent. Two signs in tensor (1.6) correspond to two geometric possibilities—left- and right-handed helices. Thus, spatial changes in the dielectric permittivity tensor of a cholesteric reduce to a harmonic rotation ($\varphi \sim z$) of two principal axes of the tensor $\hat{\epsilon}(r)$ with the variation of the coordinate along the cholesteric axis (see Fig. 1.3).

1.2.2 Eigenwaves

In the absence of external currents and charges, the Maxwell equation for a cholesteric has the form

$$\begin{aligned} \text{rot}\,\mathbf{E} &= c^{-1}\partial\mathbf{B}/\partial t, \quad \text{div}\,\mathbf{D} = 0, \quad \mathbf{D} = \hat{\epsilon}\mathbf{E} \\ \text{rot}\,\mathbf{H} &= c^{-1}\partial\mathbf{D}/\partial t, \quad \text{div}\,\mathbf{B} = 0, \quad \mathbf{B} = \hat{\mu}\mathbf{H}. \end{aligned} \tag{1.7}$$

Assuming the magnetic permittivity tensor, $\mu = 1$, for a plane wave propagating along the optical axis of a cholesteric, axis Z, the field is independent of coordinates x and y and (1.7) for the electric field of the wave yields

$$\partial^2\mathbf{E}/\partial z^2 = \hat{\epsilon}(z)c^{-2} \cdot \partial^2\mathbf{E}/\partial t^{-2}. \tag{1.8}$$

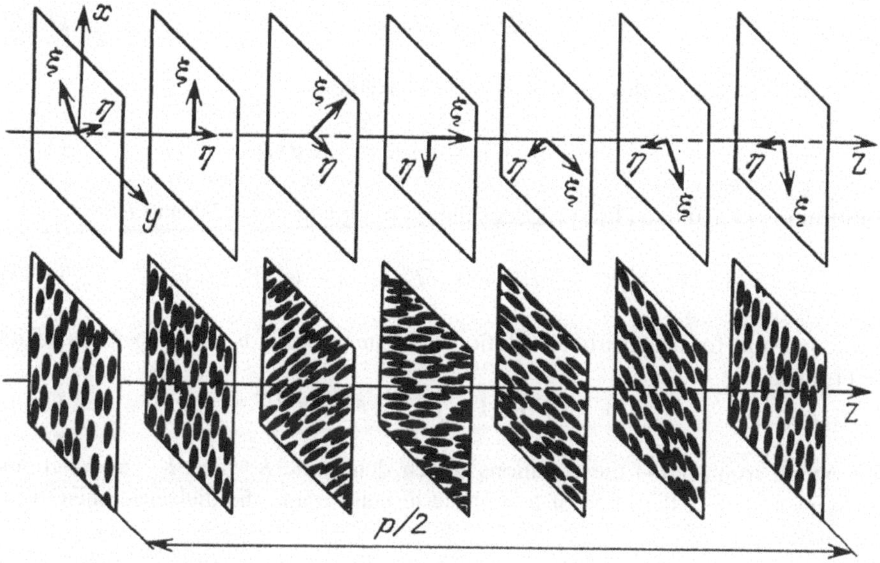

Fig. 1.3 Schematic representation of a cholesteric structure (below) and the corresponding change of spatial orientations of the principal axes of the dielectric tensor (above)

Notice that for propagation direction parallel to the helical axis, an electromagnetic wave is strictly transverse, i.e., vector **E** is normal to axis Z.

We seek the solution of (1.8) for fixed wave frequency ω. As shown in Sect. 1.1, in a spatially periodic medium (as in a cholesteric), the simplest solution of the Maxwell equation is not a plane wave, e^{ikz}, as in a homogeneous medium, but a Bloch wave of the form

$$\mathbf{E}(z,\ t) = \sum_s \mathbf{E}_s e^{i(\mathbf{K}+s\tau)z} e^{-i\omega t}. \tag{1.9}$$

Thus, the problem reduces to a search for amplitude E_s and wave vectors K in the superposition (1.9). As shown in Sect. 1.1, in the general case of a medium with one-dimensional periodicity, an infinite number of amplitudes E are nonzero.

In a cholesteric, specific polarization properties of scattering, determined by the dielectric permittivity tensor (1.6), result in the fact that if a wave propagates along the optical axis, only two amplitudes E_s in the exact solution (1.9) are nonzero while all the other amplitudes vanish. Then, the solution of (1.8) is sought in the form

$$\mathbf{E}(z, t) = e^{-i\omega t}(E^+\mathbf{n}_+ e^{i\mathbf{K}^+ z} + E^-\mathbf{n}_- e^{i\mathbf{K}^- z}), \tag{1.10}$$

where $\mathbf{n}_\pm = (\hat{\mathbf{x}} \pm i\hat{\mathbf{y}})/\sqrt{2}$ are the unit vectors of circular polarizations (hereinafter $\hat{\mathbf{a}} = \mathbf{a}/|\mathbf{a}|$) and K^+ and K^- are related by the Bragg condition

$$K^{+} - K^{-} = \tau. \tag{1.11}$$

We introduce the following notation for convenience

$$K^{+} = \eta^{+} q, \; K^{-} = \eta^{-} q, \; q = \omega\sqrt{\bar{\epsilon}}/c. \tag{1.12}$$

Substituting (1.10) into (1.8) produces the set of equations for E^{+} and E^{-}

$$[1 - (\eta^{+})^2]E^{+} + \delta E^{-} = 0, \; \delta E^{+} + [1 - (\eta^{-})^2]E^{-} = 0. \tag{1.13}$$

The condition for a non-trivial solution of system (1.13) is the zero value of its determinant

$$[1 - (\eta^{+})^2][1 - (\eta^{-})^2] - \delta^2 = 0. \tag{1.14}$$

We obtain from (1.14) the equations which determine K^{+} and K^{-} as functions of frequency or q, the pitch of a cholesteric helix p and the dielectric anisotropy parameter δ:

$$\begin{aligned} K_j^{+}/q = \eta_j^{+} = q^{-1}(\tau/2 \pm k^{\pm}), \; \eta_j^{-} = \eta_j^{+} - 2\tilde{\lambda}, \\ k^{\pm} = q[1 + \tilde{\lambda}^2 \pm (4\tilde{\lambda}^2 + \delta^2)^{1/2}]^{1/2}, \; \tilde{\lambda} = \tau/2q. \end{aligned} \tag{1.15}$$

Expression (1.15) for the fixed value of frequency ω determines four values of $\eta_j^{\pm}$, i.e., four values of wave vectors $K_j^{\pm}$. Four corresponding solutions of system (1.13) and expression (1.10) determine eigenwaves in a cholesteric for the propagation direction coinciding with its optical axis.

Let us number the eigensolutions in the following way: $j = 1, 4$ for signs "+" and "–" before K^{+} in the first equation in (1.15) and $j = 2, 3$ for signs "+" and "–" before K^{-} in the same expression.

The amplitude ratio, E^{-}/E^{+}, in the jth eigensolution is determined by the following expression (Fig. 1.4)

$$\xi_j = (E^{-}/E^{+}) = \delta/[(\eta_j^{-})^2 - 1]. \tag{1.16}$$

1.2.3 Properties of Eigensolutions

It follows from (1.10) that eigensolutions are superpositions of two circularly polarized plane waves. Depending on the sign of K^{+} and K^{-} in (1.10), these are either two waves with opposite circular polarizations propagating in the same direction (same signs of K^{+} and K^{-}) or two waves with the same circular polarization propagating in opposite directions (different signs of K^{+} and K^{-}). We should like to remind here that the changes in the sign of the wave vector in the expression describing the circular wave is equivalent to the simultaneous change of propagation direction and sense

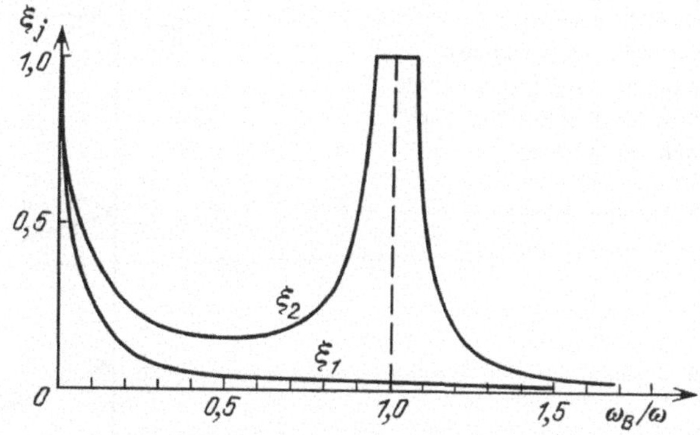

Fig. 1.4 Amplitude ratio $\xi_j = |E^-/E^+|$ in eigensolutions versus frequency

of the circular polarization of the wave. The existence of eigensolutions in which the constituent plane waves propagate in opposite directions reflects the possibility that light diffracts from the spatial structure of cholesterics.

From (1.10), (1.12), and (1.15), it follows that solutions 1 and 4 containing k^+ in (1.15) correspond to two non-diffracting modes in the periodic structure of a cholesteric: one propagating along the Z-axis, the other in the opposite direction. Solutions 2 and 3 relate to modes which experience diffraction in a cholesteric.

We focus on the physical characteristics of eigensolutions and their frequency dependence in the case when the dielectric anisotropy parameter of a cholesteric $\delta \ll 1$. In solution 1, K^+ and K^- have the same signs for all frequencies whereas the ratio of one amplitude (E^+ or E^-) to the other is of the order of δ (Fig. 1.4) everywhere except the high-frequency limit $\omega/c \gg \tau/\delta$ where both amplitudes are equal. Therefore, solution 1 corresponds to a wave circularly polarized in the direction opposite the screw sense of the cholesteric helix, within the accuracy of δ, for all the frequencies except $\omega/c > \tau/\delta$. For $\omega/\tau \gg \tau/\delta$, i.e., in the limit of very short waves, solution 1 corresponds to a plane-polarized wave, the polarization plane of which is rotated, with period p, matching the screw sense of the helix and the variation in the z coordinate. The polarization plane is parallel to the local orientation of the principal axes of the dielectric permittivity tensor $\hat{\epsilon}$ which corresponds to the least principal value, ϵ_2. The wave vector of the eigensolution in this limit is $q(1 - \delta/2)$. As mentioned previously, solution 4 is analogous to solution 1 but describes wave propagation in the opposite direction. Depending on frequency, wave vectors K^+ and K^- in solution 2 may be real with the same signs, real with opposite signs, or complex quantities. The region of complex K^+ and K^+ is close to the Bragg frequency $\omega_B = \tau c/2\sqrt{\bar{\epsilon}}$ in the range $\omega_B/\sqrt{1+\delta} < \omega < \omega_B\sqrt{1-\delta}$; it corresponds to a "forbidden band" and pure imaginary values of K^-. That is, in a cholesteric, no waves propagate in this frequency range if their circular polarization matches the screw sense of the cholesteric helix. Independently of δ, in the forbidden band $|E^-/E^+| = 1$, and selective reflection of light associated with diffraction occurs in cholesterics. Outside the forbidden band, except in the high-frequency limit, the ratio of E^- or E^+ to the other amplitude is of the order of δ and solution 2 corresponds, with

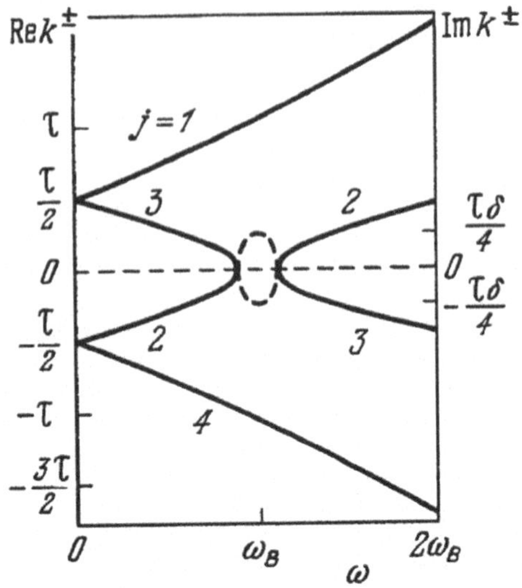

Fig. 1.5 Roots of the dispersion (1.14) versus frequency (solid lines denote Re K, dashed lines indicate Im K [see (1.15)]

sufficient accuracy, to the wave circularly polarized in the sense of the cholesteric helix. In the limit of very short waves $\omega c \gg \tau/\delta$, this solution becomes a plane-polarized wave with the polarization plane normal to that of solution 1 which rotates with period p and varies with z. In this limit, the wave vector of solution 2 is equal to $q(1+\delta/2)$. Solution 3 has the same properties as solution 2 with one exception: outside the forbidden band, it describes wave propagation in the opposite direction. The frequency dependence of k^+ and k^- in eigensolutions is illustrated in Fig. 1.5 which shows that the dispersion law $K(\omega)$ for solutions 1 and 4 is close to that for a wave propagating in a homogeneous medium. The dispersion law for solutions 2 and 3 is diffractive in nature and has a forbidden band (cf. Fig. 1.2). In summary, the features of light propagation in a cholesteric: selective diffraction reflection is expressed by light propagating along the optical axis whose circular polarization matches the screw sense of the cholesteric helix; it is essential that only first-order diffraction reflection occurs and no higher-order reflection is possible which corresponds to $|s| = 1$ in relationships of type (1.5). The absence of higher-order reflections for light propagation along the optical axis of a cholesteric is also illustrated by dispersion curves in Fig. 1.5 with only one forbidden band for each of solutions 2 and 3.

1.3 Solution of Boundary Problem

The description of experimentally observed optical characteristics in the interaction of light with periodic media requires that we know the properties of the eigenwaves in the corresponding media and find the solution of the Maxwell equations that satisfy certain boundary conditions, i.e., to solve the boundary problem. The solutions of the boundary problem for periodic media are illustrated with the following example of a cholesteric.

1.3.1 Planar Cholesteric Texture

Solution of the Boundary Problem. Using the eigensolutions obtained previously, we solve the problem of light reflection and transmission in a cholesteric sample of finite thickness. Consider the case of normal incidence onto a plane-parallel plate of thickness L with optical axis normal to the sample surface. That is, we consider the optical properties of an ideal planar texture for normal incidence of light.

We must determine which and to what degree eigensolutions of system (1.13) are "excited" by the incident wave to find the wave amplitudes of light reflected from and transmitted through the cholesteric. For normal incidence of light, the boundary conditions (continuous tangential components of electric and magnetic fields at the boundary) reduce to continuous electric and magnetic fields on sample surfaces since the wave in a cholesteric is strictly transverse. Therefore, the boundary conditions are written in the form

$$\begin{gathered}\mathbf{E}^e(0)+\mathbf{E}^r(0)=\mathbf{E}(0)\ ,\ \ \mathbf{E}^t(L)=\mathbf{E}(L)\\ \text{rot}\ [\mathbf{E}^e(z)+\mathbf{E}^r(z)]_{z=0}=[\text{rot}\ \mathbf{E}(z)]_{z=0}\\ \text{rot}\ \mathbf{E}(z)_{z=L}=\text{rot}\ \mathbf{E}(Z)_{z=L}.\end{gathered}\tag{1.17}$$

The electric fields of the incident $\mathbf{E}^e$, reflected $\mathbf{E}^r$ and transmitted $\mathbf{E}^t$ waves and the wave $\mathbf{E}$ propagating in the sample are expressed in the form

$$\begin{aligned}\mathbf{E}^e&=\mathrm{e}^i(q_e z-\omega t)(E_e^+\mathbf{n}_++E_e^-\mathbf{n}_-)\\ \mathbf{E}^r&=\mathrm{e}^{-i(q_e z+\omega t)}(E_r^+\mathbf{n}_-+E_r^-\mathbf{n}_-)\\ \mathbf{E}^t&=\mathrm{e}^{i(q_e z-\omega t)}(E_t^+\mathbf{n}_++E_t^-\mathbf{n}_-)\\ \mathbf{E}&=\mathrm{e}^{-i\omega t}\sum_{j=1}^{4}E_j^+(\mathrm{e}^{iK_j^+z}\mathbf{n}_++\xi_j\mathrm{e}^{iK_j^-z}\mathbf{n}_-)\end{aligned}\tag{1.18}$$

where $\mathbf{n}_+$, $\mathbf{n}_-$ are the unit vectors of the circular polarizations and q_e is the wave vector of light outside the sample. When waves propagate in the positive direction of the axis, (particularly incident waves), $\mathbf{n}_+$ and $\mathbf{n}_-$ describe the left- and right-handed circular polarizations, respectively, whereas $\mathbf{n}_+$ and $\mathbf{n}_-$ describe the right- and left-handed circular polarizations, respectively, when reflecting waves propagate in the opposite direction. This fact is reflected in expression for $\mathbf{E}^r$ in (1.18).

Substituting (1.18) into (1.17), equating each factor before $\mathbf{n}_+$ and $\mathbf{n}_-$ to zero, and transforming, we find that reflected and transmitted amplitudes are expressed in terms of the amplitudes of eigenwaves E_j excited in the sample

$$\begin{aligned}
E_r^+ &= (1/2)\sum_j \xi_j(1 - K_j^-/q_e)E_j^+, \\
E_t^+ &= (1/2)\sum_j \mathrm{e}^{i(K_j^+ - q_e)L}(1 + K_j^+/q_e)E_j^+, \\
E_r^- &= (1/2)\sum_J (1 - K_j^+/q_e)E_j^+, \\
E_t^- &= (1/2)\sum_j \xi_j \mathrm{e}^{i(K_j^- - q_e)L}(1 + K_j^-/q_e)E_j^+.
\end{aligned} \tag{1.19}$$

The amplitudes of eigenwaves E_j^+ are given by the following set of equations

$$\begin{aligned}
&\sum_j (1 + K_j^+/q_e)E_j^+ = 2E_e^+, \sum_j \mathrm{e}^{iK_j^+L}(1 - K_j^+/q_e)E_j^+ = 0, \\
&\sum_J \xi_j(1 + K_j^-/q_e)E_j^+ = 2E_e^-, \sum_j \xi_j \mathrm{e}^{iK_j^-L}(1 - K_j^-/q_e) = 0.
\end{aligned} \tag{1.20}$$

Thus, the case of light transmission and reflection for normal incidence of a beam onto a planar cholesteric texture reduces to the solution of a non-uniform set of four linear equations. Its solution and (1.15) and (1.16) for parameters ξ_j and $K_j^{\pm}$ provide the exact description of optical characteristics in the general case of a planar layer. In particular, it describes their dependence on frequency, light polarization and the dielectric properties of the medium outside the cholesteric.

Although the exact expressions for reflected and transmitted wave amplitudes are complicated, their structures are simple. Thus, the expression for the amplitude E_r^+ of the diffracted, circularly polarized wave is written as the following determinant ratio

$$E_r^+ = [\det D]^{-1} \left| \begin{array}{ccccc} & & D & & |E_e^+ \\ & & & & |E_e^- \\ & & & & |0 \\ -- & -- & -- & -- & |0 \\ {}^r\alpha_{51}^+ & {}^r\alpha_{52}^+ & {}^r\alpha_{53}^+ & {}^r\alpha_{54}^+ & 0 \end{array} \right| \tag{1.21}$$

where D is the matrix of system (1.20) and the elements of the fifth row are given by the following equation

$$ {}^r\alpha_{5j}^+ = \xi_j(1 - r\eta_j^-) \tag{1.22}$$

where $r = \sqrt{\bar{\epsilon}/\epsilon}$, ϵ is the dielectric permittivity outside the cholesteric. Expressions for E_r^-, E_t^+, and E_t^- are also derived from (1.21) if the elements of the fifth row are substituted by

$$^{r}\alpha_{5j}^{-} = (1 - r\eta_j^+)\,,\ ^{t}\alpha_{5j}^{+} = \mathrm{e}^{i(K_j^+ - q_e)L}(1 + r\eta_j^+)\,^{t}\alpha_{5j}^{-} = \xi_j \mathrm{e}^{i(K_j^- - q_e)L}(1 + r\eta_j^-), \tag{1.23}$$

respectively. In the general case, the exact solutions of the boundary problem (1.19)–(1.23) are cumbersome and it is more convenient to analyze them by numerical methods. The form of solution (1.21) is most convenient since it is used directly in computer calculations.

Analytical expressions are obtained directly from the general solution for some particular cases or by using simplifying assumptions. We present such cases to clarify the analysis and physical aspects of the problem.

1.3.2 Reflection from Thick Layers

Assume that thickness L of a planar cholesteric layer is very large, so that in solving the boundary problem, we may say that the liquid crystal occupies a half-space. The physical condition for such a simplification is the small distance over which light is absorbed in the cholesteric relative to layer thickness. The obtained results are applicable, however, over a wider range. Thus, for light frequencies in the selective reflection band, the corresponding condition becomes a small value of extinction length relative to layer thickness L. Extinction length is the distance over which the diffracted wave is attenuated by a factor of e due to diffraction reflection.

If light is incident from the outside half-space onto the half-space filled with the cholesteric, only two eigensolutions of those described by (1.13)–(1.15) are "excited" in the sample, which correspond to wave propagation into the cholesteric depth, i.e., to solutions 1 and 2. Solutions 3 and 4 which describe light propagation in the opposite direction cannot be "excited". This is proved by assuming that there is an arbitrary weak absorption in a cholesteric—a small imaginary addition to $\bar{\epsilon}$. It can be shown, using (1.20), that the amplitudes of the corresponding solutions decrease exponentially with thickness L. The physical explanation for the absence of excitation in the half-space of solutions 3 and 4 is that if the excitation did occur, the amplitudes of the corresponding fields would have increased infinitely toward the cholesteric depth; this is inconsistent with the finite values of the incident wave amplitude.

Thus taking into account that amplitudes E_j^+ and E_j^- for a half-space have nonzero values in (1.20) and (1.19) for solutions 1 and 2 only, i.e., $j = 1, 2$ and the cited fourth-order systems reduce to second-order systems. These are (1.19) and (1.20) without exponential factors

$$\sum_{j=1,2}(1+r\eta_j^+)E_j^+ = 2E_e^+,\ \sum_{j=1,2}\xi_j(1+r\eta_j^-)E_j^+ = 2E_e^-$$
$$E_r^+ = 1/2\sum_{j=1,2}\xi_j(1-r\eta_j^-)E_j^+, E_r^- = 1/2\sum_{j=1,2}(1-r\eta_j^+)E_j^+. \tag{1.24}$$

The solution of this system is represented in another form (1.21). Simplification occurs in the reduction of determinant rank by two orders (1.21). The corresponding expressions are obtained by canceling the third and fourth columns and rows of the determinant (1.21). For example, the amplitude of reflected wave is described as

$$E_r^+ = \begin{vmatrix} 1+\tilde{\eta}_1^+ & 1+\tilde{\eta}_2^+ & E_e^+ \\ \xi_1(1+\tilde{\eta}_1^-) & \xi_2(1+\tilde{\eta}_2^-) & E_e^- \\ \xi_1(1-\tilde{\eta}_1^-) & \xi_2(1-\tilde{\eta}_2) & 0 \end{vmatrix}[\det D]^{-1} \tag{1.25}$$

where

$$\tilde{\eta}_j^+ = r\eta_j^+,$$
$$\det D = \begin{vmatrix} 1+\tilde{\eta}_1^+ & 1+\tilde{\eta}_2^+ \\ \xi_1(1+\tilde{\eta}_1^-) & \xi_2(1+\tilde{\eta}_2^-) \end{vmatrix}.$$

Similarly, using (1.21), we obtain the expression for amplitude E_r^-:

$$E_r^- = [\det D]^{-1}\begin{vmatrix} 1+\tilde{\eta}_1^+ & 1+\tilde{\eta}_2^+ & E_e^+ \\ \xi_1(1+\tilde{\eta}_1^-) & \xi_2(1+\tilde{\eta}_2^-) & E_e^- \\ 1-\tilde{\eta}_1^+ & 1-\tilde{\eta}_2^+ & 0 \end{vmatrix}. \tag{1.26}$$

Equations (1.25) and (1.26) determine the reflection coefficient and polarization of the reflected wave for an arbitrary polarization of incident light. Since the eigenwaves in a cholesteric are superpositions of circularly polarized waves, we consider first the reflection of circularly polarized waves from a cholesteric, i.e., assume that either E_e^+ or E_e^- is zero. It is possible to find four amplitudes from (1.25) and (1.26), two E_r^{++} and E_r^{+-} which describe light reflected from diffracting and non-diffracting polarizations into a wave with diffracting polarization in (1.25) when E_e^- and E_e^+ are set equal to zero in (1.25). Two other amplitudes, E_r^{-+} and E_r^{--}, describe the reflection of waves with the same polarization into waves with non-diffracting polarization given by (1.26) when E_e^- and E_e^+ are set to zero.

The reflection coefficients R^+ for circularly polarized diffracting waves and R^- for circularly polarized non-diffracting waves are expressed in terms of the introduced amplitudes

$$R^+ = |E_e^+|^{-2}(|E_r^{++}|^2 + |E_r^{-+}|^2),\ R^- = |E_e^-|^{-2}(|E_r^{+-}|^2 + |E_r^{--}|^2). \tag{1.27}$$

Expressions (1.25)–(1.27) produce the exact solution to the problem of light reflected from a half-space filled with a cholesteric for an arbitrary relationship of cholesteric dielectric properties to surrounding medium.

Generally, reflection from the sample occurs according to two mechanisms—diffraction and dielectric (associated with the jump in the dielectric constant at the dielectric-medium interface). But in order to reveal the diffraction properties of reflection typical of periodic media, particularly in cholesterics, the analysis of obtained results should begin with the simulation where the averaged dielectric constant $\overline{\epsilon}$ of the cholesteric coincides with that of the surrounding medium, ϵ.

1.3.3 The Case Where Medium and Cholesteric Have Equal Dielectric Constants

All formulae are simplified by substitution of $\tilde{\eta}$ by η.

Assuming that δ is small, (1.25) and (1.26), the reflected wave amplitudes, in the zeroth approximation with respect to δ, are:

$$E_r^{++} = -\xi_2 E_e^+, \; E_r^{-+} = E_r^{+-} = E_r^{--} = 0. \tag{1.28}$$

The results show that, in the zeroth approximation with respect to δ, reflection from a cholesteric is experienced only by light with left or right-handed circular polarization. The corresponding reflection coefficient R^+ is

$$R^+ = |\xi_2|^2 = |E_2^-/E_2^+|^2 \tag{1.29}$$

and is governed by the amplitude ratio of waves E^- and E^+ in the second eigensolution. In the forbidden band, $|\xi_2| = 1$ we find that, within the considered approximation with respect to δ, only the wave with diffracting polarization experiences reflection from a cholesteric in the frequency range $\omega_B\sqrt{1+\delta} < \omega < \omega_B\sqrt{1-\delta}$, with reflection coefficient $R^+ = 1$. If a cholesteric reflects a wave with diffracting polarization (i.e., left or right-handed circular polarization), the reflected and incident waves share polarization of the same sense. However, reflection from the dielectric boundary changes the sense of the circular polarization, i.e., the left-handed circular polarization changes to right-handed and vice versa.

In summary, we conclude that, in the zeroth approximation with respect to δ, reflection from a cholesteric occurs inside the selective reflection band; independent of incident wave polarization, reflected light is of circular polarization and matches the sense of cholesteric helix. The value of the reflection coefficient in the selective reflection band is independent of frequency and determined by the fraction of diffraction polarization present in the incident beam, i.e.,

$$R = R^+[1/2(1-P) + P|\mathbf{e}\mathbf{n}_+^*|^2] \tag{1.30}$$

where e is the incident wave polarization vector and P is the degree of polarization.

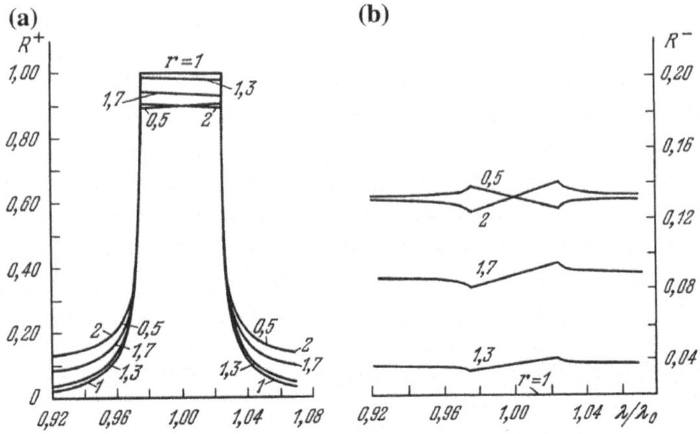

Fig. 1.6 Calculated reflection spectra of a cholesteric for normal incidence of light with the jump in dielectric permittivity (different r) at the sample boundary for $\delta = 0.05$. **a** Right-handed circular polarization of the incident light; **b** left-handed circular polarization of the incident light [7]

The obtained formulae describe light reflection from thick cholesteric layers, when $\bar{\epsilon} = \epsilon$ and the anisotropy parameter is small (Fig. 1.6, curve for $r = 1$). The corresponding expressions do not account for light reflection from the cholesteric boundary. We emphasize that light reflection from the boundary occurs also in the case where the averaged dielectric constant $\bar{\epsilon}$ of the cholesteric coincides with that of the external medium but is insignificant for small values of δ; such reflections from the boundary are described by (1.25) and (1.26). If we consider first-order terms with respect to δ, the amplitudes of circularly polarized reflected waves are given by

$$\begin{aligned} E_r^{++}/E_e^{+} &= -\xi_2(1 - \Delta\omega/\omega_B),\ E_r^{+-}/E_e^{-} = \delta\xi_2/4 \\ E_r^{-+}/E_e^{+} &= \delta\xi_2/4,\ E_r^{--}/E_e^{-} = \delta/8(1 + (\Delta\omega)/2\omega_B)) \end{aligned} \tag{1.31}$$

where $\Delta\omega = \omega - \omega_B$.

It follows from (1.31) that in the selective reflection band $R^- \approx 5\delta^2/64$, i.e., the reflection coefficient of non-diffracting circular polarization is independent practically of frequency and is of the order of δ^2. For diffracting polarization, the reflection coefficient differs from unity (by a value of the order of δ) and its proportional to δ^2 outside the selective reflection band when the frequency of light deviates from the boundary of the selective reflection band by $\Delta\omega > \delta\omega_B$ (Fig. 1.6, curve for $r = 1.$). Thus, the diffractive nature of reflection in a cholesteric appears in the frequency range of the selective reflection band. Outside it, the reflection coefficient is determined by the difference in dielectric properties on both sides of the interface. It is independent of the polarization and is of the order of δ^2, i.e., it is proportional to the difference between dielectric constants of the cholesteric and the medium squared.

We emphasize that the properties of diffraction reflection studied on a cholesteric are common to all periodic media. In particular, they are also characteristic of the previously described periodic medium with scalar dielectric permittivity. Thus, the curve for $r = 1$ in Fig. 1.6 qualitatively describes reflection from a half-space with the sole difference that now reflection coefficient is independent of wave polarization and coincides with R^+, the reflection coefficient of light with diffracting polarization in a cholesteric.

1.3.4 Dependence of Reflection on Polarization

In previous sections, we considered light reflection with circular polarization which is independent of cholesteric surface anisotropy axis orientation, i.e., of the director **n** orientation. The reflection characteristics of non-circularly polarized reflected light, however, depend on cholesteric surface director **n**. This principle is illustrated best by plane-polarized incident light.

Expressions (1.25) and (1.26) give plane-polarized light reflection coefficient dependence on director orientation at cholesteric surface:

$$R(\xi) = R_{un} - |E_e|^{-2}[\sin 2\xi \mathrm{Re}(E^{++}E^{+-} + E^{-+}E^{--}) + \cos 2\xi \mathrm{Im}(E^{++}E^{+-} + E^{-+}E^{--})] \tag{1.32}$$

where ξ is the angle between the director orientation on the cholesteric surface and the polarization plane of light and $R_{un} = |E_e|^{-2}(|E^{++}|^2 + |E^{--}|^2 + |E^{+-}|^2 + |E^{-+}|^2)/2$ is the reflection coefficient of an unpolarized beam.

It follows from (1.32) [see, e.g., (1.31)] that reflection coefficient modulation amplitude depends on the orientation of the polarization plane of light and may reach a value of the order of dielectric anisotropy δ whereas the reflection coefficient maximum and the minimum are determined by the director orientation on the sample surface. Hence, the possibility exists to determine both molecule (director) orientation on the sample surface and the value for the dielectric anisotropy δ of the cholesteric using polarization dependence on reflection coefficient. Thus, within the linear approximation with respect to δ, orientations of the polarization plane corresponding to maximum and minimum reflection are independent of frequency in the selective reflection band for $\delta > 0$ and are given by the expressions $\xi = -\pi/4$ and $\xi = \pi/4$ for the right-handed cholesteric helix and $\xi = \pi/4$ and $\xi = -\pi/4$ for the left-handed helix. Expressions (1.31) and (1.32) produce the formula for the plane-polarized light reflection coefficient in the selective reflection band

$$R(\xi) = (1/2)(1 - 2\Delta\omega/\omega_B) \mp (\delta/4)\sin 2\xi \tag{1.33}$$

where the upper sign corresponds to the right-handed and the lower to the left-handed sense of the cholesteric helix.

1.3.5 Effect of Dielectric Boundaries

We considered the situation where cholesteric boundary effects on reflected light are minimal and reflection from a cholesteric is determined by diffraction from the periodic structure alone (i.e., boundary effect vanishes if $\delta \to 0$). This analysis facilitates studying diffraction properties of reflection from cholesterics which manifest themselves most clearly in the polarization properties and frequency characteristics of reflection.

Experimental measurements must meet the condition $\overline{\epsilon} = \epsilon$. The common and clearest case is one in which the mean dielectric constant $\overline{\epsilon}$ of a cholesteric differs from the dielectric constant of the external medium. Large differences between medium ϵ and mean cholesteric dielectric constant $\overline{\epsilon}$, ($|\overline{\epsilon} - \epsilon| > \delta$), affect reflection characteristics due to the strong frequency dependence of diffraction reflection phase in the selective reflection band and the interference between it and dielectric reflection. Diffraction reflection phase changes by π over the frequency range of the selective reflection band (1.16) and this explains strong interference effects.

A qualitative illustration of the variation in the reflection characteristics from a half-space filled with a cholesteric when $\overline{\epsilon} \neq \epsilon$ follows (see [4] for more details).

Analysis of general formulae (1.25–1.27) indicates that light reflection from a cholesteric with non-diffracting circular polarization (R^-) is due to the difference between $\overline{\epsilon}$ and ϵ alone (within the zeroth approximation with respect to δ). In this approximation, reflection coefficients R^+ and R^- are independent of frequency in the selective reflection band and their dependence on frequency is described by terms of nonzero order in δ. If anisotropy δ is small, the approximation of reflection coefficients of circularly polarized waves as independent of frequency in the selective reflection band is satisfactory [7–9]. In this case, the simplified equations for R^+ and R^- apply:

$$\begin{aligned} R^+ &= \frac{1 + 14r^2 + r^4}{(1+r)^4}, \quad R^- = \left(\frac{1-r}{1+r}\right)^2 \\ r &= \sqrt{\overline{\epsilon}/\epsilon}. \end{aligned} \tag{1.34}$$

The reflection coefficient of unpolarized light

$$R_{wn} = R^+/2 + R^-/2 = (1 + 6r^2 + r^4)(1+r)^{-4} \tag{1.35}$$

is independent of frequency as well.

Within the same approximation, however, reflection of plane-polarized light shows strong reflection coefficient frequency dependence in the selective reflection band as r deviates from unity (Fig. 1.7)

$$\begin{aligned} R_L = 2|(2r\xi_2)(1+r)^{-2} \pm 1/2(1-r)(1+r)^{-1}|^2 \\ + 1/2(1-r)^2(1+r)^{-2} \end{aligned} \tag{1.36}$$

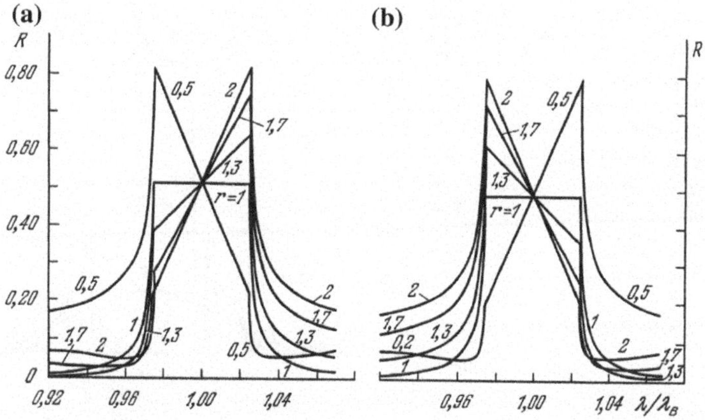

Fig. 1.7 Calculated reflection spectra analogous to those in Fig. 1.6 for linear polarizations of the incident light. **a** Light polarization along director orientation; **b** light polarization normal to director orientation on the cholesteric surface [7]

where ξ_2 is determined by (1.16), the upper sign relates to light polarized along the director on the cholesteric surface and lower sign to the polarization normal to the director (assuming anisotropy in δ is positive).

Since the previously mentioned frequency dependence is due to the phase frequency dependence of the wave experiencing diffraction reflection [see (1.6) for ξ_2], it corresponds to the following physical picture: if there is an interface, the waves reflected from it interfere with those experiencing diffraction reflection. The phase of the wave experiencing diffraction reflection depends on frequency, whereas the phase of the wave reflected from the dielectric boundary is constant. Such interference is most prominent for plane-polarized light. (Recall that within the zeroth approximation with respect to δ, waves reflected from the dielectric boundary and those which experienced diffraction reflection have opposite circular polarizations and cannot interfere.)

As follows from (1.36), the reflection coefficient for the plane-polarized wave depends on the polarization orientation with respect to the director. This is similar to the case where $\overline{\epsilon} = \epsilon$ and this fact is used to determine the director orientation on the surface. In this case, however, polarization orientation at maximum and minimum reflection depends on light frequency (in the selective reflection band). The amplitude of modulation is determined not only by δ but, in the zeroth approximation in δ, also by $r - 1$, i.e., it is less sensitive to δ than in the case where $\overline{\epsilon} = \epsilon$ (i.e., $r = 1$) [7].

The following numerical calculations [7, 9] (Figs. 1.6 and 1.7) illustrate this analysis: in the case of reflection from a half-space filled with a cholesteric, Fig. 1.6 shows the computed reflection coefficients for light of diffracting R^+ and non-diffracting R^- circular polarizations versus frequency for different values of medium dielectric constant outside the cholesteric. In particular, Fig. 1.6 shows how well equation (1.34) describes reflection from a cholesteric for small δ. For the value $\delta = 0.05$ used in the calculation, the deviation of R^+ from unity for $\overline{\epsilon} = \epsilon$ (i.e., $r = 1$) cannot be shown

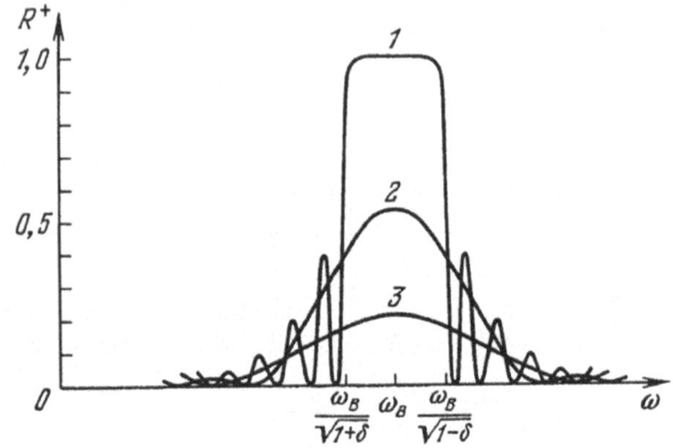

Fig. 1.8 Reflection coefficient of light with diffracting polarization for normal incidence versus frequency in samples of different thickness. (1) Thick sample ($L\delta/p \gg 1$); (2) sample of intermediate thickness ($L\delta/p \sim 1$); (3) thin sample ($L\delta/p \ll 1$)

on the scale of the drawing in the selective reflection band since this difference is less than 2×10^{-4}. The corresponding calculated value of the reflection coefficient for non-diffracting polarization of all the wavelengths is also very small, less than 3×10^{-4}. Figure 1.6 also demonstrates that the constant reflection coefficient for circular polarizations in the selective reflection band is a good approximation for all reasonable values of r.

Figure 1.7 represents the calculated reflection coefficients of plane-polarized light versus frequency for two different orientations of the molecule on the cholesteric surface and shows that the constant reflection coefficient in the selective reflection band is a good approximation only if $\overline{\epsilon} = \epsilon$.

For a layer of finite thickness, the reflection coefficient R and transmission coefficient, $T = 1 - R$ are described by general formulae (1.21)–(1.23) and not discussed here, since they are cumbersome (see [4, 10]). Although the general character and polarization properties of reflection are the same, reflection intensity decreases and Pendellösung beats and beats due to finite sample thickness appear in reflection (transmission) (Fig. 1.8).

1.3.6 Method of Characteristic Matrices

Another method of solving the boundary problem for a layered medium is to apply one procedure of characteristic matrices [1]. The method permits expression of reflection and transmission characteristics of multilayered system using the characteristics of one layer and this reveals how transmission and reflection depend on the number of layers and the properties inherent in each.

Although the structure of expressions derived within this approach is simple, generally, quantitative results are obtained only after numerical calculations. Therefore, the analytical results obtained by this method apply to the simplest situations in periodic media with scalar dielectric permittivity [1]. Nevertheless, this method applies to media with complicated structure [11] as well since it can be used in computer calculations, the accuracy of which is adjustable, in principle, since it is not limited by the method itself.

We briefly consider the fundamental idea of the method of characteristic matrices and refer the reader to the references in [11–13] for details.

If the electric (or magnetic) components of the transverse electromagnetic wave (TE or TM) at the entrance and exit surfaces of a plane-parallel layer are given by

$$Q = \begin{bmatrix} U(z) \\ V(z) \end{bmatrix}, \quad Q_0 = \begin{bmatrix} U_e \\ V_e \end{bmatrix} \tag{1.37}$$

then, without restriction to normal incidence, we obtain the following relationship between Q and Q_0 via the characteristic 2 by 2 matrix M for the layer:

$$Q_0 = MQ. \tag{1.38}$$

If a layered medium consists of N different layers (in the general case), it is also described by the characteristic matrix M_N related to the characteristic matrices of individual layers M_i as

$$\begin{aligned} Q_0 &= M_N Q(z_n) \\ M_N &= M_1 M_2 \ldots M_N \end{aligned} \tag{1.39}$$

where $Q(Zn)$ determines the wave amplitude on the exit surface of the Nth layer. If all layers possess the same properties, the characteristic matrix has the form

$$M_N = (M_1)^N. \tag{1.40}$$

It is easy to extract the amplitude coefficients of reflection and transmission (for the TE-type waves) using elements m_{ik} of the characteristic matrix and the boundary conditions for the electric and magnetic component vectors of the wave

$$\begin{aligned} r = R/A &= [m_{11} + m_{12}p_e)p_1 - (m_{21} + m_{22}p_e)][(m_{11} + m_{12}p_e)p_1 \\ &\quad + (m_{21} + m_{22}p_e)]^{-1} \\ &= T/A = 2p_1[(m_{11} + m_{12}p_e) + (m_{21} + m_{22}p_e)]^{-1} \end{aligned} \tag{1.41}$$

where A, R, and T are the amplitudes of the incident, reflected, and refracted waves

$$p_1 = \cos\theta_1\sqrt{\epsilon_1/\mu_1}, \quad p_e = \cos\theta_e\sqrt{\epsilon_e/\mu_e}.$$

Here ϵ_1, μ_1, ϵ_e, and μ_e are dielectric and magnetic permittivities of homogeneous media which limit the layered structure under consideration (see Fig. 1.9) and θ_1

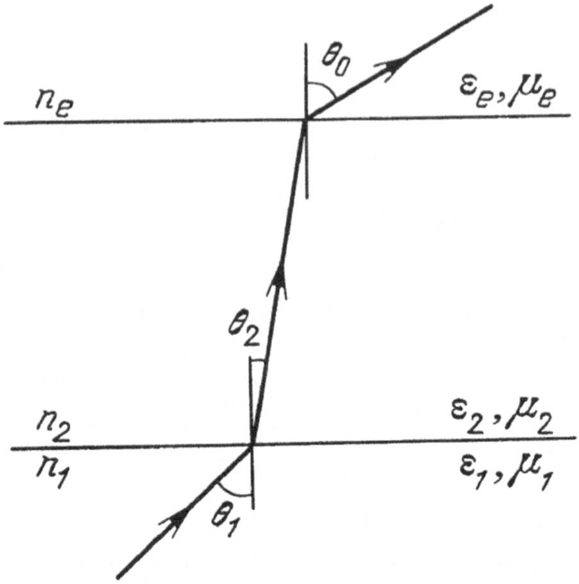

Fig. 1.9 Schematic representation of light propagation through a layer with the jump in dielectric permittivity at its boundaries

and θ_e are the angles between the direction of stratification (z-axis) and direction of propagation of incident and transmitted waves, respectively.

Thus, the main idea of the approach is to seek the characteristic matrix of the layer, (1.38), and structure as a whole, (1.39). In case a, complicated coordinate dependence of layer characteristics (even for scalar dielectric permittivity), the search for the characteristic matrix for the layer may be complicated. It is simplified, however, if a layered structure is considered as a sequence of different layers of dielectric properties. Thus, for a structure formed by an array of two-layered units with different properties, the characteristic matrix of the layered structure as a whole is expressed via the characteristics of individual layers and Chebyshev polynomials of order $N - 1$ and $N - 2$, which are functions of a dimensionless argument determined by the characteristics of the layers, frequency, and incidence angle of the wave [1].

Since the simplified formulae are cumbersome, we give the exact expressions of the characteristic matrix for the structure built by N double layers and normal incidence which the condition that each of them is a quarter-wave layer, i.e., $h_1\sqrt{\epsilon_1}(\omega/c) = h_2\sqrt{\epsilon_2}(\omega/c) = \pi/2$ (the latter condition corresponds to the fulfillment of the Bragg condition):

$$M_N = \begin{vmatrix} (-\epsilon_2/\epsilon_1)^{N/2} & 0 \\ 0 & (-\epsilon_1/\epsilon_2)^{N/2} \end{vmatrix}. \tag{1.42}$$

Therefore, the reflection coefficients for the layered medium of N layers must be written assuming the above conditions, as

$$R_N = \left\{ \left[1 - \sqrt{\epsilon_{e2}/\epsilon_{e1}} (\epsilon_1/\epsilon_2)^N \right] \left[1 + \sqrt{\epsilon_{e2}/\epsilon_{e1}} (\epsilon_1/\epsilon_2)^N \right]^{-1} \right\}^2 . \tag{1.43}$$

Equation (1.43) demonstrates that reflection coefficient R_N with the number of layers caused by light diffraction increases and tends to unity for $N \to \infty$ even for an arbitrary small difference between ϵ_1 and ϵ_2. In our case, where absorption in the layers is absent, the transmission coefficient T is found from (1.41) or directly from the law of conservation of energy, $T = 1 - R$.

We stated the general principles of the method of characteristic matrices and illustrated its use with the simplest example. This method, combined with numerical computer calculations, solves the boundary problem in more complicated systems, such as optics of liquid crystals [13–15], X-ray optics of layered crystals [16, 17], etc.

The method, although exact in principle, loses its clarity with complicated structure and makes analysis of the interaction between radiation and periodic media difficult. Therefore, the results obtained by this numerical method should be verified and compared with the results obtained by the approximate analytical solutions of corresponding problems. Moreover, if the accuracy of the approximate analytical method is sufficient, it is the method of choice due to its simplicity.

The examples given here almost exhaust the known exact analytical solutions of the Maxwell equations and the corresponding boundary problems for periodic media. The following chapters deal with more complicated structure; however, we always refer the reader to the simple results. On the other hand, in many complicated situations, the physics of the effect is studied using approximate methods since, in many important cases, the accuracy is sufficient and they provide the correct physical interpretation of the effects of radiation interaction with periodic media.

The examples given here almost exhaust the known exact analytical solutions of the Maxwell equations and the corresponding boundary problems for periodic media (see, e.g., [18]). It also worth to be mentioned that the recent researches [19–21] showed that the polarization dependencies of light reflection and transmission on the director orientation at the surfaces for layers of finite thickness can be successfully used for studies of the director distribution in CLC samples and, particular, for an experimental restoration of the actual surface anchoring potential. The following chapters deal with more complicated situations; however, we always refer the reader to the simple results. On the other hand, in many complicated situations, the physics of the effect is studied using approximate methods since, in many important cases, the accuracy is sufficient and they provide the correct physical interpretation of the effects of radiation interaction with periodic media.

References

1. M. Born, E. Wolf, *Principals of Optics* (Pergamon Press, Oxford, 1965)
2. C. Elachi, Proc. IEEE **64**, 1666 (1976)
3. L. Brilluen, M. Parodi, *Propagation des Ondes dans les Milieux Periodicues* (Paris, 1956)
4. V.A. Belyakov, A.S. Sonin, *Optics of Cholesteric Liquid Crystals* (Nauka, Moscow, 1982). (in Russian)
5. H. De Vries, Acta Crystallogr. **4**, 219 (1951)
6. E.I. Kats, ZhTF **59**, 1854 (1970) [Sov. Phys. JETP **32**, 104 (1971)]
7. M. Tur, Mol. Cryst. Liquid Cryst. **29**, 345 (1975)
8. V.A. Belyakov, V.E. Dmitrienko, Pis'ma v. Zh. Tekh. Fiz. **7**, 19 (1981). (in Russian)
9. A.V. Tolmachev, A.S. Sonin, Fiz. Tverd. Tela **17**, 3096 (1975); Kristallografiya **21**, 794 (1976) (in Russian)
10. V.A. Belyakov, V.E. Dmitrienko, V.P. Orlov, UFN **172**, 221 (1979) [Sov. Phys. Usp. **22**, 63 (1979)]
11. R.M.A. Azzam, N.M. Bashara, *Ellipsometry and Polarized Light* (North Holland Publ. Comp, Amsterdam, New York, Oxford, 1977)
12. W. Shercliff, *Polarized Light* (Harvard University Press, Cambridge, 1962)
13. S. Chandrasekhar, *Liquid Crystals* (Cambridge University Press, 1976)
14. D.W. Berreman, T.J. Scheffer, Phys. Rev. Lett. **25**, 577 (1970); Mol. Cryst. Liquid Cryst. **11**, 395 (1970); Phys. Rev. **5A**, 1397 (1971)
15. V.A. Belyakov, S.M. Osadchii, V.A. Korotkov, Kristallografiya **31**, 522 (1986). (in Russian)
16. A.V. Kolpakov, Yu.N. Belyaev, Vestn. MGU (Ser. 3, Fizika, Astronomiya **26**, 91 (1985) (in Russian)
17. M.A. Andreeva, K. Rocete, Yu.P. Khapachev, Phys. Stat. Sol (a) **88**, 455 (1985)
18. V.A. Belyakov Differaction Optics of Complicatedly Structured Periodic Media (Moscow, Nauka, 1988) in Russian; V.A. Belyakov Differaction Optics of Complex Structured Periodic Media (Springer Verlag New York, 1992) (Revised translation from Russian)
19. V.A. Belyakov, Mol. Cryst. Liquid Cryst. **489**, 54 (2008)
20. V.A. Belyakov, S.V. Semenov, D.V. Shmeliova, Mol. Cryst. Liquid Cryst. **559**, 31 (2012)
21. K. Nowicka, D. Dardas, W. Kuczynski, V.A. Belyakov, D.V. Shmeliova, Liquid Crystals **41**, 1448 (2014)

Chapter 2
Approximate Description of Interaction of Radiation with Regular Media

In the previous chapter considering the interaction of an electromagnetic radiation with regular media, we have made no assumption that the modulation of their dielectric properties is small. Neither was made an assumption that for finite samples, the intensity of the scattered radiation is low with respect to that of the incident radiation. The description of the interaction of an electromagnetic wave even with simplest periodic media is rather complicated and, as a rule, the problem stated in such a general form cannot be solved in an analytical form. But in many instances important for practice the problem has a small parameter both for electromagnetic and other radiations which provides a sufficiently accurate analytical solution of the problem. The physical conditions for a small parameter are: (i) weak interaction between radiation and individual scatterer or a subunit of the matter playing the role of such a scatterer; (ii) weak radiation scattering from a microscopic sample, i.e., small wave intensity relative to the scattered wave incident. If the first condition is met, the second one is necessarily fulfilled if the size of a scattering sample is sufficiently small. The first condition is related to the interaction between radiation and medium. Thus, a small parameter exists if X-rays interact with atoms [1], or neutrons with atoms and atomic nuclei [2] (scattering amplitudes for individual centers are much smaller than $(n\lambda)^{-1/2}$ where λ is radiation wavelength and n is scatterer density —in the continuum medium, amplitude of the spatial modulation of scatterer density. On the other hand, there is no such a small parameter, e.g., for low-energy electron scattering by atoms [3]. And, in resonance scattering of Mössbauer radiation by nuclei [4], the corresponding parameter may be sufficiently small to give the accuracy necessary to apply approximate methods.

A description of the interaction between radiation and regular media with complicated structure assuming the second condition (the so-called kinematical approximation of scattering theory) follows.

V. Belyakov, *Diffraction Optics of Complex-Structured Periodic Media*, Springer Series in Optical Sciences 203,
https://doi.org/10.1007/978-3-319-43482-7_2

2.1 Kinematical Approximation

Let a plane monochromatic wave be incident onto a sample of regularly arranged scatterers of finite dimensions. Let scattered wave intensity I_s be much lower than that of incident wave I such that, in the first approximation, transmitted beam attenuation may be neglected relative to incident beam, i.e., it is possible to assume that incident wave intensity is constant within the sample.

For the time being, we consider only coherent elastic scattering of a wave from scatterers where $f_i(k_0e_0,\ k_1e_1)$ is the coherent scattering amplitude by an individual scatterer i, subscript i with discrete values and k_0e_0 and k_1e_1, wave and polarization vectors of incident and scattered waves, respectively. We reserve consideration of coherent scattering amplitude form until the sections which deal with specific regular media and various types of radiations [1–4].

2.1.1 Scattering Cross Section

In the kinematical approximation, radiational scattering is usually described by cross sections $\mathrm{d}\sigma/\mathrm{d}\Omega$ defined as

$$\mathrm{d}\sigma(\mathbf{k}_0,\ \mathbf{e}_0;\ \mathbf{k}_1,\ \mathbf{e}_1)/\mathrm{d}\Omega_{\mathbf{k}_1} = \left|\sum_i f_i(\mathbf{k}_0,\ \mathbf{e}_0;\ \mathbf{k}_1,\ \mathbf{e}_1)\mathrm{e}^{i(\mathbf{k}_0-\mathbf{k}_1)\mathbf{r}_i}\right|^2. \tag{2.1}$$

Summation is over all sample scatterers, and r_i is the ith scatterer coordinate. Expression (2.1) is transformed to the form

$$\mathrm{d}\sigma(\mathbf{k}_0,\ \mathbf{e}_0;\ \mathbf{k}_1\mathbf{e}_1)/\mathrm{d}\Omega_{\mathbf{k}_1} = \left|\sum_i f_i\mathrm{e}^{i(\mathbf{k}_0-\mathbf{k}_1)\mathbf{r}_i}\right|^2 \cdot \left|\sum_n \mathrm{e}^{i(\mathbf{k}_0-\mathbf{k}_1)\mathbf{r}_n}\right|^2 \tag{2.2}$$

where summation is over all scatterers i within the unit cell of the periodic structure (in the first factor) and over all sample unit cells (in the second factor); r_n is the radius-vector of the nth unit cell. The first factor in (2.2),

$$F(\mathbf{k}_0,\ \mathbf{e}_0;\ \mathbf{k}_1,\ \mathbf{e}_1) = \sum_i f_i(\mathbf{k}_0,\ \mathbf{e}_0;\ \mathbf{k}_1,\ \mathbf{e}_1)\mathrm{e}^{(\mathbf{k}_0-\mathbf{k}_1)\mathbf{r}_i} \tag{2.3}$$

is structure amplitude. It relates scattering and symmetry characteristics and scatterer type in the crystal unit cell. The second factor in (2.2) accounts for strong cross-sectional dependence on $k_0 - k_1$ and leads to the Bragg condition for scattering maxima

$$\mathbf{k}_0 - \mathbf{k}_1 = \boldsymbol{\tau},\ |\mathbf{k}_0| = |\mathbf{k}_1| \tag{2.4}$$

where τ is the structure reciprocal lattice vector.

In the limit of infinite scatterer number, scattering cross section (2.2) is related to one unit cell and takes the form known in the theory of X-ray diffraction

$$\begin{aligned} d\sigma/d\Omega &= \lim_{N\to\infty}(|F|^2/N)\left|\sum_{n=1}^{N} e^{i(\mathbf{k}_0-\mathbf{k}_1)\mathbf{r}_n}\right|^2 \\ &= (2\pi)^3/V|F|^2\delta(\mathbf{k}_0-\mathbf{k}_1-\boldsymbol{\tau}) \end{aligned} \quad (2.5)$$

where V is unit cell volume.

Media scattering differs from X-ray diffraction in the complicated polarization and angular dependence of structure amplitude; however, as in X-ray and thermal-neutron diffraction, symmetry analysis of structure amplitude permits one to find reciprocal lattice vectors τ in relationships (2.2)–(2.5) for which F is zero, that is, diffraction reflection experiences extinction.

2.1.2 Polarization Characteristics

Thus, structure amplitude reflects features typical of periodic media with complicated structure. Structure complexity is associated microscopically with the vector nature of scattering from individual scatterers and the complex spatial structure of these scatterers in the medium.

The consequences of scatterer non-scalar nature are different extinction rules for such structures (see below) compared with those for structures formed by scalar scatterers and complicated scattering polarization characteristics.

We focus on the relation between scattering polarization characteristics and structure amplitude, give the general relationships between the polarization properties of both incident and scattered radiation and crystal structure, and present the polarization-dependent scattering cross sections for monochromatic radiation. Consider the differential scattering cross section of polarized radiation

$$d\sigma(\mathbf{k}_0,\ \mathbf{e}_0;\mathbf{k}_1,\ \mathbf{e}_1)/d\Omega_{\mathbf{k}_1} = (2\pi)^3/V|F_\tau(\mathbf{k}_0,\ \mathbf{e}_0;\mathbf{k}_1,\ \mathbf{e}_1)|^2\delta(\mathbf{k}_0-\mathbf{k}_1-\boldsymbol{\tau}). \quad (2.6)$$

(Notation here is as earlier.) Expression (2.6), a function of e, reaches its maximum for values of $e_1 \equiv n_0'$. Vector $n_0'(e_0)$ is a polarization vector of the scattered radiation and corresponding cross section,

$$d\sigma(\mathbf{k}_0,\ \mathbf{e}_0;\mathbf{k}_1,\ \mathbf{n}_0)/d\Omega_{\mathbf{k}_1} \equiv d\sigma((\mathbf{k}_0,\ \mathbf{e}_0;\mathbf{k}_1)/d\Omega_{\mathbf{k}_1}$$

is the differential scattering cross section for a quantum with polarization vector e_0. Introducing a coherent scattering tensor T_{ik} for the unit cell such that

$$F(\mathbf{k}_0,\mathbf{e}_0;\mathbf{k}_1,\ \mathbf{e}_1) = \sum_{ik} e_{1i}^* T_{ik} e_{0k} = \mathbf{e}_1^*\hat{T}\mathbf{e}_0 \quad (2.7)$$

yields the following formula for $\mathbf{n}_0'(\mathbf{e}_0)$

$$\mathbf{n}_0'(\mathbf{e}_0) = \hat{T}\mathbf{e}_0|\hat{T}\mathbf{e}_0|^{-1} \tag{2.8}$$

where $\hat{T}\mathbf{e}_0$ is a vector, the kth component of which equals $\sum_i T_{ki}e_{0i}$.

The explicit expressions for vector n_0 are given for some particular cases. The scattering cross section of unpolarized radiation is expressed by $\mathrm{d}\sigma(\mathbf{k}_0,\ \mathbf{e}_0;\mathbf{k}_1)$ as

$$\mathrm{d}\sigma/\mathrm{d}\Omega_{\mathbf{k}_1} = 1/2\sum_{i=1,2}\mathrm{d}\sigma(\mathbf{k}_0,\ \mathbf{e}_0;\mathbf{k}_1)/\mathrm{d}\Omega_{\mathbf{k}_1}. \tag{2.9}$$

Summation is over two mutually orthogonal polarization vectors e_{0i}. The scattered radiation is partially polarized with polarization matrix

$$\rho = \sum_{i=1,2}\rho(\mathbf{n}_0'(\mathbf{e}_{0i}))(\mathrm{d}\sigma(\mathbf{k}_0,\ \mathbf{e}_{0i};\mathbf{k}_1)/\mathrm{d}\Omega_{\mathbf{k}_1}) \times\left[\sum_i \mathrm{d}\sigma(\mathbf{k}_0,\ \mathbf{e}_{0i},\ \mathbf{k}_1)/\mathrm{d}\Omega_{\mathbf{k}_1}\right]^{-1}. \tag{2.10}$$

In the case of electromagnetic radiation, $\rho(\mathbf{e})$ is the polarization density matrix for a photon with polarization vector e [5]. Its elements are given by

$$\rho_{ik} = e_i e_k^* \tag{2.11}$$

where e_i are the coefficients in the expansion of the polarization vector in polarization unit vectors.

The scattering cross sections of partially polarized radiation with polarization degree P have the form

$$(\mathrm{d}\sigma/\mathrm{d}\Omega_{\mathbf{k}_1})_P = (1-P)\mathrm{d}\sigma/\mathrm{d}\Omega_{\mathbf{k}_1} + P\mathrm{d}\sigma(\mathbf{k}_0,\ \mathbf{e}_0;\mathbf{k}_1/\mathrm{d}\Omega_{\mathbf{k}_1}) \tag{2.12}$$

where $\mathbf{e}_0$ is the vector of polarization presented in the radiation. The polarization density matrix for the scattered radiation is equal to

$$\rho_P = \left[(1-P)(\mathrm{d}\sigma/\mathrm{d}\Omega_{\mathbf{k}_1})\rho + P\frac{\mathrm{d}\sigma(\mathbf{k}_0,\mathbf{e}_0;\mathbf{k}_1)\rho(\mathbf{n}_0')}{\mathrm{d}\Omega_{\mathbf{k}_1}}\right] \times\left[(1-P)\frac{\mathrm{d}\sigma}{\mathrm{d}\Omega_{\mathbf{k}_1}} + P\frac{\mathrm{d}\sigma(\mathbf{k}_0,\mathbf{e}_0;\mathbf{k}_1)}{\mathrm{d}\Omega_{\mathbf{k}_1}}\right]^{-1}. \tag{2.13}$$

Note that the above expressions for polarization and scattering cross sections are obtained for monochromatic radiations.

2.1.3 *Scattering of Light in Cholesterics*

Consider selective (diffraction) light scattering by cholesterics as an illustration of the general formulae in the previous section. Unlike discrete scatterers, a cholesteric corresponds to continuous scatterer distribution (dielectric tensor).

Scattering Cross Section. Examine light transmission and reflection for a uniform plane-parallel plate of a cholesteric with the optical axis normal to the surface (Fig. 2.1). To simplify the presentation and neglect light reflection and refraction from the boundary due to different refraction indices of the sample and the outer medium, we assume that the medium outside the crystal is homogeneous and its dielectric constant is equal to the mean dielectric constant $\bar{\epsilon}$ of the cholesteric. In our approximation, we neglect primary beam intensity variations and polarization characteristics in the sample. Thus, the problem reduces to the determination of scattered beam intensity and polarization characteristics as a function of the direction of primary beam propagation and polarization. These characteristics, for a monochromatic primary beam, are described by the differential (with respect to angle) scattering cross section.

In the kinematical approximation, we obtain from (2.1) and (2.2) that the scattering cross section of light for a cholesteric sample is described by [6]

$$\frac{d\sigma(\mathbf{k}_0, \mathbf{e}_0; \mathbf{k}_1, \mathbf{e}_1)}{d\Omega_{\mathbf{k}_1}} = \left(\frac{\omega^2}{4\pi c^2}\right)^2 \left| \int [\mathbf{e}_1^*(\hat{\epsilon} - \bar{\epsilon})\mathbf{e}_0] \exp[i(\mathbf{k}_0 - \mathbf{k}_1)\mathbf{r}] d\mathbf{r} \right|^2 \qquad (2.14)$$

where $\hat{\epsilon}(\mathbf{r})$ is the dielectric permittivity tensor of the cholesteric and $\mathbf{k}_0$, $\mathbf{k}_1$, $\mathbf{e}_0$, $\mathbf{e}_1$ are the wave and polarization vectors of the incident and scattered waves. Integration is over sample volume. Expression (2.14) may be transformed using the Fourier expansion of $\hat{\epsilon}(\mathbf{r})$

$$\hat{\epsilon}(\mathbf{r}) = \sum_{s=0,\pm 1} \hat{\epsilon}_s \exp[is\tau\mathbf{r}] \qquad (2.15)$$

where

$$\hat{\epsilon}_0 = \begin{pmatrix} \bar{\epsilon} & 0 & 0 \\ 0 & \bar{\epsilon} & 0 \\ 0 & 0 & \epsilon_3 \end{pmatrix}, \quad \hat{\epsilon}_1 = \epsilon_{-1}^* = \begin{pmatrix} 1 & \mp i & 0 \\ \mp i & -1 & 0 \\ 0 & 0 & 0 \end{pmatrix} \cdot \frac{\bar{\epsilon}\delta}{2}.$$

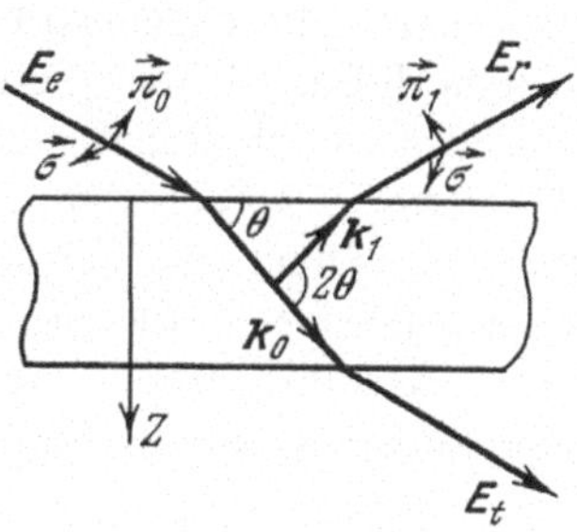

Fig. 2.1 Geometry of diffraction scattering for a planar texture of a cholesteric

Here τ is the reciprocal lattice vector of the cholesteric. Substituting (2.15) into (2.14) produces

$$\frac{d\sigma(\mathbf{k}_0, \mathbf{e}_0; \mathbf{k}_1, \mathbf{e}_1)}{d\Omega_{\mathbf{k}_1}} = \left(\frac{\omega^2}{4\pi c^2}\right)^2 \left|\sum_s \mathbf{e}_1^* \hat{\epsilon}_s \mathbf{e}_0 \int \exp[i(\mathbf{k}_0 - \mathbf{k}_1 + s\tau] \mathrm{d}r\right|^2 . \quad (2.16)$$

In the limit of an infinite sample, the integral in (2.16) is proportional to the delta function $\delta(\mathbf{k}_0 - \mathbf{k}_1 + s\tau)$ [cf. (2.5)] and the scattering cross section acquires the form known in the theory of X-ray and neutron scattering [1, 2]. In particular, scattering directions are determined by the kinematical Bragg condition (2.4) which gives no information about the intensity or polarization properties of scattering. This condition is also written in the form

$$\sin\theta = \frac{s\lambda}{p} \quad (2.17)$$

where 2θ is scattering angle (Fig. 2.1) and λ is light wavelength. If the Bragg condition is violated, the scattered wave has zero intensity—there is no scattering. The intensity and polarization characteristics of scattering depend on the details of sample structure and a structure factor, the coefficient before the integral in expression (2.16), whereas X-ray scattering amplitude dependence on crystal structure is reflected in structure amplitude. Physically, structure amplitude refers to the amplitude of scattering by the crystal unit cell. For cholesterics, an analogue of the X-ray, structure amplitude is the quantity

$$F(\mathbf{k}_0,\ \mathbf{e}_0; \mathbf{k}_1,\ \mathbf{e}_1) = \mathbf{e}_1^* \hat{\epsilon}_s \mathbf{e}_0 \quad (2.18)$$

which describes the scattering amplitude from a cholesteric layer with thickness equal to a half-pitch of the cholesteric helix.

The simple expressions given above are the basis for a series of qualitative features observed in optics of cholesterics [7]: Expression (2.16) represents cholesteric color dependence on the angle of observation and its change with the temperature; there are diffraction-reflected waves of different frequencies by different observation angles. In particular, for normal incidence, diffraction reflection is experienced by light with wavelength $\lambda = p$. Figure 2.2 (normal incidence) illustrates the situation where k_0 and k_1 are of opposite direction, that is, selective reflection of normally incident light is a particular case of diffraction reflection. The temperature-related variations in cholesteric color explained by (2.16) correspond to the change in the wavelength of light diffracted from the cholesteric which, in turn, relates to the temperature dependence of cholesteric pitch.

Polarization Characteristics. Expression (2.16) also delineates the polarization dependence of light reflection and transmission in a cholesteric. To see this, we analyze the dependence of structure amplitude $F(\mathbf{k}_0, \mathbf{e}_0, \mathbf{k}_1, \mathbf{e}_1)$ on polarization vectors $\mathbf{e}_0$ and $\mathbf{e}_1$. As follows from expression (2.15), $\hat{\epsilon}_s = 0$ for $|s| \geq 1$; thus, in the kinematical approximation, only first-order reflection occurs (it follows from (2.15) and (2.18)

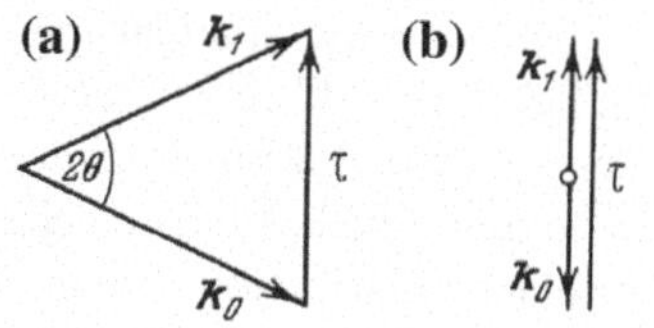

Fig. 2.2 Illustration of the diffractive nature of selective reflection of light in a cholesteric. The Bragg condition for oblique (**a**) and normal (**b**) angles of incidence

that for higher-order reflection $F \equiv 0$). Amplitude $F(\mathbf{k}_0, \mathbf{e}_0, \mathbf{k}_1, \mathbf{e}_1)$ depends on incident wave polarization vector $\mathbf{e}_0$ for a certain polarization $\mathbf{e}_0^{\perp}$ it goes to zero. This implies that polarization $\mathbf{e}_0^{\perp}$ does not experience diffraction scattering, whereas polarization orthogonal to it $\mathbf{e}_0^s$ experiences the strongest diffraction scattering. Similarly, expression (2.18) for $F(\mathbf{k}_0, \mathbf{e}_0, \mathbf{k}_1, \mathbf{e}_1)$ depends on polarization vector $\mathbf{e}_1$. Amplitude, a function of $\mathbf{e}_1$, reaches its maximum for a certain polarization $\mathbf{e}_1^s$ and goes to zero for the orthogonal polarization $\mathbf{e}_1^s$, whereas the values of $\mathbf{e}_1^s$ and $\mathbf{e}_1^{\perp}$ are independent of incident wave polarization, $\mathbf{e}_0$. Therefore, scattered wave polarization $\mathbf{e}_1^s$ is independent of incident wave polarization.

To determine $\mathbf{e}_0^s$ and $\mathbf{e}_1^s$, expression (2.18) should be rewritten in explicit form. Let us choose as one of the polarization unit vectors σ in the direction normal to the scattering plane and write polarization vectors $\mathbf{e}_0$ and $\mathbf{e}_1$ in the form[1]

$$\mathbf{e}_0 = \boldsymbol{\sigma} \cos \alpha_0 + i\boldsymbol{\pi}_0 \sin \alpha_0; \quad \mathbf{e}_1 = \boldsymbol{\sigma} \cos \alpha_1 + i\boldsymbol{\pi}_1 \sin \alpha_1 \tag{2.19}$$

where vectors $\boldsymbol{\sigma}, \boldsymbol{\pi}_0, \mathbf{k}_0(\boldsymbol{\sigma}, \boldsymbol{\pi}, \mathbf{k}_1)$ form the right-handed triad. Then expression (2.18) becomes

$$F(\mathbf{k}_0,\ \mathbf{e}_0;\ \mathbf{k}_1,\ \mathbf{e}_1) = -\frac{\bar{\epsilon}\delta}{2}(\cos \alpha_0 \pm \sin \alpha_0 \sin \theta) \times (\cos \alpha_1 \pm \sin \alpha_1 \sin \theta). \tag{2.20}$$

Minimizing expression (2.20) with respect to α_0 shows that the cholesteric provides the diffraction scattering for elliptically polarized light. Parameter $\alpha_0 = \alpha_0^s$ for that polarization [see (2.19)] depends on angle θ and is given by the expression

$$\alpha_0^s = \pm \text{ arctg } (\sin \theta). \tag{2.21}$$

Light with orthogonal polarization $\alpha_0^{\perp} = \mp$ arc tag(sin θ) does not interact with the cholesteric. Scattered wave polarization also depends on the angle of incidence and corresponds to parameter $\alpha_1^s = \alpha_0^s$, i.e., is also determined by expression (2.21).

[1]Representation of polarization vectors in form (2.19) implies that, in the most general form of the polarization vector, $\mathbf{e} = (\sigma \cos \alpha + \pi e^{i\beta} \sin \alpha)e^{i\eta}$, we assume that the phase, η, which is not important here, is zero and $\beta = \pi/2$. This corresponds to a certain choice of orientation for the axes of the polarization ellipse. The choice is determined by the symmetry of the problem, namely one axis should be in the scattering plane while the other is normal to it.

Thus, the kinematical approach accounts for selective reflection and transmission of light with different polarization and its dependence on the angle of incidence; in the special case of normal incidence, $(\theta = \pi/2)$, circularly polarized light is selectively scattered. The left-handed cholesteric helix produces scattered light with left-handed circular polarization, for the right-handed helix—right-hand circular polarization. In the limit of scattering angle tending to zero, light with linear σ polarization is scattered with linear polarization, whereas that with π polarization is not scattered. In the general case, $0 < \theta < \pi/2$, elliptically polarized light experiences selective scattering and the scattered radiation is also elliptically polarized. In the kinematical approximation, polarization of scattered radiation is determined by angle θ and is independent of either the degree or type of primary beam polarization.

2.1.4 Kinematical Approximation Limitations

As stated, the kinematical approximation gives the clear physical picture of diffraction scattering and some useful relationships to describe the experimental results. However, the quantitative description is possible within this approximation, only in the situations where coherently scattering volumes are small to justify the neglect of radiation attenuation within coherently scattering volumes.

The limitations of the kinematical approximations may be illustrated on the case of a cholesteric. It is applicable either to very thin planar samples ($L\delta/p \ll 1$ where L is the sample thickness) or to mosaic samples with small perfect regions.

In thick crystals, multiple Bragg scattering becomes important and the kinematical approximation does not apply in quantitative analyses and even cannot be used for the explanation of some qualitative effects. Thus, the kinematical approximation accounts for the rotation of polarization plane associated with the difference in interaction between left- and right-handed polarization in cholesterics and the consequent differences in refraction index for each polarization. But the kinematical approach fails to explain the experimentally observed change in polarization plane rotation sign with frequency [7]. In addition, it does not rationalize higher-order diffraction reflection (for higher-order diffraction reflection the structure amplitude (2.18) goes to zero), frequency or angular width of the selective reflection bands for light in cholesterics and existence of frequency or angular range for light reflection of any polarization for the case of oblique incidence, etc. [7]. The above-mentioned phenomena cannot be explained within the kinematical theory because it accounts for only single scattering and these phenomena require more rigorous consideration of the interaction of light with cholesterics [8, 9]. The kinematical approximation also fails in the description of the interaction of other types of radiations (Mössbauer gamma radiation, neutrons, fast charged particles in crystals) with media. In each of these cases, it is necessary to find the exact or numerical solutions of the Maxwell equations (see Chap. 1) or to find approximate analytical solutions based on the approach similar to the X-ray dynamical theory.

2.2 Dynamical Theory

Now consider the interaction of an electromagnetic wave with a sample of three-dimensional periodic structure in the case where the system does not meet the conditions necessary to apply the kinematical approximation. Assume that the dielectric permittivity tensor $\hat{\epsilon}(\mathbf{r})$ is a three-dimensional periodic function of coordinates and that the magnetic permittivity of the medium is unity, $\hat{\mu} \equiv 1$. As we seek the approximate analytical solutions of the Maxwell equations, we assume that all amplitudes $\hat{\epsilon}_{\boldsymbol{\tau}}$ for $\boldsymbol{\tau} \neq 0$ ($\boldsymbol{\tau}$ is the reciprocal lattice vector) in the Fourier expansion of the dielectric permittivity tensor

$$\hat{\epsilon}(\mathbf{r}) = \sum_{\boldsymbol{\tau}} \hat{\epsilon}_{\boldsymbol{\tau}} \mathrm{e}^{i\boldsymbol{\tau}\mathbf{r}} \tag{2.22}$$

are small relative to amplitude $\hat{\epsilon}_0$. Such an assumption corresponds to many practical situations. The small parameter of the problem is $\sim 10^{-2}$ for optics of liquid crystals and $\sim 10^{-4}$–10^{-5} for Mössbauer and X-ray crystal optics. Thus, in most instances, the accuracy of this approximation satisfies the practical requirements of the experiment.

2.2.1 *Set of Dynamical Equations*

If the conditions of the previous section are met, it is possible to find from the Maxwell equations that an electric field vector for an arbitrary direction of light propagation in a periodic medium satisfies the following equation

$$\hat{\epsilon} \frac{\partial^2 \mathbf{E}}{\partial t^2} = -c^2 \mathrm{rot}\,\mathrm{rot}\,\mathbf{E} \tag{2.23}$$

where (2.22) defines $\hat{\epsilon}$. Since the medium is periodic, the solution of (2.23) is a Bloch wave [10]

$$\mathbf{E}(\mathbf{r},\, t) = \mathrm{e}^{i\mathbf{k}_0\mathbf{r} - i\omega t} \sum_{\boldsymbol{\tau}} \mathbf{E}_{\boldsymbol{\tau}} \mathrm{e}^{i\boldsymbol{\tau}\mathbf{r}}. \tag{2.24}$$

Substituting (2.24) into (2.23) produces the set of homogeneous equations for $E_{\boldsymbol{\tau}}$

$$- k_{\tau}^2 \mathbf{E}_{\boldsymbol{\tau}} + \frac{\omega^2}{c^2} \sum_{\boldsymbol{\tau}'}{}' \hat{\epsilon}_{\boldsymbol{\tau}-\boldsymbol{\tau}'} \mathbf{E}_{\boldsymbol{\tau}'} + (\mathbf{k}_{\boldsymbol{\tau}} \mathbf{E}_{\boldsymbol{\tau}}) \mathbf{k}_{\boldsymbol{\tau}} = 0 \tag{2.25}$$

where (2.22) defines $\mathbf{k}_{\boldsymbol{\tau}} = \mathbf{k}_0 + \boldsymbol{\tau}$ and $\hat{\epsilon}_{\boldsymbol{\tau}}$. The general properties of (2.23) are analyzed in numerous works (see, e.g., [10, 11]). Using Floquet's theorem, it is shown [12] that, for fixed frequency and angle of incidence, there are four k_0 values in a crystal for which (2.23) has non-trivial solutions of form (2.24).

Numerical methods used for solving system (2.25) provide any desired accuracy. However, for solution analysis, it is preferable to have analytical expressions, even approximate ones. Thus, (2.25) is solved below in the two-wave approximation of X-ray dynamical diffraction theory [4, 8, 9]. This approximation is based on the fact that if the Bragg condition (2.4) is met, the set of (2.25) has, as a rule, only two amplitudes $\mathbf{E}_0$ and $\mathbf{E}_\tau$ whose values are as large as that of incident wave; all remaining amplitudes are smaller by a factor of at least $|\epsilon_\tau|/|\epsilon_0|$.

The explicit form of the corresponding equations is given in the two-wave approximation, and we isolate from system (2.25) two vector equations for waves $\mathbf{E}_1(\mathbf{E}_\tau)$ and $\mathbf{E}_0$ with wave vectors $\mathbf{k}_0$ and $\mathbf{k}_1 = \mathbf{k}_\tau = \mathbf{k}_0 + \boldsymbol{\tau}$

$$\left(\hat{\epsilon}_0 - \frac{k_0^2 c^2}{\omega^2}\right)\mathbf{E}_0 + \hat{\epsilon}_\tau \mathbf{E}_1 = 0$$

$$\hat{\epsilon}_{-\tau}\mathbf{E}_0 + \left(\hat{\epsilon}_0 - \frac{k_1^2 c^2}{\omega^2}\right)\mathbf{E}_1 = 0. \qquad (2.26)$$

The accuracy of the approximation permits neglect of the small (order of $\frac{|\epsilon_\tau|}{|\epsilon_0|}$) non-transversality of the waves $\mathbf{E}_0$ and $\mathbf{E}_1$ in (2.26). Thus, it is possible to assume that they are orthogonal to $\mathbf{k}_0$ and $\mathbf{k}_1$, respectively.

We do not specify at this point, the form of (2.26) by writing the explicit form of Fourier-harmonics $\hat{\epsilon}(r)$ for the interaction of radiation with concrete periodic medium. Notice that (2.26) and the solutions of the corresponding boundary problems have been studied extensively (see, e.g., [7, 8]) for the simplest case of scalar dielectric permittivity $\epsilon(r)$ which corresponds, e.g., to X-ray diffraction from crystals. In this instance, polarization separation occurs and vector (2.26) for E_0 and E_1 separate into two uncoupled sets of two equations for scalar amplitudes E_0 and E_1. In the case where the medium is characterized by tensor dielectric permittivity $\hat{\epsilon}(r)$, the solutions of (2.26) and the corresponding boundary conditions are complicated and have non-trivial polarization properties. This results in qualitatively new effects for media with structure complexity unobserved in diffraction optics of simple periodic media, e.g., the complicated polarization structure of forbidden bands [7] and some specific features of the Borrmann effects in such media [4].

2.2.2 *Dispersion Surfaces*

We describe in general form the solution of the set of dynamical equations and boundary problems, in the two-wave approximation, of dynamical diffraction theory. The solution of the set of (2.26) determines the eigenwaves, 1.4., amplitudes $\mathbf{E}_0$ and $\mathbf{E}_1$ in expansion (2.24) and their dependence on the parameters of the problem—frequency ω, the direction of wave propagation, etc., and the dependence of wave vectors $\mathbf{k}_0$ and $\mathbf{k}_1$ of two plane waves in this expansion on those parameters.

Possible values of wave vectors $\mathbf{k}_0$ and $\mathbf{k}_1$ are determined from the solvability condition (2.26), i.e., the zero value of its determinant. Different values of amplitudes $\mathbf{E}_0$ and $\mathbf{E}_1$ in the eigensolution correspond to different allowed values of $\mathbf{k}_0$ and $\mathbf{k}_1$. Since (2.26) is a set of four linear homogeneous equations, the condition of its solvability and the relation between $\mathbf{k}_0$ and $\mathbf{k}_1$ established by the Bragg condition (2.4) yield, in the general case, four different relationships between $\mathbf{k}_0$ and $\mathbf{k}_1$ which correspond to four different eigensolutions of form (2.24). In contrast to the kinematical approach, which under diffraction conditions [the Bragg condition (2.4) plus condition $|\mathbf{k}_0| = |\mathbf{k}_1|$] uniquely determines vectors $\mathbf{k}_0$ and $\mathbf{k}_1$, the dynamical approach permits some variations for $\mathbf{k}_0$ and $\mathbf{k}_1$ under diffraction conditions not yielding single values of k_0 and k_1 as in the kinematical approximation. The general solution of the set of (2.26) is a linear combination of all four eigensolutions of the system (2.26)

$$\mathbf{E}(\mathbf{r},\ t)_j = (\mathbf{E}_{0j}\mathrm{e}^{i\mathbf{k}_{0j}\mathbf{r}} + \mathbf{E}_{1j}\mathrm{e}^{i\mathbf{k}_{1j}\mathbf{r}})\mathrm{e}^{-i\omega t}$$
$$\mathbf{E}(\mathbf{r}, t) = \sum_{j=1} C_j\mathbf{E}(\mathbf{r},\ t)_j \tag{2.27}$$

where coefficients C_j and quantities $\mathbf{K}_{0j}$ and $\mathbf{k}_{1j}$ in each eigenwave are determined by the boundary conditions.

It is usually said that the values of $\mathbf{k}_0$ and $\mathbf{k}_1$ (consistent with the Bragg condition and Maxwell equations) determine the dispersion surfaces [8, 9] which are the geometric loci of the ends of vectors $\mathbf{k}_0$ and $\mathbf{k}_1$ (Fig. 2.3). These surfaces are used to solve the following boundary problem. Dispersion surfaces are shown in Fig. 2.3 in the simplest case. There are four branches of the dispersion surface since the condition of zero determinant for the set (2.26) reduces to an equation of degree four. However, two of the four branches are shown for clarity. If any point of the dispersion curves is connected with the ends of the reciprocal lattice vector $\boldsymbol{\tau}$ (Fig. 2.3), the values of wave vectors $\mathbf{k}_0$ and $\mathbf{k}_1$ in expansion (2.24) satisfy both the Bragg condition and Maxwell equation.

Dispersion curves describe propagating modes. This means that wave vectors on the dispersion curve are real. In the space between the branches of the dispersion

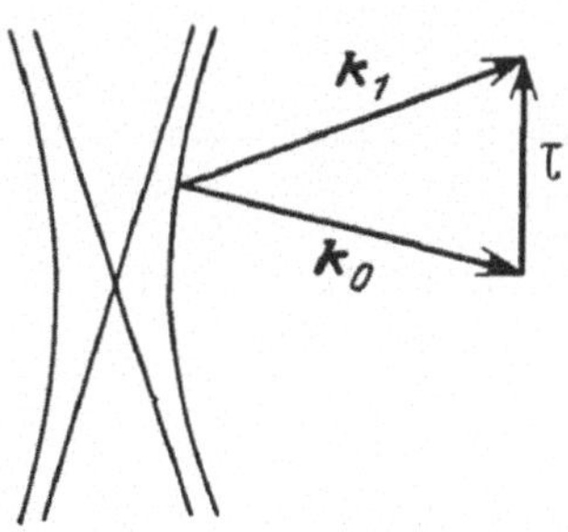

Fig. 2.3 Dispersion surface

surfaces, there are no propagating modes. The corresponding regions of k-space represent forbidden bands. In these regions, wave vectors $\mathbf{k}_0$ and $\mathbf{k}_1$ of the Maxwell equation solution are complex quantities.

We determine wave vectors $\mathbf{k}_0$ and $\mathbf{k}_1$ in the eigensolution "excited" in a sample by an incident electromagnetic wave with dispersion equations and graphical construction. According to the wave vector boundary conditions, their tangential components on the sample surface are continuous. Thus, the difference between the wave vectors inside and outside the sample is directed along the surface normal. Hence, the following method is developed to seek eigensolutions "excited" in the sample for a given direction of wave propagation outside the sample (Fig. 2.4). Draw a straight line normal to sample surface S from the end of vector q corresponding to the wave vector outside the sample. The points of line intersection with the dispersion surfaces determine the values of wave vectors $\mathbf{k}_0$ and $\mathbf{k}_1$ of the eigensolutions "excited" in the sample.

In special cases (e.g., for surface normals to $\boldsymbol{\tau}$), it is possible that the surface normal drawn through the end of vector q does not intersect the dispersion surface—the incident wave cannot "excite" propagating modes. This case corresponds to diffraction reflection of light from the sample. This is an illustration of the search for eigensolutions "excited" in a crystal with specific boundary conditions.

2.2.3 *Solution of the Boundary Problem*

In practice, an analytical method that does not use explicitly dispersion surfaces is more convenient when solving the boundary problem. This approach uses boundary conditions to establish relationships between wave vector $\mathbf{q}$ of the wave outside the sample and wave vectors $\mathbf{k}_0$ and $\mathbf{k}_1$ inside it. Then the solution of the dispersion equation is sought for a fixed value of q, on the condition that the determinant of system (2.26) is zero. The result describes the superposition of eigensolutions "excited" in the sample as a function of the parameters of the wave outside the sample, e.g., its angle of incidence or frequency (for the fixed angle of incidence).

Let a plane monochromatic wave $\mathbf{E}^e e^{i\mathbf{qr}}$ be incident onto a crystal surface in the shape of a plate of finite thickness. For simplicity, we neglect light refraction and reflection at the sample boundary associated with different refraction indices of two

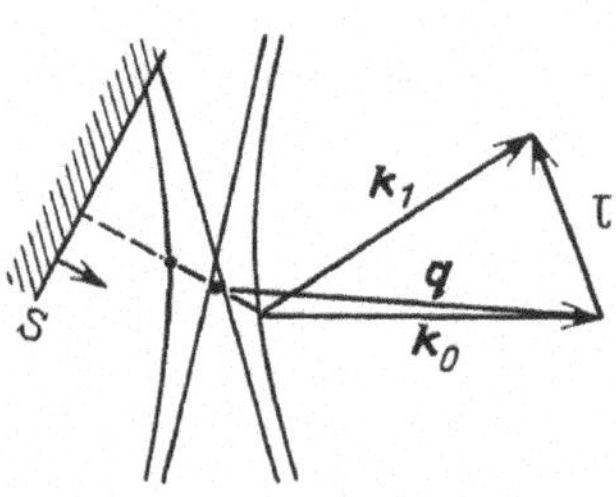

Fig. 2.4 Dispersion surface and solution of the boundary problem (**S** is the surface normal, **q** is the wave vector outside the sample)

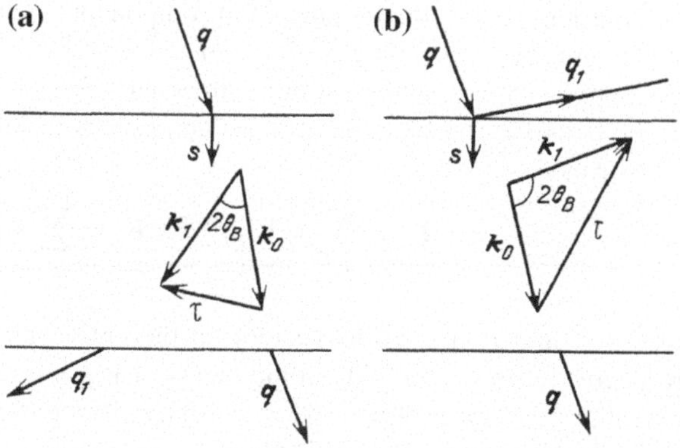

Fig. 2.5 Diffraction geometry for the Bragg (**b**) and Laue (**a**) cases

media and assume that the wave falls from a medium with dielectric permittivity that coincides with the mean dielectric permittivity $\bar{\epsilon}$ of the crystal.

Again, we assume that the sample is a plane-parallel plate. Accounting for the continuity of the tangential components of wave vectors k_0 and k_1 at the boundary, we write (see Fig. 2.5)

$$\mathbf{k}_0 = \mathbf{q} + \mathbf{q}\xi s, \ \mathbf{k}_1 = \mathbf{k}_0 + \boldsymbol{\tau} \tag{2.28}$$

where s is the inward surface normal and ξ is the small quantity of the order of ϵ_τ. We determine the relation between the absolute values of wave vectors $\mathbf{k}_0$ and $\mathbf{k}_1$ and the angle of incidence $\gamma_0 = \hat{\mathbf{q}, \mathbf{s}} \approx \hat{\mathbf{k}_0, \mathbf{s}}$. We represent the vectors in the form

$$k_0 = \mathbf{q}(1 + \eta_0) \ , \ k_1 = \mathbf{q}(1 + \eta_1). \tag{2.29}$$

From formulae (2.28), (2.29), and linear terms in small parameter ξ, we obtain the relation between η_1 and η_0 as

$$\eta_1 = 0/2 + \eta_0/b \tag{2.30}$$

where $b = \cos\gamma_0/\cos\gamma_1$, $\gamma_1 = \hat{\mathbf{k}_1 \mathbf{s}}$, $\alpha = \boldsymbol{\tau}(\boldsymbol{\tau} + 2\mathbf{q})/\mathbf{q}^2$ and parameter ξ obeys the equation $\eta_0 = \xi \cos\gamma_0$. Notice that parameter α is the measure of angular (frequency) deviation from the Bragg condition. Substituting (2.29) and (2.30) into the coefficients of system (2.26) and equating its determinant to zero yield the dispersion equation which determines η_0 and therefore $\mathbf{k}_0$ and $\mathbf{k}_1$ as a function of parameter α. In the general case [for an arbitrary dielectric permittivity tensor $\hat{\epsilon}(\mathbf{r})$], the dispersion equation is a fourth-degree equation. For particular forms of $\hat{\epsilon}(\mathbf{r})$ [4, 8, 9, 13] and specific scattering geometries [4, 7], the equation becomes biquadratic. We present cases such as these below.

Upon finding the eigenvalues "excited" in the sample as functions of α, from the solutions of dispersion equations, solution of the boundary problem reduces to the determination of combined coefficients C_j in (2.27). They are determined from the continuity condition for tangential components of the electric and magnetic fields at the boundary (procedure similar to that used for solution of the boundary problem in Chap. 1—corresponding solution in general form is given in [4, 7]). The solution of the boundary problem determines amplitudes of diffracted, $\mathbf{E}^r$, and transmitted, $\mathbf{E}^t$, waves

$$\mathbf{E}^r = \sum_{j=1}^{4} C_j \mathbf{E}_{1j} e^{i\mathbf{k}_{1j}\mathbf{r}_e}, \ \mathbf{E}^t = \sum_{j=1}^{4} C_j \mathbf{E}_{0j} e^{i\mathbf{k}_{0j}\mathbf{r}_t} \tag{2.31}$$

where $\mathbf{r}_e$ and $\mathbf{r}_t$ are radius-vectors on the entry and exit surfaces of the sample, respectively. In (2.31), only components of $\mathbf{k}_{1j}$ and $\mathbf{k}_{0j}$, normal to the surface, are essential in scalar products, $\mathbf{k}_{1j}\mathbf{r}_e$ and $\mathbf{k}_{0j}\mathbf{r}_t$, since their tangential components are equal for all eigenwaves (all j values).

The general scheme of the solution of the dynamical system (2.26) and boundary problem presented here is extended in Chap. 3 to the description of the optics of chiral liquid crystals for an arbitrary light propagation direction and is extended in [13] to interaction between various types of radiation including neutrons and Mossbauer radiation for specific media with complicated spatial structure (magnetically ordered structures, some liquid crystals, etc.). One approach to various problems demonstrates diffraction scattering property dependence (in different media) on features of radiation interaction with the media, on the one hand, and the common physical nature of that interaction for different types of radiation, on the other. It should be noted that recently a lot of attention was paid, especially, to the periodic media with the periodicity scale in the range of optical wavelength value. The new term "photonic crystals" was commonly accepted for these media to distinguish them from the conventional crystals with the periodicity scale in the range of X-Ray wavelength value. The numerous published on the subject books (see, for example [14–17]) demonstrate that the common for all periodical media effects of radiation interaction with the media reveal themselves especially strong in the photonic crystals (as an illustrative example may be named the flourishing nonlinear optics of photonic crystals in contrast to the X-Ray nonlinear optics, where up to now practically there were no observations of the nonlinear effects).

References

1. R.W. James, *The Optical Principles of the Diffraction of X-rays* (Bell and Sons, London, 1967)
2. I.I. Gurevich, L.V. Tarasov, *Physics of Low-Energy Neutrons* (North-Holland, Amsterdam, 1968)
3. R.B. Hirsch, A. Howie, B.B. Nicholson, D.W. Pashley, M.J. Whelan, *Electron Microscopy of Thin Crystals* (Butterworth, London, 1965)
4. V.A. Belyakov, UFN **115**, 553 (1975) [Sov. Phys.-Usp. **18**, 267 (1975)]
5. V.B. Berestetskii, E.M. Lifshitz, L.P. Pitaevskii, *Relativistic Quantum Theory, Part 1* (Nauka, Moscow, 1968). (in Russian)
6. L.D. Landau, E.M. Lifshitz, *Course of Theoretical Physics, vol. 8, Electrodynamics of Continuous Media* (Pergamon Press, Oxford, New York, 1960)

7. V.A. Belyakov, A.S. Sonin, *Optics of Cholesteric Liquid Crystals* (Nauka, Moscow, 1982). (in Russian)
8. Z.G. Pinsker, *Dynamical Scattering of X-rays in Crystals* (Springer Verlag, Berlin, 1978)
9. B.W. Batterman, H. Cole, Rev. Mod. Phys. **36**, 681 (1964)
10. L. Brilluen, M. Parodi, *Propagation des Ondes dans les Milieux Periodicues* (Paris, 1956)
11. C. Elachi, O. Yeh, J. Opt. Soc. Am. **63**, 840 (1973)
12. V.A. Belyakov, V.E. Dmitrienko, UFN **146**, 369 (1985) [Sov. Phys.-Usp. **28**, 535 (1985)]
13. V.A. Belyakov, Diffraction Optics of Complicatedly Structured Periodic Media. (Moscow, Nauka, 1988) in Russian; V.A. Belyakov Diffraction Optics of Complex Structured Periodic Media (Spirnger, New York, 1992) (Revised translation from Russian)
14. J.D. Joannopoulus, R.D. Meade, J.N. Winn, Photonic Crystals: Moulding the Flow of Light (Prinston University Press, 1995)
15. V.F. Shabanov, S. Ya. Vetrov, A.V. Shabanov, Optics of Real Photonic Crystals. 2005, Novosibirsk, SD RAN Publishing House (in Russian)
16. B.I. Mantsysov, *Coherent and Nonlinear Optics of Photonic Crystals* (FIZMATLIT, Moscow, 2009). (in Russian)
17. Liquid Crystal Microlasers, in *Transwold Research Network*, ed. by L.M. Blinov, R. Bartolino (Kerala, India, 2010)

Chapter 3
Optics of Chiral Liquid Crystals

In Chap. 2, we considered the interaction between a short wave radiation and crystals within the framework of diffraction theory. Here, we demonstrate that the same approach describes the interaction between optical (long wave) radiation and periodic media of complex structure, specifically chiral liquid crystals.

Different types of liquid crystals have very interesting and useful applications which exploit their optical properties [1–4]. The most unusual and sophisticated optical properties are found in chiral liquid crystals [1–4] the main representatives of which are cholesteric liquid crystals (or cholesterics), chiral smectic liquid crystals, and the blue phases of liquid crystals.

Although these types of liquid crystals have different structures and properties, all consist of mirror-asymmetric (chiral) molecules and exhibit spatial periodicity with periods generally in the range of visible light wavelengths. This explains the observation that the periodicity of these liquid crystals manifests itself most prominently in their optical properties, typically those due to light diffraction.

3.1 Optics of Cholesteric Liquid Crystals (CLC)

In Chaps. 1 and 2, we presented the optical properties of CLCs in the general discussion of periodic media optics. Those properties were considered in the context of the exact solution of the Maxwell equations for light propagating along the optical axis and within the kinematical approximation for an arbitrary propagation direction.

Since the purpose of those considerations was to emphasize the general features of periodic media optics, only some of the interesting properties of CLC optics were mentioned. Here, we present a detailed consideration of the problem and begin with a description of the salient facts of CLC optics.

V. Belyakov, *Diffraction Optics of Complex-Structured Periodic Media*, Springer Series in Optical Sciences 203,
https://doi.org/10.1007/978-3-319-43482-7_3

Cholesteric liquid crystals selectively reflect light of definite polarization and wavelength. For example, when a light beam is incident on a CLC along its optical axis, the beam component with some circular polarization is reflected in a relatively narrow frequency bandwidth while the other (orthogonal) polarization penetrates into the crystal without reflection. The color of a CLC depends on the angle at which one looks at the crystal. The optical peculiarities of CLCs are connected with their structure shown in Fig. 1.3 for a single crystal sample; the long axes of cholesteric molecules are shown as oblong marks.

It was observed that the light polarization planes rotate by several thousand degrees per millimeter in thin layers; this is far above the natural optical activity of molecules. The wavelength at which these anomalous events occur changes in the presence of an external field (electric or magnetic) or temperature variance.

In any plane perpendicular to the optical axis Z, all molecules are similarly oriented, however, but there is no long-range order at the molecular centers of gravity. The orientation of the molecule's axes varies with coordinate Z as

$$\varphi(z) = \frac{2\pi z}{P} \tag{3.1}$$

where $\varphi(z)$ is the angle of molecular rotation around the Z-axis. The quantity P is the period or pitch of a cholesteric helix—at that distance, the molecular axis rotates by 2π. The value of P is usually several thousand angstreams.

The structure shown in Fig. 1.3 of cholesteric crystals is idealized. Specifically, for a fixed value of Z, the orientations of individual molecules may slightly differ from that determined by (3.1), which gives the Z dependence of the mean orientation of molecules. This direction is usually described by unit vector $\mathbf{n}$, the director. The degree of orientational ordering of molecules is characterized by the order parameter [3].

$$S = (3\langle\cos^2\theta\rangle - 1)/2 \tag{3.2}$$

where $\langle\cos^2\theta\rangle$ is the mean square cosine of the angle of deviation of the molecular long axis from the director. Hence, (3.1) describes the spatial variation of the director's orientation which, at $S \neq 1$, coincides with average molecular orientation.

The principal values of the CLC dielectric permittivity tensor $\hat{\varepsilon}(\mathbf{r})$ (1.6) are determined by molecular dielectric anisotropy and depend on the order parameter, so that in an isotropic liquid with $S = 0$ the parameter δ (dielectric anisotropy) vanishes [5]. Development of a quantitative theory of CLC dielectric permittivity and other types of liquid crystals encounters difficulties [5]. Usually, it is assumed that the anisotropy δ is proportional to S and molecular anisotropy, while the proportionality factor remains without precise determination [3].

3.1.1 The Fundamental Equations

Consider light propagation in a CLC at an arbitrary angle to the CLC optical axis. This case is difficult for analysis because there is no exact solution to the Maxwell equations. Thus, in a series of papers, the solutions for light propagation in an arbitrary direction with respect to cholesteric axis were analyzed in general, whereas the solutions were obtained by numerical calculations for specific values of the parameters (see [1]). However, there is a small parameter, the anisotropy of dielectric properties δ, which makes it possible to use the simple and exact theory described in Chap. 2.

Oblique incidence in CLC optics qualitatively differs from that of normal incidence (propagation along the optical axis); there are higher-order reflections at frequencies which are multiples of the Bragg frequency; also, first-order reflection differs from that of normal incidence: A region appears in which any polarization is reflected and the polarization properties of solutions become more complex. The physical cause for this is that when light propagates at an angle to the helical axis we must account for both the diffractive scattering of light with the polarization given by (2.21) and birefringence which affects the polarization properties of the solutions. Hence, the results of kinematic theory, according to which diffraction scattering occurs for light whose polarization is defined by (2.21), become invalid for thick crystals and the diffraction pattern of light obliquely incident on a CLC becomes more complicated.

The more exact description of CLC light diffraction in the range of the first-order reflexes (considered within the kinematic theory in Chap. 2) requires that we use the system of dynamic equations in the two-wave approximation (2.26) and substitute into it the explicit form of the CLC dielectric tensor (1.6). To solve the system (2.26), it is convenient to reduce it to a form which is used routinely in the dynamic theory of diffraction:

$$
\begin{aligned}
&[1 - k_0^2/q^2 + (\delta/2)\cos^2\theta]E_0^\sigma - (\delta/2)E_1^\sigma + (i\delta/2)\sin\theta E_1^\pi = 0\\
&[1 - k_0^2/q^2 - (\delta/2)\cos^2\theta]E_0^\pi - (i\delta/2)\sin\theta E_1^\sigma - (\delta/2)\sin^2\theta \cdot E_1^\pi = 0\\
&-(\delta/2)E_0^\sigma + (i\delta/2)\sin\theta \cdot E_0^\pi + [1 - k_1^2/q^2 + (\delta/2)\cos^2\theta]\cdot E_1^\sigma = 0\\
&-(i\delta/2)\sin\theta \cdot E_0^\sigma - (\delta/2)\sin^2\theta \cdot E_0^\pi + [1 - k_1^2/q^2 - (\delta/2)\cos^2\theta]\cdot E_1^\pi = 0
\end{aligned}
\tag{3.3}
$$

where E^σ and E^π are the σ and π components of amplitudes $\mathbf{E}_0$ and $\mathbf{E}_1$, respectively, and $q = (\omega/c)[\bar{\epsilon}(1 - \frac{\delta}{2}\cos^2\theta)]^{1/2}$ is the mean value of the vector of the wave propagating at an angle $\frac{\pi}{2} - \theta$ to the optical axis.

Equating the determinant of (3.3) to zero yields the conditions for the system to be solvable in a form similar to (1.14):

$$
t^4 - (2\Delta^2 + 2m^2 - 1)t^2 + \Delta^4 - (2m^2 + 1)\Delta^2 + 2m^2\Delta = 0 \tag{3.4}
$$

where we assume

$$t = (k_1^2 - k_0^2)[q^2\delta(1+\sin^2\theta)]^{-1},\ m = \cos^2\theta(1+\sin^2\theta)^{-1}$$
$$\Delta = (2q^2 - k_0^2 - k_1^2)[q^2\delta(1+\sin^2\theta)]^{-1}. \quad (3.5)$$

3.1.2 Eigensolutions

We use the method described in Sect. 2.2 of Chap. 2 to solve the boundary problem, and we consider the sample a plane-parallel layer as above. When the optic axis is perpendicular to the CLC surface, expressions (2.28)–(2.30) show that the parameter in (3.4) coincides within α, to within a factor, and is associated with the deviation of the incidence angle (or wavelength) from the value given by the Bragg condition (2.4):

$$\Delta = 2\sin\theta(2q\sin\theta - \tau)[\delta q(1+\sin^2\theta)]^{-1}. \quad (3.6)$$

The quantity Δ is a convenient parameter to describe the boundary problem and the CLC optical properties. We must find the dependence of all optical parameters, particularly eigenvalue dependence on Δ. The dependence of Δ on t is given by (3.4) as

$$t_j = \pm\left\{\Delta^2 + m^2 - 1/2 \pm [(\Delta^2 + m^2 - 1/2)^2 + (2m^2+1)\cdot\Delta^2 - 2m^2\Delta - \Delta^4]^{1/2}\right\}^{1/2} \quad (3.7)$$

Solutions t_j are enumerated as in the case of normal incidence [see (1.15)]; that is, if the sign in the square brackets is "plus," then $j = 1, 4$, respectively, for signs "plus" and "minus" in the braces; if the sign in the square brackets is "minus," then $j = 2$, 3, respectively, for signs "plus" and "minus" in the braces. The value of t given by (3.7) is similar to η of (1.14).

The eigensolutions of (3.3) are expressed by t_j

$$\mathbf{E}_j(\mathbf{r}, t) = e^{-i\omega t}(\mathbf{E}_{0j}e^{i\mathbf{k}_{0j}\mathbf{r}} + \mathbf{E}_{1j}e^{i\mathbf{k}_{1j}\mathbf{r}}) \quad (3.8)$$

where $\mathbf{E}_{0j} = E^{\sigma}_{0j}.\boldsymbol{\sigma} + E^{\pi}_{0j}\boldsymbol{\pi}$, $\mathbf{E}_{1j} = E^{\sigma}_{1j}\boldsymbol{\sigma} + E^{\pi}_{1j}\boldsymbol{\pi}_1$,

$$E^{\sigma}_{0j} = a_{1j} = (\Delta - m + t_j)[(\Delta - t_j)^2 - m^2],$$
$$E^{\pi}_{0j} = a_{2j} = -i\sin\theta(\Delta + m + t_j)[(\Delta - t_i)^2 - m^2] \quad (3.9)$$
$$E^{\sigma}_{1j} = a_{3j} = (\Delta - m - t_j)(\Delta + t_j - m^2)$$
$$E^{\pi}_{1j} = a_{4j} = -i\sin\theta(\Delta + m - t_j)(\Delta + t_j - m^2)$$
$$\mathbf{k}_{0j} = \mathbf{q} + qs\delta(\Delta + m + t_j)[2(1+m)\sin\theta]^{-1}, \mathbf{k}_{1j} = \mathbf{k}_{0j} + \boldsymbol{\tau}. \quad (3.10)$$

We note that (3.10) was obtained using the boundary conditions for wave vectors (the continuity of their tangential components for the case when the CLC surface is perpendicular to the CLC optical axis). Here, **q** is the incident wave vector, and s is the vector normal to the crystal surface.

The eigensolutions of (3.8) are the superpositions of two waves with wave vectors $\mathbf{k}_{0j}$ and $\mathbf{k}_{1j}$, the polarization of which is generally elliptical.

The frequency dependence of polarization parameters for waves 1 and 2 is presented in Fig. 3.1. In contrast to normal incidence in which eigenpolarizations are circular, eigenpolarizations in oblique incidence are elliptical and vary noticeably within the selective reflection band range [6].

The rocking curves (angular dependence of the reflection coefficient) for linear polarizations have plateau-like shapes as in X-ray diffraction, and the asymmetry is a manifestation of eigenpolarization angular (frequency dependence) which exhibits maximal diffractive reflection (see also the analogous measured frequency dependences at Fig. 3.11).

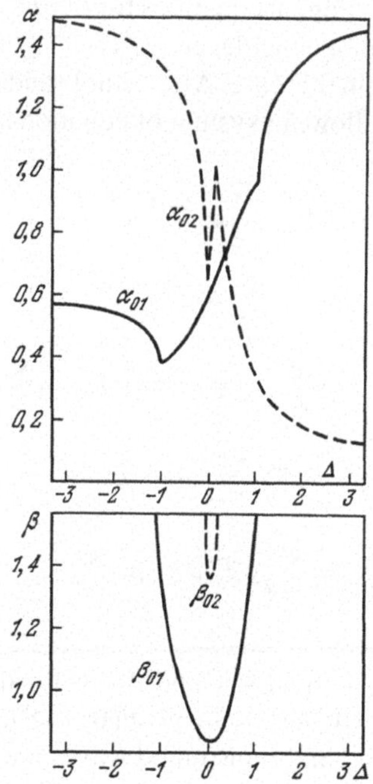

Fig. 3.1 Frequency (angular) dependences of eigensolution polarization parameters for the case of oblique light propagation with respect to the cholesteric axis. The parameters α_j and β_j describe the eigenpolarization vectors in the representation $n_j = \bar{\sigma}\cos\alpha_j + \bar{\pi}\sin\alpha_j\exp(i\beta_j)$. The orientation of the polarization ellipse with respect to $\bar{\sigma}$ is given by ψ, and the ratio of its axes is given by $b = tg\alpha'$ where $tg\,2\psi_j = tg\,2\alpha_j\cos\beta_i$; $\sin 2\alpha' = \sin 2\alpha_j \sin\beta_j$. The calculations were done for the following values of the parameters: $\theta = \pi/4, \bar{\epsilon} = 2.445, \delta = 0.05, p = 0.78\,\mu\text{m}$

3.1.3 The Solution to the Boundary Problem

The general solution to (3.3) is the superposition of eigensolutions of type (2.27). If the dielectric constant of the medium adjacent to the CLC boundary is equal to the CLC mean dielectric constant, the problem is solved as above.

First, we find the amplitudes of the reflected and transmitted waves. Let a wave of amplitude E^{e} and polarization **e** be incident from outside the sample surface located at $z = z_1$ and

$$\mathbf{E}^{\mathrm{e}} = \mathbf{e}E^{\mathrm{e}}\mathrm{e}^{i(\mathbf{qr}-\omega t)} \tag{3.11}$$

where

$$\mathbf{e} = \boldsymbol{\sigma}\cos\alpha + \boldsymbol{\pi}_0 \sin\alpha \mathrm{e}^{i\beta}. \tag{3.12}$$

We seek the amplitudes of the reflected and transmitted waves in the form $\mathbf{E}^{\mathrm{r}} = E^{\mathrm{r}}_{\sigma}\boldsymbol{\sigma} + E^{\mathrm{e}}_{\pi}\boldsymbol{\pi}_1$ and $\mathbf{E}^{\mathrm{t}} = E^{\mathrm{t}}_{\sigma} + E^{\mathrm{t}}_{\pi}\boldsymbol{\pi}_0$, respectively, while the field within the sample has the form (2.27). The continuity condition of components of **E** and **H** on the crystal's surface produces a system of equations for E^{r}_{σ}, E^{r}_{π}, E^{t}_{σ} and E^{t}_{π}. The boundary conditions for the fields in oblique incidence are as (1.17), except that only tangential components must be equated in this case. After simple and cumbersome calculations, we obtain, from (1.17) the following system of equations:

$$\sum_j a_{1j}C'_j = E^{\mathrm{e}}_{\sigma},\ \sum_j a_{2j}C'_j = E^{\mathrm{e}}_{\pi}$$

$$\sum_j a_{3j}C'_j\mathrm{e}^{i\Delta k_j L} = 0,\ \sum_j a_{4j}C'_j\mathrm{e}^{i\Delta k_j L} = 0$$

$$E^{\mathrm{r}}_{\sigma} = \sum_j a_{3j}C'_j\mathrm{e}^{i\tau z_1},\ E^{\mathrm{r}}_{\pi} = \sum_j a_{4j}C'_j\mathrm{e}^{i\tau z},$$

$$E^{\mathrm{t}}_{\sigma} = \mathrm{e}^{i\mathbf{qse}}\sum_j a_{1j}C'_j\mathrm{e}^{i\Delta k_j L};$$

$$E^{\mathrm{t}}_{\pi} = \sum_j \mathrm{e}^{i\mathbf{qs}L}a_{2j}C'_j\mathrm{e}^{i\Delta k_j L} \tag{3.13}$$

where $C'_j = C_j\mathrm{e}^{i\mathbf{k}_{0j}}$ **s** and z_1 of equal size and Δk_j is the diffractive correction to the wave vector as defined by (3.10) and a_{tq} are given by (3.9). The following expressions for the amplitudes of reflected and transmitted waves are obtained from (3.13).

$$E^{\mathrm{r}}_{\sigma} = (E^{\mathrm{e}}/D)D^{\mathrm{r}}_{\sigma} = (E^{\mathrm{e}}/D)\begin{vmatrix} a_{11} & a_{12} & a_{13} & a_{14} & \cos\alpha \\ a_{21} & a_{22} & a_{23} & a_{24} & \mathrm{e}^{i\beta}\sin\alpha \\ \gamma_1 a_{31} & \gamma_2 a_{32} & \gamma_3 a_{33} & \gamma_4 a_{34} & 0 \\ \gamma_1 a_{41} & \gamma_2 a_{42} & \gamma_3 a_{43} & \gamma_4 a_{44} & 0 \\ a_{31} & a_{32} & a_{33} & a_{34} & 0 \end{vmatrix} \tag{3.14}$$

$$D = \begin{vmatrix} a_{11} & a_{12} & a_{13} & a_{14} \\ a_{21} & a_{22} & a_{23} & a_{24} \\ \gamma_1 a_{31} & \gamma_2 a_{32} & \gamma_3 a_{33} & \gamma_4 a_{34} \\ \gamma_1 a_{41} & \gamma_2 a_{42} & \gamma_3 a_{43} & \gamma_4 a_{44} \end{vmatrix}$$

$$\gamma_j = e^{i\ell(\Delta + t_j + m)}, \quad \ell = \delta q L[2(1+m)\sin\theta_B]^{-1} \tag{3.15}$$

where L is crystal length.

The other components of $\mathbf{E}^t$ and $\mathbf{E}^r$ are determined by (3.14) if we substitute the first four elements of the last row in matrix D^t_σ by a_{4j} for D^t_π, $\gamma_j a_{1j}$ for D^r_σ, and $\gamma_j a_{2j}$ for D^r_π where $j = 1, 2, 3, 4$. The parameters α^t, β^t, and α^r, β^r which determine, as in (3.12), $\mathbf{e}^t$ and $\mathbf{e}^r$, the polarization vectors of the transmitted and reflected waves, are given by

$$tg\alpha^t e^{i\beta^t} = D^t_\pi / D^t_\sigma; \quad tg\alpha^r e^{i\beta^r} = D^t_\pi / D^r_\sigma \tag{3.16}$$

Hence, (3.14) and (3.16) yield a solution to the boundary problem and describe light transmission and reflection for arbitrary incidence angles and sample thickness.

Before analyzing expression (3.14) in the general case, we introduce some special cases which result in simpler expressions than those produced in the analysis of expression (3.14).

Thin Crystals. Consider scattering by thin crystals that satisfies condition $\ell \ll 1$ [see (3.15)]. The determinants of (3.14) show that wave polarization with maximal reflectivity (3.12) is determined by

$$tg\alpha e^{i\beta} = -i\sin\theta_B. \tag{3.17}$$

The wave reflected from a thin crystal is always completely polarized, and its parameters α and β are given by (3.17) no matter what the polarization of the incident wave. Concerning the first approximation, the wave whose polarization is orthogonal to that defined by (3.17) is not reflected at all. This result agrees with the kinematic consideration of Chap. 2.

Thick Crystals. For thick crystals, $\ell \gg 1$, the solution to the boundary problem is much simpler than that for $\ell \sim 1$ because only eigensolutions with nonzero coefficients in (2.27) are those which correspond to wave damping inside the crystal depth. The damping of eigensolutions for real $\hat{\epsilon}$ is due to diffraction scattering and the fact that parameters $\mathbf{k}_0$ and $\mathbf{k}_1$ in (3.10) have nonzero imaginary parts in a certain range of Δ; that is, there are zones in k-space where wave propagation is forbidden. As a result, diffraction reflection bands are present [7–10] where boundaries are determined from (3.4).

From the solution of (3.4), three different cases of light reflection are distinguished in the dependence of parameter Δ [see (3.5)], i.e., on deviation of the incidence angle (or frequency) from that given by the Bragg condition (2.4). If Δ is such that all the solutions t_j of (3.4) are real, the wave vectors in (3.10) are also real for all solutions of (3.8). Such waves pass through a crystal without attenuation, and the reflectivity in this range is small. The other specific case is when Δ is such that two solutions, $t_{2,3}$, are imaginary (with different signs) and the two others, $t_{1,4}$, real. There, one of the eigenwaves (say $\mathbf{E}_2$) is attenuated exponentially inside the crystal, and the other, $\mathbf{E}_3$, increases. Then, the incident wave with elliptical polarization corresponding to $\mathbf{E}_{02}$ is completely reflected, while the orthogonally polarized wave excites non-damping waves $\mathbf{E}_1$ and $\mathbf{E}_4$ in the crystal and only slightly contributes to reflection. In that region of Δ, reflectivity for unpolarized light is practically 1/2 (the region Δ_p). There is one more region of Δ in which the four roots t_j are complex and complex-conjugated in pairs. There, light with any polarization is almost completely reflected and the reflectivity for unpolarized light is unity (the region Δ_T).

Equation (3.4) determines boundaries of the regions Δ_p and Δ_T abut for different values of parameter $m = \cos^2\theta(1+\sin^2\theta)^{-1}$. For instance, at $0 < m < (1+\sqrt{13})/18$,

$$-(1+\sqrt{1+8m^2})/2 < \Delta_p < 0 < \Delta_T < (\sqrt{1+8m^2}-1)/2 \qquad (3.18)$$
$$(\sqrt{1+8m^2}-1)/2 < \Delta_p < 1.$$

The variations of Δ at reflection band boundaries over the range of θ are presented in Fig. 3.2.

Consider now the results of dynamic diffraction theory in different regions. When the roots are real, the formulae obtained from (3.14) are simple only for incidence angles far from the exact Bragg condition at $|\Delta| \gg 1$

$$E^{\mathrm{r}}_{\sigma} = \Big\{-\sin\theta(1+m)(1-\mathrm{e}^{2i\tilde{t}_1\ell})\cos\alpha - i(1-m)[1-\mathrm{e}^{i(\tilde{t}_1+\tilde{t}_2)\ell}]$$
$$\times\, \mathrm{e}^{i\beta}\sin\alpha\Big\}(2\Delta\sin\theta)^{-1}$$
$$E^{\mathrm{r}}_{\pi} = (1/2\Delta)\Big\{-(1-m)(1-\mathrm{e}^{2i\tilde{t}_2\ell})\mathrm{e}^{i\beta}\sin\alpha + i(1+m)\sin\theta$$
$$\times\,[1-\mathrm{e}^{i(\tilde{t}_1+\tilde{t}_2)\ell}]\cos\alpha\Big\} \qquad (3.19)$$

where $\tilde{t}_{1,2} = (\Delta/|\Delta|)t_{1,2}$ and $t_{1,2}$ is given by (3.7).

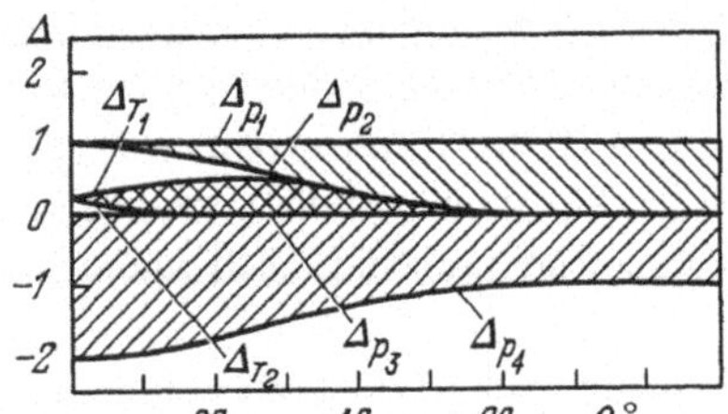

Fig. 3.2 Diffraction scattering band dependence on the incidence angle of light in a cholesteric liquid crystal. The parameter Δ characterizes deviation from the Bragg condition. The selective reflection band of an individual polarization is hatched, and the reflection bands of any polarization are double-hatched

Equation (3.19) shows that the intensity of reflected waves decreases as Δ^{-2} and the polarization, determined by $E^{\mathrm{r}}_{\pi}/E^{\mathrm{r}}_{\sigma}$, depends on Δ ad L in a complicated way. In the region Δ_T, light with any polarization is reflected and all t_j are complex and complex-conjugated in pairs. For $\ell \gg 1$, two parameters $\gamma_j = \exp[i\ell(\Delta + m + t_j)]$ in determinant (3.14) (e.g., at $j = 1, 2$) are exponentially small, and the two others exponentially large. As follows from (3.14), $E^{\mathrm{t}}_{\sigma} = 0$ and $E^{\mathrm{t}}_{\pi} = 0$ because the columns in D^T_{σ} and D^{t}_{π} are linearly dependent but $D \neq 0$. Thus, in region Δ_T the intensity of the transmitted wave is zero; hence, a wave with any polarization is completely reflected since there is no absorption in CLCs.

Expression (3.16) determines the polarization of the reflected wave in Δ_T and is simplified as compared with the general case:

$$tg\alpha^{\mathrm{r}}e^{i\beta} = \begin{vmatrix} \cos\alpha & a_{11} & a_{12} \\ e^{i\beta}\sin\alpha & a_{21} & a_{22} \\ 0 & a_{41} & a_{42} \end{vmatrix} \times \begin{vmatrix} \cos\alpha & a_{11} & a_{12} \\ e^{i\beta}\sin\alpha & a_{21} & a_{22} \\ 0 & a_{31} & a_{32} \end{vmatrix}^{-1} \tag{3.20}$$

where a_{ij} are given by (3.9) for complex t_1 and t_2 which correspond to the solutions that damp within the crystal. Figure 3.3 shows the dependence of $\Delta\theta_T$ (the angular width of the total reflection band Δ_T) on incidence angle. The frequency width of this band is given by

$$\Delta\omega_T/\omega = \Delta\theta_T ctg\theta. \tag{3.21}$$

Now consider the intermediate region in which t_1 and t_2 are imaginary and t_3, t_4 real. The amplitude of transmitted light is, neglecting exponentially small corrections,

$$E^{\mathrm{t}}_{\sigma} = D^{-1}(A_{21}\cos\alpha - a_{11}e^{i\beta}\sin\alpha) \begin{vmatrix} a_{32} & a_{33} & a_{34} \\ a_{42} & a_{43} & a_{44} \\ a_{12} & a_{13} & a_{14} \end{vmatrix}$$

$$E^{\mathrm{t}}_{\pi} = D^{-1}(a_{21}\cos\alpha - a_{11}e^{i\beta}\sin\alpha) \begin{vmatrix} a_{32} & a_{33} & a_{34} \\ a_{42} & a_{43} & a_{44} \\ a_{22} & a_{23} & a_{24} \end{vmatrix} \tag{3.22}$$

$$D = \begin{vmatrix} a_{11} & 0 & a_{13} & a_{14} \\ a_{21} & 0 & a_{23} & a_{24} \\ 0 & a_{32} & \gamma_3 a_{33} & \gamma_4 a_{34} \\ 0 & a_{42} & \gamma_3 a_{43} & \gamma_4 a_{44} \end{vmatrix}.$$

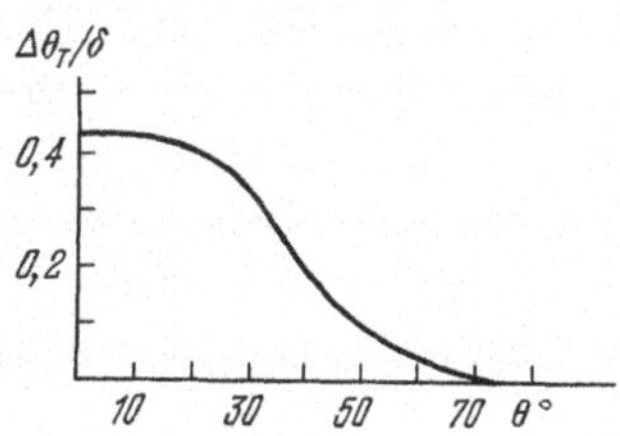

Fig. 3.3 Total reflection band angular width $\Delta\theta_T$ dependence on Bragg's angle θ [10]

Here, t_1 corresponds to the solution which attenuates inside the crystal. Equation (3.22) shows that transmitted wave polarization is determined by E_1^t/E_σ^t and depends neither on incident polarization nor on crystal thickness. It also follows from the formulae that there is always a polarization which is reflected completely, for this polarization $tg\alpha_1 \exp(i\beta_1) = a_{21}/a_{11}$. The polarization of the reflected wave is, in this case, given by

$$tg\alpha_1^r \exp(i\beta_1^r) = a_{41}/a_{31}. \tag{3.23}$$

Also, (3.22) shows that the transmission is maximal if the incident wave is polarized so that $tg\alpha_2 \exp(i\beta_2) = \frac{a_{21}^*}{a_{11}^*}$, i.e., if it is orthogonal to that of the wave which experiences total reflection.

The results of this section permit analysis of the interaction between light and a cholesteric liquid crystal for arbitrary incidence angles. The optical properties become anomalous in the vicinity of the angle which satisfies the Bragg condition, and the corresponding region is of the order of δ, i.e., determined by the local anisotropy of the dielectric constant.

Although the solutions corresponding to the diffraction reflection band in CLCs are similar to those for X-ray diffraction, they are more complicated than those of X-ray diffraction theory. However, polarization properties of eigensolutions are more complex; therefore, optical polarization properties of cholesteric crystals are unusual. The most interesting results are those of dynamic theory, related to oblique incidence of light on a sample, because exact solutions are still unknown for this case; the corresponding light scattering properties differ considerably from those in the case of light propagating along the axis.

When light is incident obliquely, in addition to the selective reflection band, there is a band of total light reflection with any polarization near the region where the Bragg condition holds. The polarization of the selectively reflected light and total reflection band depend on incidence angle. In normal incidence, the angular and frequency widths of the reflection band are zero and the totally reflected polarization is circular. As seen in Fig. 3.3, the total reflection band Δ_T is at a maximum when incidence angle $\theta \lesssim 25°$; in this region, $\Delta\theta_T \approx 0.48\delta$, or at $\delta = 0.01$, it is $14'$. Therefore, the dimension of the diffraction region is such that the above relation is verifiable experimentally.

According to a detailed analysis for small incidence angles $\theta \leq 30°$, weak reflection occurs far from the Bragg condition as well as between the total and the selective reflection bands. Thus, in the region of diffraction reflection of obliquely incident light, there is a complicated and interesting dependence of reflected light polarization properties and reflection coefficient on incidence angle at a fixed frequency or on frequency at a fixed angle (Fig. 3.4).

It was shown in [7–10] that the angular (frequency) width of the bands of polarization selective reflection and total reflection may be either adjacent to or separated from the regions where there is no diffraction reflection. The behavior of unpolarized light reflection as a function of frequency (angle) is shown in Fig. 3.5. The value, 1/2, of the reflection coefficient corresponds to the selective reflection band for a definite

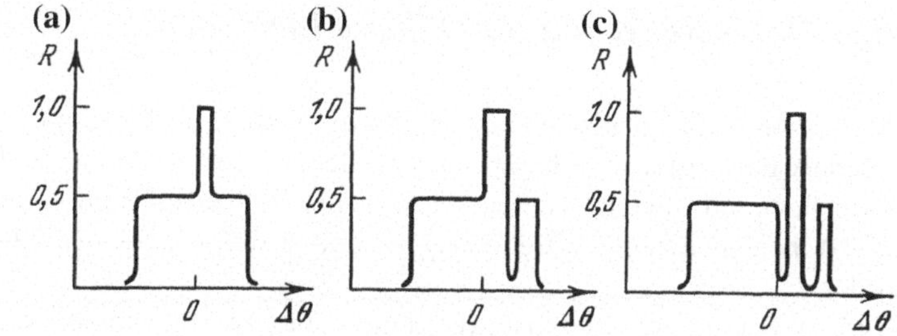

Fig. 3.4 Qualitative behavior of the reflection coefficient of unpolarized light as a function of deviation from the Bragg condition. Depending on the incidence angle θ, there are either one ($90° > \theta \geq 32°$) (**a**), two ($32° \geq \theta \geq 25°$) (**b**), or three ($25° \geq \theta \geq 0°$) (**c**) selective reflection bands

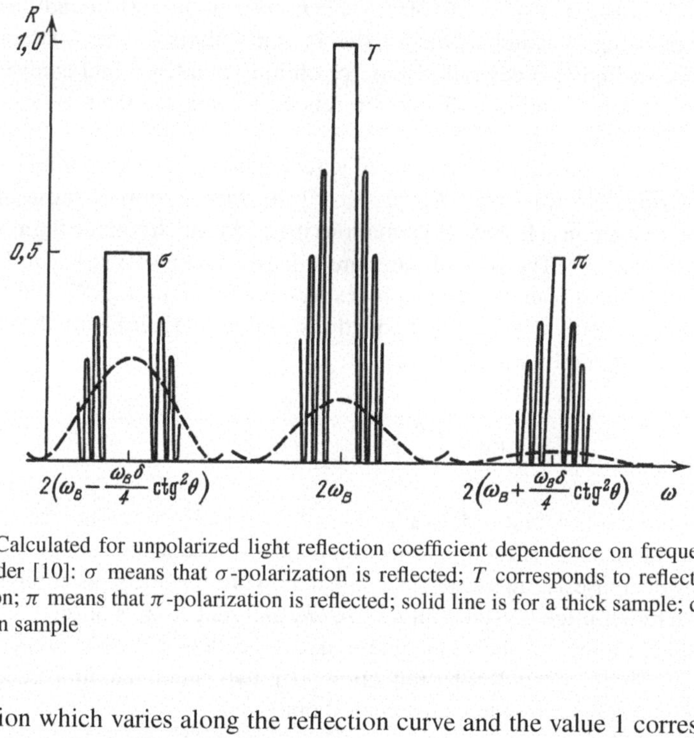

Fig. 3.5 Calculated for unpolarized light reflection coefficient dependence on frequency in the second order [10]: σ means that σ-polarization is reflected; T corresponds to reflection of any polarization; π means that π-polarization is reflected; solid line is for a thick sample; dashed line is for a thin sample

polarization which varies along the reflection curve and the value 1 corresponds to total reflection band. The total and selective reflection bands for small θ may not be adjacent but separated by intervals of weak reflection. This fact follows from Fig. 3.4 which shows the structure of reflection regions as a function of wavelength and incidence angle as obtained from (3.4). It is seen that for normal incidence ($\theta = 90°$) the width of the polarization-independent band becomes zero in agreement with the results of the exact solution.

3.1.4 High-Order Reflection for Oblique Incidence

General Considerations. Consider one more specific feature of the case of oblique incidence, namely the existence of higher-order reflection corresponding to the case of diffraction reflection from CLCs at $|s| > 1$ in (2.17). That is, diffraction reflection exists at a fixed incidence angle at frequency ω_B and at frequencies $s\omega_B$ where s is reflection order.

Higher-order reflection was first discovered theoretically while numerically solving the Maxwell equations [8, 11] and was later observed experimentally [11]. As mentioned above, there is no reflection at multiple frequencies $s\omega_0$ in the case of normal incidence. In addition, according to the results of kinematic theory for oblique incidence, no direct scattering occurs with changes in the wave vector by $s\tau(|s| > 1)$ because the Fourier expansion (1.6) of tensor $\hat{\epsilon}(\mathbf{r})$ contains no corresponding harmonics. However, using dynamic diffraction theory, particularly (2.25), we demonstrate the existence of higher-order reflections for oblique incidence and analyze their s dependence. It was found that diffraction reflection intensity decreases rapidly as reflection order s increases.

Although there is no direct scattering of waves E_0 to E_t, such scattering is possible through intermediate waves $E_1, E_2 \ldots, E_{s-1}$ [10]. Description of these processes demands that we retain at least $S + 1$ equations in (2.25) which contain the amplitudes of intermediate waves. These amplitudes are at least a factor of δ less than E_0 and E_s and may be excluded from the system by expressing them through E_0 and E_s. Thus, we obtain for E_0 and E_s the system of equations, similar to (2.26), with the exception that $\hat{\epsilon}_{\pm\tau}$ is substituted by $F_s\hat{\epsilon}_{\pm\tau}$, where

$$F_s = \left(\frac{\delta}{8}\right)^{s-1} \frac{s^{2s}}{(s!)^2} ctg^{2s-2}\theta. \tag{3.24}$$

As δ in the case of first-order reflection, the value δF_s describes characteristic angular (or frequency) diffraction reflection bands, whose widths are of the order of δ^s; i.e., they decrease as s increases. We note that to observe the s-order reflection we must have a rather thick crystal with $L \sim p/\delta F_s$ and reflection is observed when s is not very large. Figure 3.5 shows the qualitative dependence of reflection coefficient on the frequency for second-order reflection [10]; two lateral maxima are located at the distance $\pm 2\omega_s \delta ctg^2\theta/4$ from the central one.

The width of the σ-maximum is $\Delta\omega_\sigma = 2\omega_B\delta^2 ctg^2\theta/4 \cdot \sin^2\theta$. In this maximum, light with linear σ-polarization is reflected as σ-polarized, while in the π-maximum, whose width is $\Delta\omega_\pi = 2\omega_B\delta^2 ctg^2\theta/4$, π-polarized light is reflected as π-polarized. In the central T-maximum, σ-polarized light becomes π-polarized after reflection and vice versa; i.e., reflection occurs for any polarization. The width of this maximum is $\Delta\omega_T = 2\omega_B\delta^2 ctg^2\theta/4 \sin\theta$. If crystal thickness is small, the maxima in Fig. 3.5 broaden (dashed line) and become much less prominent.

A similar picture of three maxima is observed with incidence angle variance at a fixed frequency. Second-order reflection takes place near the Bragg angle, $\theta_B =$

$\arcsin(\tau/q)$, and expression (3.25) defines angular widths of the σ-, π-, and T-maxima, respectively:

$$\frac{\Delta\theta_\sigma}{\Delta\omega_\sigma} = \frac{\Delta\theta_\pi}{\Delta\omega_\pi} = \frac{\Delta\theta_T}{\Delta\omega_T} = \frac{1}{\omega} tg\theta. \tag{3.25}$$

Studies of higher-order reflection (particularly second order) are useful for CLCs with large pitch whose first-order reflection occurs in the infrared. Notice that for these crystals Bragg angles are small and hence more easily observed, high reflection orders. This is demonstrated in the formulae above where reflection band frequency and angular width decrease and crystal thickness (necessary for reliable observation) increases.

Second-Order Selective Reflection. We will use the general results for higher-order reflection in CLCs presented above for the case of second-order reflection within the framework of multiwave dynamic diffraction theory. Second-order reflections are due to multiple light scattering events and are, to a certain extent, analogous to the Reninger reflections in X-ray scattering [12]. No such reflections appear in kinematic diffraction theory since the corresponding scattering amplitudes are strictly zero. The phenomenon in question is more complicated in the case of CLCs because of the specific light polarization properties in CLC.

Assume that the direction of a monochromatic wave incident on a CLC is close to or coincides with the Bragg direction for second-order reflection, i.e., $s = 2$ in (2.17) (Fig. 3.6). This means that the selectively reflected light frequency is $2\omega_B$ if the corresponding first-order frequency for the same incidence angle is ω_B.

We consider second-order Bragg reflexes using expressions (2.25) and (2.26). Since the two-wave approximation is not sufficient to obtain second-order reflection, we retain in the infinite system (2.25) the equations for three waves: E_0, E_1, and E_2. Neglecting higher-order terms in δ (usually, δ is about 10^{-2}), we obtain

$$\begin{aligned}
&\left(\hat{\epsilon}_0/\bar{\epsilon} - \frac{k_0^2}{q^2}\right)\mathbf{E}_0 + (\hat{\epsilon}_{-\tau}/\bar{\epsilon})\mathbf{E}_1 = 0\\
&(\hat{\epsilon}_\tau/\bar{\epsilon})\mathbf{E}_0 + (1 - k_1^2/q^2)\mathbf{E}_1 + (\mathbf{k}_1\mathbf{E}_1)\mathbf{E}_1/q^2 + (\hat{\epsilon}_{-\tau}/\bar{\epsilon})\mathbf{E}_2 = 0\\
&(\hat{\epsilon}_\tau/\bar{\epsilon})\mathbf{E}_1 + (\hat{\epsilon}_0/\bar{\epsilon} - k_2^2/q^2)\mathbf{E}_2 = 0.
\end{aligned} \tag{3.26}$$

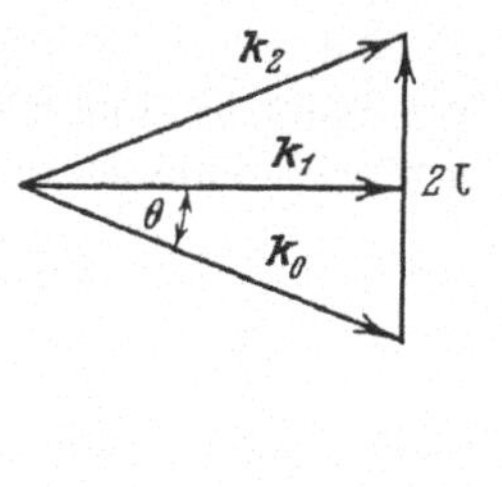

Fig. 3.6 Bragg condition in vector form for second-order selective reflection

From the second equation, we find

$$\mathbf{E}_1 = [q^2\hat{\epsilon}_\tau \mathbf{E}_0 - \mathbf{k}_1(\mathbf{k}_1\hat{\epsilon}_\tau \mathbf{E}_0) + q^2\hat{\epsilon}_{-\tau}\mathbf{E}_2 \\ - \mathbf{k}_1(\mathbf{k}_1\hat{\epsilon}_{-\tau}\mathbf{E}_2)][(k_1^2 - q^2)2\bar{\epsilon}]^{-1} \tag{3.27}$$

and substitute $\mathbf{E}_1$ into the third and first equations. When retaining only terms with the lowest orders in δ, we obtain

$$(\hat{\epsilon}_0/\bar{\epsilon} - k_0^2/q^2)\mathbf{E}_0 + \frac{\delta}{2}ctg^2\theta\hat{\epsilon}_{-\tau}\mathbf{E}_2 = 0 \\ \frac{\delta}{2}ctg^2\theta\hat{\epsilon}_\tau \mathbf{E}_0 + (\hat{\epsilon}_0/\bar{\epsilon} - k_2^2/q^2)\mathbf{E}_2 = 0. \tag{3.28}$$

It was taken into account that the component of $\mathbf{k}_1$ along the Z-axis was zero, while that perpendicular to Z was $q\cos\theta$ (see Fig. 3.6). Note that the coefficients of $\hat{\epsilon}_{\pm\tau}$ in (3.28) are nonzero because intermediate wave $\mathbf{E}_1$ is not transversal.

Let us analyze the solutions for system (3.28). This system differs from system (2.26), (3.3) in the coefficients of $\hat{\epsilon}_{\pm\tau}$; however, its solution is less complex than that of (3.3) because (3.28) is reduced to two equations which separately describe π- and σ-polarized light scattering. Physically, this separation is due to CLC birefringence. For second- and higher-order diffraction scattering, the difference between wave vectors of eigenwaves is proportional to birefringence values, i.e., to δ, and greater than the diffraction corrections to wave vectors; these corrections are proportional to δ^s. This means that polarization properties of eigenwaves in the presence of diffraction are the same, to within the order of δ, as those in the case of direct transmission in the absence of diffraction. That is, higher-order diffractive scattering is too weak to appreciably change the polarization properties and wave vectors of eigenwaves. This means that, in the presence of diffraction, the polarizations of $\mathbf{E}_0$ and $\mathbf{E}_2$ practically coincide with σ- and π-polarizations, the eigenpolarizations for direct transmission without diffraction. possible polarization combinations: $\mathbf{E}_0$ and $\mathbf{E}_2$ are either both σ- or π-polarized, π- and σ- or σ- and π-polarized, respectively. Since the birefringence is strong compared with δ^2, the angular (or frequency) regions in which the Bragg condition is satisfied for the combinations above are presented at Fig. 3.7. Thus, in corresponding angular (or frequency) intervals, whose widths are of the order of δ^2, one wave with eigenpolarization is subjected to diffractive scattering, while that of the other eigenpolarization does not satisfy the Bragg condition and experiences negligible diffractive scattering. This is the physical reason for simplification of (3.28) which is reduced to a system of two equations. Naturally, the same result is obtained by solving the complete system and neglecting small polarization corrections of the order of δ. Thus, system (3.28) is reduced to the system of scalar equations:

$$(1 - k_0^2/q_p^2)E_0^p + F_{pq}E_2^q = 0 \\ F_{qp}E_p^r + (1 - k_2^2/q_q^2)E_2^q = 0 \tag{3.29}$$

Fig. 3.7 Bragg condition in vector form for a CLC with strong birefringence

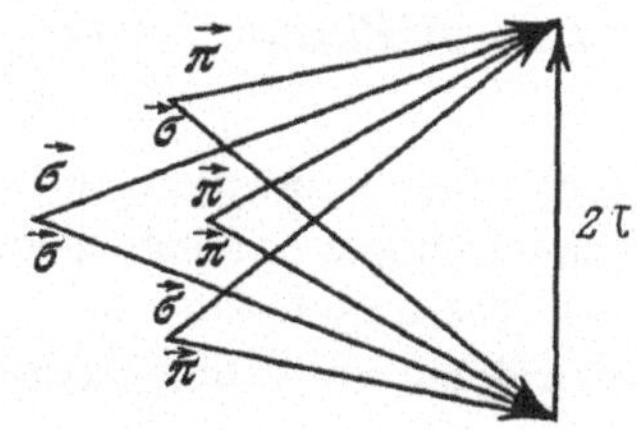

where indices p and q at the scalar amplitudes E_0 and E_2 and q indicate corresponding wave polarizations (these indices may be π or σ as well)

$$q_q = \frac{\omega\sqrt{\epsilon_q}}{c}; \quad \epsilon_\pi = \bar{\epsilon}\left(1 - \frac{\delta}{2}\cos^2\theta\right), \quad \epsilon_\sigma = \bar{\epsilon}\left(1 + \frac{\delta}{2}\cos^2\theta\right)$$

$$F_{\pi\sigma} = F_{\sigma\pi} = \frac{\delta^2}{4}\frac{\cos^2\theta}{\sin\theta}; \quad F_{\pi\pi} = \frac{\delta^2}{4}\cos^2\theta, \quad F_{\sigma\sigma} = \frac{\delta^2}{4}ctg^2\theta.$$

System (3.29) is identical with the system of equations from X-ray diffraction theory [13, 14].

The Bragg angle θ_B corresponding to the polarization combination above is:

- for scattering $\sigma - \pi$ and $\pi - \sigma$: $\theta_B^{\pi\sigma} = \theta_B^{\sigma\pi} = \theta_B = \arcsin\frac{\tau c}{\omega\sqrt{\bar{\epsilon}}}$,
- for scattering $\sigma - \sigma$ and $\pi - \pi$: $\theta_B^{qq} = \theta_B + \Delta\theta^{qq}$,

where $\Delta\theta^{\sigma\sigma} = -\Delta\theta^{\pi\pi} = \delta ctg\theta_B$. If we assume that the incidence angle is close to some of the θ_B^{pq} angles and solve the boundary problem (as for first-order diffraction reflection) or use the results of X-ray diffraction theory [13, 14], we obtain the reflection coefficient R and transmission coefficient T dependence on incidence angle and thickness L of a planar cholesteric texture [1, 10].

We assumed above that, in an experiment the incidence angle is varied at fixed frequency ω. Conversely, if angle θ is fixed and frequency varied, diffraction scattering occurs in the vicinity of Bragg's frequencies ω_B^{pq}. For polarization-changing scattering, $\omega_B^{\sigma\pi} = \omega_B^{\pi\sigma} = \omega_B = c\tau/(\sqrt{\bar{\epsilon}}\sin\theta)$ and polarization-conserving scattering $\omega_B^{qq} = \omega_B + \Delta\omega^{qq}$, $\frac{\Delta\omega^{\pi\pi}}{\omega_B} = \frac{-\Delta\omega^{\sigma\sigma}}{\omega_B} = -\delta ctg\theta/4$.

We note that the mentioned angular (or frequency) positions of diffraction maxima (Fig. 3.5) are given, within the approximate approximation, to the accuracy of δ and widths, and to the accuracy of δ^2; i.e., to a first nonvanishing approximation, a more exact solution requires that we use a more accurate approximation. For example, positions of maxima can be found to the accuracy of δ^2 by retaining, in (3.28), higher-order terms in δ on the basis of (3.26).

Thus, higher-order diffraction scattering and corresponding polarization properties are simpler than those of first order. This analysis indicates that the complex polarization properties of first-order diffraction scattering at oblique incidence are due to birefringence and diffraction scattering of the same strength.

3.1.5 A Planar Layer

The explicit expressions related to light propagation in CLCS are given for an unlimited sample or a half-space filled with a CLC in Chap. 1 and in this chapter as well. These expressions are used occasionally to describe experiments, but the most typical and interesting experimental cases are those on samples that are not thick ($\delta L/p \sim 1$).

Notice that, in many cases, experiments involved normal or near normal incidence of light on a planar texture. The analytical solution to this problem may be used for any layer's thickness. In the general case, however, it is complicated and it is necessary to invoke numerical methods. Thus, we analyze the optical properties of a layer of arbitrary thickness L for normal incidence assuming no reflection at dielectric boundaries. The analytic description of this case is simple; for a layer of finite thickness, we study the polarization and frequency properties of both transmitted light and reflected light. Thus, we neglect light reflection at CLC boundaries and assume that anisotropy δ is small and $\bar{\epsilon} = \epsilon$ or $r = (\bar{\epsilon}/\epsilon)^{1/2}$ differs from unity by less than δ. Under these conditions, three eigensolutions are excited in a sample; two correspond to diffracting polarization, and the third to the non-diffracting wave propagating along the direction of incidence. Systems (2.18) and (2.19), which describe the solution to the boundary problem in a general case, become more simple and take the form (1.24) except that $r = 1$ and index j assumes values 1, 2, and 3. Solutions of the corresponding systems are described by expressions of form (1.21) if the fourth row and the fourth column are canceled in the corresponding matrices. Thus, the problem reduces to solving a system of three linear equations.

The assumption of no boundary reflection permits other simplifications which reduce the problem to a system of two linear equations. This is so because, in this case, the polarizations in system (1.20) become separated since the incident wave with diffracting circular polarization excites only solutions No. 2 and No. 3, whereas the wave with non-diffracting polarization excites only solution No. 1. This means that the amplitude and phase of the eigenwave at the CLC surface are determined by $E_{\rm e}^-$; i.e., it is identical with corresponding values for the non-diffracting component of the incident wave.

Accounting for the approximation, i.e., neglect of boundary reflection, which introduces an error of the order of δ, we obtain the equation for the amplitudes of diffracting eigensolutions E_2 and E_3 retaining only the terms of lowest order in δ. We do this by assuming that in system (1.20) $\eta_2^+ = \eta_3^+ = 1$, $\eta_2^- = \eta_3^- = -1$. This yields

$$E_2^+ + E_3^+ = E_{\rm e}^+, \quad {\rm e}^{iK_2^- L}\xi_2 E_2^+ + {\rm e}^{iK_3^- L}\xi_3 E_3^+ = 0 \tag{3.30}$$

From the expression for the amplitudes of reflected and transmitted waves with diffracting polarization (1.19), we find:

$$E_{\rm r}^+ = \sum_{j=2,3} \xi_j E_j^+, \quad E_{\rm t}^+ = \sum_{j=2,3} {\rm e}^{iK_j^+ L} E_j^+. \tag{3.31}$$

And from (3.30) and (3.31), or from the corresponding simplified form of (1.21), we obtain the following expressions for the amplitudes of reflected and transmitted waves of diffracting polarization:

$$\begin{aligned} E_r^+ &= i\delta \sin k^- L\Big\{(4k^-\tilde{\lambda}^2/\tau)\cos k^- L + i[\tilde{\lambda}^2 + (2k^-\tilde{\lambda}/\tau)^2 - 1] \\ &\quad \sin k^- L\Big\}^{-1} E_e^+ \\ E_t^+ &= 4e^{iqL}(k^-\tilde{\lambda}^2/\tau)\Big\{(\tilde{\lambda}/\tau)\cos k^- L + (i/4)[\tilde{\lambda}^2 + (2k^-\tilde{\lambda}/\tau)^2 - 1] \\ &\quad \sin k^- L\Big\}^{-1} E_e^+ \end{aligned} \tag{3.32}$$

where the notation is as in Chap. 1. Under these conditions, the reflection coefficient R^+ and transmission coefficient T^+ for a wave with diffracting polarization are:

$$\begin{aligned} R^+ &= (\delta^2 \sin^2 k^- L)[16(k^-\tilde{\lambda}^2\tau)^2 + \delta^2 \sin^2 k^- L]^{-1} \\ T^+ &= 1 - R^+ = (4k^-\tilde{\lambda}^2/\tau)^2[16(k^-\tilde{\lambda}^2/\tau)^2 + \delta^2 \sin k^- L]^{-1}. \end{aligned} \tag{3.33}$$

Thus, the wave with non-diffracting polarization has no reflection, $R^- = 0$ and $T^- = 1$, within this approximation.

In (3.33), in more detail, within the frequency range $\omega_B(1+\delta)^{-1/2} < \omega < \omega_3(1-\delta)^{-1/2}$ (i.e., when $|q^2 - \tau^2/4| < q^2\delta$) the value k^- becomes imaginary and $\sin k^- L$ in (3.33) becomes $sh|k^-|L$. When the crystal is sufficiently thick ($L > p/\delta$) $sh|k^-|L \gg 1$ and $R^+ \approx 1$, i.e., light with right-handed circular polarization is totally selectively reflected (Fig. 1.8). The frequency width of the reflection band is $\Delta\omega = \omega_B[(1-\delta)^{-1/2} - (1+\delta)^{-1/2}] \approx \delta\omega_B$; i.e., it is defined by the anisotropy δ of dielectric permittivity. When the frequency moves away from the region of strong reflection outside the selective reflection band, the reflection does not decrease smoothly—it oscillates. The oscillations are due to light diffraction within a limited volume and unassociated with boundary reflection. These oscillations are not observed, as a rule, in experiments, because samples are not sufficiently perfect.

In thin crystals ($L \ll p/\delta$),

$$R^+ \simeq \delta^2 \frac{\sin^2[(q - \tau/2)L]}{(2q/\tau - 1)^2} \tag{3.34}$$

i.e., reflected intensity is proportional to δ^2, while there is prominent broadening of the reflection curve (see Fig. 1.8); this result is also obtained from (2.16) within the kinematic approximation. The intermediate case where $L \sim p/\delta$ is also shown in Fig. 1.8.

Experimentalists often measure the difference between transmissions for opposite circular polarizations, i.e., the so-called circular dichroism $D = (T^- - T^+)/(T^- + T^+)$. This value for the non-absorbing cholesteric crystals can be expressed in terms of T^+ and R^+ as

$$D = \frac{1 - T^+}{1 + T^+} = \frac{R^+}{2 - R^+}. \tag{3.35}$$

3.1.6 Optical Rotation

General Features. It is well known [3, 15] that optical rotation in a medium is due to the difference in phase velocities of waves with right- and left-handed circular polarizations [16]. In plane cholesteric texture, the phase difference for certain wavelengths may be large since the wave with circular polarization is subjected to strong diffraction reflection while the other polarization interacts weakly with the crystal. Optical rotation in the selective reflection band of CLCs is stronger than intrinsic optical rotation (associated with molecular optical activity) due to diffraction effects and has some peculiarities. One of these peculiarities is that the wave with diffracting circular polarization is strongly attenuated in the CLC due to diffraction reflection. When light is transmitted through a crystal, its initial linear polarization changes to elliptical polarization which depends on crystal thickness. Thus, we consider the optical rotation of, say, the major axis of the polarization ellipse. In the case of light propagation along the optical axis, if incident polarization is linear, the ratio of ellipse axes is given by the following expression:

$$b = \frac{1 - \sqrt{T^+}}{1 + \sqrt{T^+}} \tag{3.36}$$

where T^+ is as in (3.33). The other peculiarity is the strong optical rotation dependence on frequency both in and outside the absorption bands. Optical rotations for frequencies at different sides of Bragg's frequency ω_B are of opposing directions, whereas those at Bragg's frequency vanish. This behavior of optical rotation seems rather strange, but it has a simple explanation: The phase velocity of one of two circular diffracting eigenwaves presented in expansion (1.19) is smaller than that of the non-diffracting wave, and the other greater. The exited diffracting wave is predominantly slow or fast depending on the sign of the difference between the frequency and ω_B. This causes frequency dependence of the optical rotation and its sign. Finally, there is one unusual property which distinguishes CLCs from other gyrotropic media—the angle of optical rotation in a CLC nonlinearly depends on sample thickness.

From solution to the boundary problem (3.33), we find that when the incident wave is plane-polarized, the amplitude of a wave leaving the crystal is

$$\mathbf{E}^{+} = \frac{E_e}{\sqrt{2}}\Big\{\mathbf{n}_{-}e^{i(k^{+}-\tau/2)L} + \mathbf{n}_{+}\frac{k^{-}\tau e^{i\tau L/2}}{k^{-}\tau\cos k^{-}L + i[(k^{-})^2+\tau^2/4-q^2]\sin k^{-}L}\Big\}. \quad (3.37)$$

The difference between two terms in (3.37) gives the rotation angle

$$\varphi = \frac{1}{2}\left[(k^{+}-\tau)L + arctg\left\{\frac{[(k^{-})^2+\tau^2/4-q^2]tgk^{-}L}{k^{-}\tau}\right\}\right]. \quad (3.38)$$

This formula is more simple for thick crystals

$$\frac{\varphi}{L} = \mathrm{Re}\frac{k^{+}-k^{-}-\tau}{2}. \quad (3.39)$$

The optical rotation frequency dependence given by (3.38) and (3.39) is presented in Fig. 3.8. Far from the selective reflection band, expressions (3.38) and (3.39) reduce to the deVries formula [17].

$$\frac{\varphi}{L} = \tau\delta^2[16\tilde{\lambda}^2(1-\tilde{\lambda}^2)]^{-1}. \quad (3.40)$$

This formula has the disadvantage that in contrast to the exact formula (3.38) it tends to $\pm\infty$ in the selective reflection band.

Optical Rotation in Thin Layers. When reflection of a CLC is not too strong (samples are used routinely to study optical rotation), we obtain, from (3.38), an approximate expression which holds both in and outside the selective reflection band

$$\frac{\varphi}{L} = \frac{\delta^2\tau}{32\tilde{\lambda}^3}\left[\left(\frac{\tau}{2\tilde{\lambda}k^{-}}\right)^2(1-\tilde{\lambda})\left(1-\frac{\sin 2k^{-}L}{2k^{-}L}\right) + \frac{1}{1+\tilde{\lambda}}\right]. \quad (3.41)$$

For a rough estimate, we assume $k^{-}/q = \tilde{\lambda} - 1$ and obtain, from (3.41), a simple formula

$$\frac{\varphi}{L} = \frac{\delta^2\tau}{16\tilde{\lambda}(1-\tilde{\lambda}^2)}\left[1 - \frac{(1+\tilde{\lambda})\sin 2q(1-\tilde{\lambda})L}{2(1-\tilde{\lambda})\tau L}\right]. \quad (3.42)$$

The first term is the deVries formula (3.40), and the second is essential in the selective reflection band at $\tilde{\lambda} \to 1$. This formula yields the correct optical rotation frequency dependence at any frequency. If the sample is thin ($L < p/\delta$), formulae (3.41) and (3.42) differ slightly from the exact (3.38) and apply if rotation angle $\varphi < 1$. We note again that in the selective reflection band formulae (3.38), (3.41), and (3.42) yield the

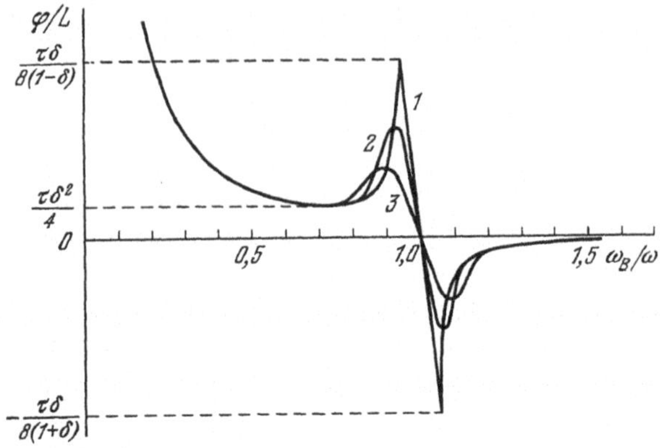

Fig. 3.8 Dependences of the angle of the polarization plane rotation on frequency for sample with different thicknesses: $1 - L\delta/p \gg 1, 2 - L\delta/p = 1, 3 - L\delta/p = 0.5$. The curves are plotted for $\delta = 0.1$

nonlinear rotation angle dependence on thickness L. Specifically, at small L values, $\varphi \sim L^2$ near the rotation maxima (see Fig. 3.8) and $\varphi \sim L^3$ near the frequency at which the rotation changes sign.

The Limit $\tilde{\lambda} = \tau/2q \ll 1$. In this case, waves 1 and 2 (or 3 and 4 depending on the direction of the incident wave) are excited in a crystal.

From (1.21), we obtain the following equation for transmitted wave amplitude [see (1.18)]

$$
\begin{aligned}
E_t^+ &= \left[E_e^+ \left(\cos\frac{(k^- - k^+)L}{2} \right) + it \sin\frac{(k^- - k^+)L}{2} \right) \\
&\quad - iE_e^- \frac{\delta t}{2\tilde{\lambda}} \sin\frac{(k^- - k^+)L}{2} \right] e^{i(k^+ + k^- + \tau)L/2} \\
E_t^- &= \left[E_e^- \left(\cos\frac{(k^- - k^+)L}{2} \right) - it \sin\frac{(k^- - k^+)L}{2} \right) \\
&\quad - iE_e^+ \frac{\delta t}{2\tilde{\lambda}} \sin\frac{(k^- - k^+)L}{2} \right] e^{i(k^+ - k^- - \tau)L/2} \qquad (3.43)
\end{aligned}
$$

where $t = (1 + \frac{\delta^2}{\tilde{\lambda}-2})^{1/2}$. Rotation angle φ nonlinearly depends on crystal thickness and generally on the direction of initial polarization.

Under the assumed limitations for q, two cases are possible. If $\tau/2q = \tilde{\lambda} \gg \delta$, (3.43) results in (3.40) and rotation angle ceases to depend on initial polarization direction.

When the wavelength is small, Mauguin's limit $\tilde{\lambda} \ll \delta$, the transmitted wave polarization, depends on molecular orientation at the CLC surface. As follows from (3.43), linearly polarized incident light becomes elliptically polarized upon leaving the crystal. If linearly polarized incident light is parallel (perpendicular) to CLC molecules at the front surface, transmitted light is also linearly polarized in the direction parallel (perpendicular) to molecules at the back surface. Hence, the polarization plane follows the orientation of CLC molecules and polarization angle is determined

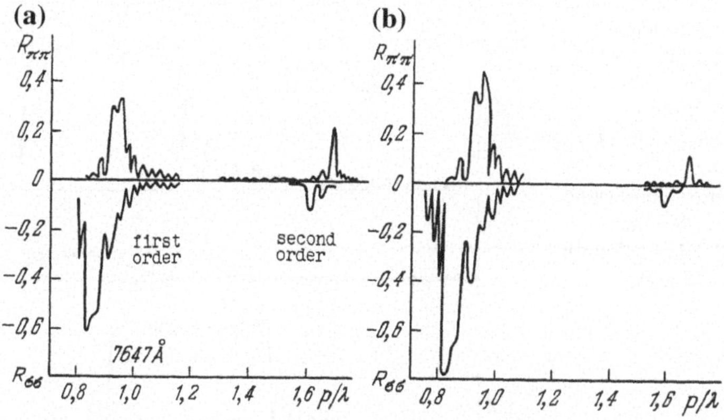

Fig. 3.9 Measured (**a**) and calculated (**b**) frequency and polarization dependence in first- and second-order reflection from a mixture of cholesterics [11]. $R_{\pi\pi}(R_{\sigma\sigma})$ is the reflection coefficient of the π- or σ-polarization into π- or σ-polarization; incidence angle is 45°; sample thickness is $L = 11.47\,\mu\text{m}$; $L/p = 15$

by the number of pitches in the CLC helix, i.e., $\varphi = 2\pi L/p$. The case $\tilde{\lambda} \ll \delta$ occurs, for example, in a mixture of right- and left-handed rotating CLC [15, 18]. This analytical description agrees with the numerical calculations and experimental results for optical parameters of CLCs in the cases of both normal and oblique incidences. Figure 3.9 shows the correlation between the numerical and the experimental data on cholesteric–nematic mixtures [11]. Specifically, it was first shown in [11] that higher-order reflections exist in the case of oblique incidence.

The optical properties of planar CLC layers were studied in detail in work done in Japan [19–22]. They showed that the dynamic theory of diffraction not only describes qualitatively CLC optical properties but also yields the quantitative results which agree with experimental data in a wide class of cholesteric liquid crystals and nematic–cholesteric mixtures (see Figs. 3.10, 3.11, and 3.12). In particular, the authors of [19–22] investigated second-order selective reflection thoroughly (Fig. 3.12) and confirmed the correspondence between second-order reflection polarized parameters and theoretical results above. We note that the experimental scattering spectra in Figs. 3.10, 3.11, 3.12, and 3.13 reproduce those calculated in every respect. Specifically, experiment confirms the features related to finite layer thickness and boundary reflection effects (the frequency beats of reflectivity depending on sample thickness). We omit the results which include reflection at dielectric boundaries for the case of oblique incidence; the analysis is similar to that described in Chap. 1 for normal incidence. Refer to [1] for more details.

Thus, we presented the use of diffraction theory approximations to describe the optical properties of cholesteric liquid crystals. Some papers [23–28] develop several approximate approaches which are outside the scope of our considered approximations. For example, the authors of [23, 27, 28] examine light propagation in a CLC whose pitch is much greater than wavelength for oblique incidence; some details

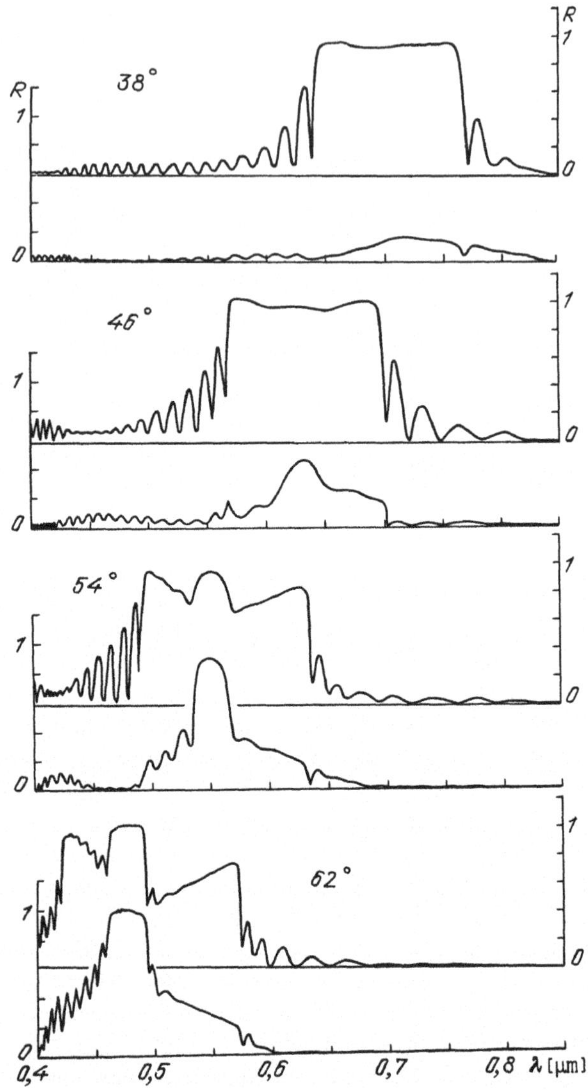

Fig. 3.10 Experimental spectra of circularly polarized light reflection from a planar cholesteric layer of thickness 6 μm for oblique incidence. The upper and lower curves correspond, respectively, to left- and right-handed polarization for every incidence angle [19]

about these approaches are found in [25, 26]. We omit the details of these complex approximate methods since it is not clear that they are useful in the presence of exact numerical methods. Nevertheless, we mention an interesting effect revealed in [27, 28] of the possible forward diffraction of light by chiral crystals with long pitches. They showed that, if diffraction conditions are satisfied, the reciprocal lattice vector can compensate for the birefringence-induced difference between two eigenwaves propagating in the same direction. Under these conditions, it is convenient to trace the transformation between orthogonal eigenpolarizations which are almost

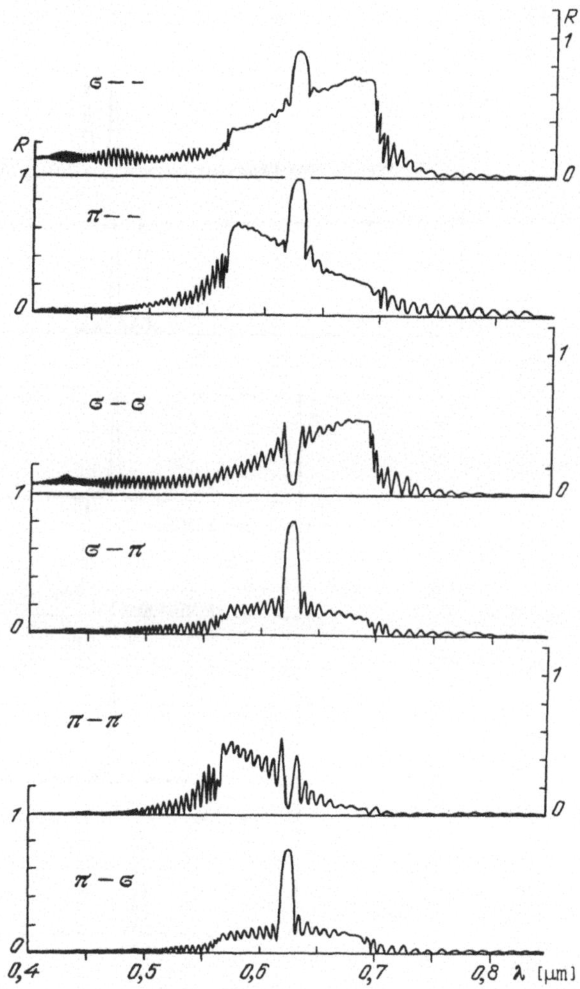

Fig. 3.11 Experimental spectra of linearly polarized light reflection from a planar cholesteric layer of 18 μm thickness and incidence angle 45°. The polarizations of incident and reflected beams (the left and right symbols) are shown. Dash instead of the right symbol corresponds to the measurements without detection of the scattered beam polarization [19]

linear [27]. We also note that the theory of linear wave transformation [29] effectively describes wave propagation in non-ideal chiral structures whose parameters vary continuously in space.

3.2 Absorbing Cholesteric Liquid Crystals

We assumed, up to this point, that cholesteric crystals do not absorb light. Actually quite often, there is considerable absorption near the absorption bands of the crystal molecules or in a special admixture (e.g., a dye). Light absorption is sensitive to CLC

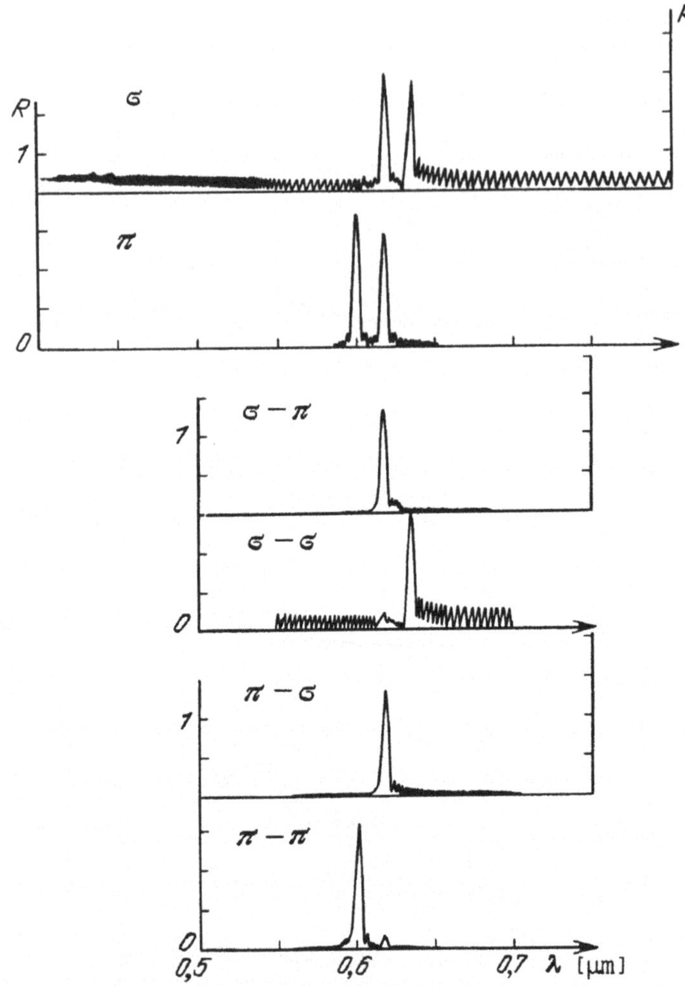

Fig. 3.12 Measured spectra of second-order reflection of linearly polarized light from a planar cholesteric layer of 38 μm thickness and incidence angle 45°. Notation is as in Fig. 3.11 [22]

helical structure such that it may vanish or there may be qualitative changes in its polarization properties. We consider here the main optical properties of absorbing CLC related to diffraction in the presence of anisotropic absorption by individual molecules and refer to [1] for an exhaustive discussion of this topic including circular dichroism in the absorption bands.

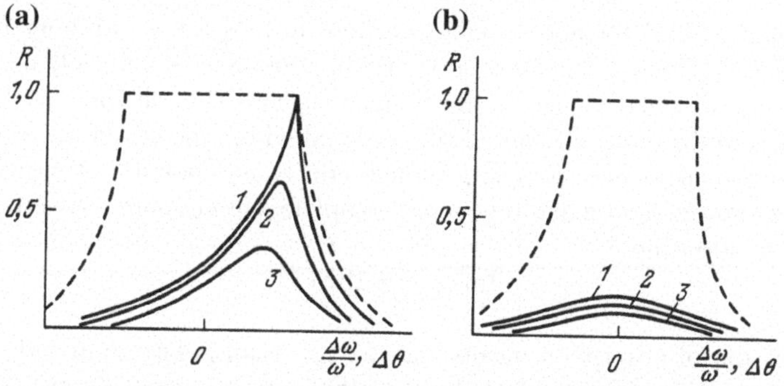

Fig. 3.13 Qualitative form of the reflection spectrum of thick cholesteric samples for different beam polarizations at oblique incidence: **a** polarization subjected to absorption suppression effect; **b** polarization not subjected to suppression effect. Curves 1–3 correspond to order parameter values $S = 1, 0.9$, and 0.7, respectively; dashed curve corresponds to the absence of absorption [30]

3.2.1 *The Dielectric Tensor of Absorbing CLCs*

Light absorption is described by the imaginary part of dielectric permittivity tensor. Hence, we use the same expression as above except that the principal values ϵ_1, ϵ_2, ϵ_3 of the tensor $\hat{\epsilon}$ are now complex. For example, if the absorption spectrum of a CLC includes several absorption bands which are linearly polarized at angles ψ_j to the molecular long axis, the imaginary parts of $\bar{\epsilon} = (\epsilon_1 + \epsilon_2)/2$ and $\epsilon_a = \epsilon_1 - \epsilon_2$ are [31]

$$\operatorname{Im}\bar{\epsilon} = \sum_j \eta_j \left[1 + 1/2\left(1 - \frac{3}{2}\sin^2\psi_j\right)S\right],$$

$$\operatorname{Im}\epsilon_a = 3S\sum_j \eta_j[1 - 3/2\sin^2\psi_j] \tag{3.44}$$

where the value of η_j is proportional to the corresponding oscillator strength and S is the order parameter.

3.2.2 *Suppression of Absorption*

Consider the effect of absorption on the optical parameters in the selective reflection band [32–35]. Absorption usually reduces the reflection from a CLC. However, since the reflection from a CLC is connected with diffraction, these reflections decrease and light absorption may be diminished considerably in the selective reflection band. The effect of absorption suppression of diffraction nature, the Borrmann effect, is known for X-rays, Mössbauer gamma quanta, and neutrons [36]. This effect in CLCs is considered in [34] for normal incidence and in [35] for an arbitrary incidence angle. It is shown therein that absorption may be suppressed completely if absorbing oscillators are directed along the molecular long axis (i.e., at $\mathrm{Im}\,\epsilon_1 \neq 0, \mathrm{Im}\,\epsilon_2 = \mathrm{Im}\,\epsilon_3 = 0$) and the molecules are ideally ordered within the crystal ($S = 1$). In the case of normal incidence, complete suppression is also possible if oscillators are oriented across the long axis. In general, complete suppression is governed by the following relationship

$$\mathrm{Im}\,\epsilon_a = \pm 2\mathrm{Im}\,\bar{\epsilon}. \tag{3.45}$$

Suppression of absorption means that, for a certain polarization and incidence angle, the intensity of light reflected from a thick crystal equals the incident intensity (Fig. 3.13). For example, in the case where absorbing oscillators lie along or across the long axis and normal incidence, reflection becomes unity at frequency $\omega = \omega_B\sqrt{2\epsilon_2/(\mathrm{Re}\,\epsilon_1 + \epsilon_2)}$ for diffracting circular polarization; this follows directly from (1.16).

The physical cause of absorption suppression is that the electric field vector of wave superposition within the crystal (1.20) is, at every point of the CLC, perpendicular to the molecular long (or short) axes (see Fig. 3.14) and, hence, to the absorbing oscillators; however, there is no absorption of these waves. For real crystals, there is partial absorption suppression; the peak of the reflection curve never reaches unity (Fig. 3.14), and its height strongly depends on order parameter S.

Absorption suppression also manifests itself in circular dichroism. Using (3.37) and (3.38), the value of circular dichroism for thick absorbing CLCs is expressed as

$$D = th[L\,\mathrm{Im}(k^- - k^+)] \tag{3.46}$$

where $k^\pm$ are defined by (1.15). At $\mathrm{Im}\,\bar{\epsilon} \gg |\delta|$, this expression becomes

$$D = th\left[\frac{q^2L}{8\tau}\frac{\mathrm{Im}\,\bar{\epsilon}(\mathrm{Re}^2\,\epsilon_a - \mathrm{Im}^2\,\epsilon_a) - \gamma Im\epsilon_a\mathrm{Re}\,\epsilon_a}{\gamma^2 + 4\mathrm{Im}^2\,\bar{\epsilon}}\right] \tag{3.47}$$

where $\gamma = 2[1 - (\omega_B/\omega)^2]\,\mathrm{Re}\,\bar{\epsilon}$.

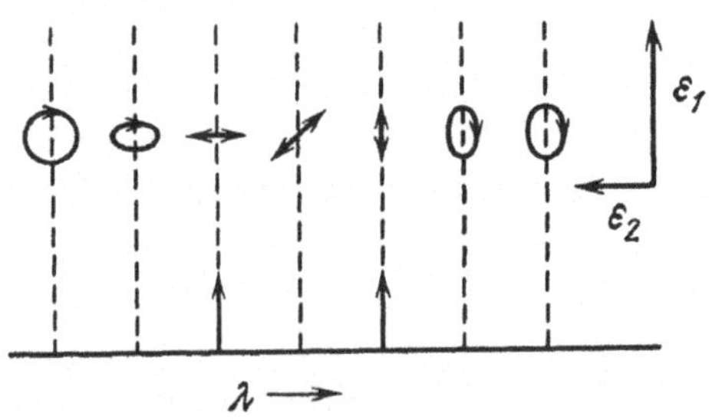

Fig. 3.14 Orientation of total field polarization in a CLC for the diffracting eigenwave as a function of wavelength [34]

It follows from (3.47) that the value of D changes sign with frequency. This means that in the diffractive reflection band the absorption not only is suppressed but also increases to more than its average value.

Absorption suppression and the frequency dependence of dichroism were observed experimentally in [37, 38] when studying light intensity changes as a function of frequency, while the light is transmitted through a CLC in the selective reflection band [37, 38].

3.2.3 Optical Rotation

Anisotropic absorption strongly affects the frequency dependence of the angle of rotation of the polarization plane. A non-absorbing crystal in the selective reflection region exhibits both right- and left-handed rotation (see Fig. 3.8), but an absorbing crystal rotates predominantly in one direction [34, 35]. This is because the absorptions of slow and fast waves in expansion (1.18) are very different; consequently, one of the waves is suppressed in CLCs and the curve of optical rotation becomes asymmetric or the sign of rotation remains unchanged.

The angle of optical rotation is determined, as for non-absorbing CLCs, by the phase difference between the addends in (3.37), except that $\hat{\epsilon}$ is now complex.

Under the assumptions of (3.47), we find the angle of polarization rotation

$$\varphi = \frac{q^2 L}{8\tau} \frac{\gamma(\mathrm{Re}^2\,\epsilon_a - \mathrm{Im}^2\,\epsilon_a) + 4\mathrm{Im}\,\bar{\epsilon}\cdot\mathrm{Im}\,\epsilon_a \mathrm{Re}\,\epsilon_a}{\gamma^2 + 4\mathrm{Im}^2\,\bar{\epsilon}}. \tag{3.48}$$

This formula produces the curves plotted in Fig. 3.15.

3.2.4 Borrmann Effect for Oblique Incidence

We briefly consider the Borrmann effect for oblique incidence light. In this case, solutions of the dynamic system (3.3) show that absorption may be suppressed completely [1, 37]; i.e., the imaginary part of the wave vector may vanish according to (3.7).

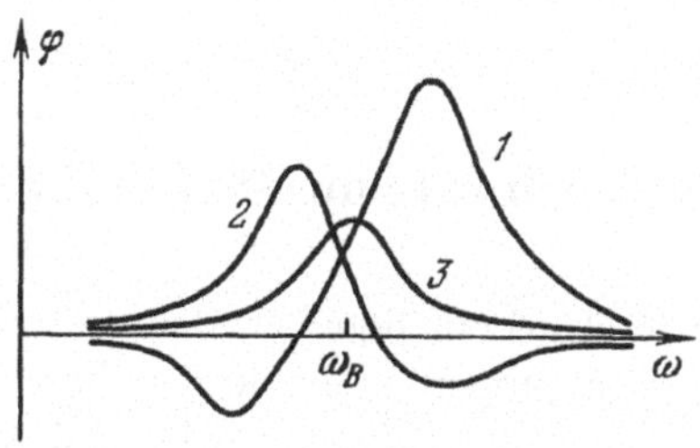

Fig. 3.15 Qualitative dependence of the angle of polarization plane rotation on frequency for different values of absorption in a CLC: 1 – $\mathrm{Re}\,\epsilon_a > \mathrm{Im}\,\epsilon_a$, 2 – $\mathrm{Re}\,\epsilon_a < \mathrm{Im}\,\epsilon_a$, 3 – $\mathrm{Re}\,\epsilon_a = \mathrm{Im}\,\epsilon_a$

As in the case of normal incidence, complete absorption occurs, i.e., reflection R is unity, when a certain relationship exists between the absorption anisotropy and its average value; this relationship is (3.45) with a positive sign. In contrast to the case of normal incidence, complete absorption suppression occurs when absorbing oscillators are oriented along the molecular long axis, i.e., $\psi = 0$ in (3.44). If $\psi = \pi/2$, and in particular, if oscillators are oriented across the long axis, it is impossible to have complete absorption suppression.

In addition, the degree of suppression strongly depends on the order parameter S; i.e., the suppression is complete only at $S = 1$ and decreases with the decrease of S. Figure 3.13 shows the calculated values of reflection for oblique incidence of light with different polarizations when absorbing oscillators lie along the molecular long axis. In Fig. 3.13a, the polarization of the incident wave coincides with that of the direct wave in the eigensolution for which complete suppression occurs. Figure 3.13b corresponds to incident polarization of the second eigensolution; this figure shows that even though the order parameter S deviates slightly from unity, the reflection value decreases sharply at its maximum. That is, if S decreases from 1 to 0.9, the reflection decreases from 1 to 0.7.

It can be shown that reflection from an absorbing CLC may reach unity for a single value of frequency or angle. Reflection is complete when incident polarization [see (3.12)] is given by

$$tg\alpha e^{i\beta} = -i/\sin\theta. \tag{3.49}$$

It was mentioned that this polarization coincides with that of the direct wave which is excited in a CLC. At the frequency value or angle in question, the total electric field of the corresponding eigensolution is, at any point in the CLC, perpendicular to the molecular long axis; i.e., the wave does not experience absorption. We note that the exact solution for non-absorbing CLC with this field configuration was first exposed by Dreher and Meier [7].

The fact that absorption cannot be completely suppressed when absorbing oscillators are perpendicular to the molecular long axis has a simple physical interpretation; it is impossible to produce a configuration such that the electric field is everywhere perpendicular to absorbing oscillators.

The Borrmann effect for oblique incidence was experimentally observed in [39], but it was clearly present only when the direction of light propagation made a small angle with the optical axis ($\pi/2 - \theta < 19°$). No effect was observed at $\frac{\pi}{2} - \theta > 19°$, perhaps because the authors made no provision for the polarization dependence of the effect.

3.3 Chiral Smectic Liquid Crystals

Chiral smectic liquid crystals are another type of liquid crystal whose structure and optical properties are close to those of CLCs. Chiral smectic liquid crystals, C^* [3], differ from CLCs in their layered structure. The C^* molecules produce mono- and bimolecular layers such that the molecular interactions within them are stronger than those between molecules of different layers. The mean orientation of all molecules within a layer is the same in a C^*, but this orientation (direction) changes from layer to layer, as in CLCs. The difference between them is that the C^* director is not orthogonal to the helical axis. Hence, C^* structure is represented as in Fig. 1.3 if all molecular axes, keeping the azimuthal angle constant, are turned so that their axes make the same indirect angle with the helical axis. Since C^* optical properties are described in the same way as those of CLCs, we superficially mention the specific features of C^*; the details are found in [1].

3.3.1 Dielectric Permittivity Tensor of a C^*

Since the structures of C^*s and CLCs differ, their dielectric permittivity tensors differ in their dependence on coordinates along the helical axis Z. Specifically, when going along Z in a CLC, only two of the CLC tensor principal axes change orientation, but in a C^* all principal axes of $\hat{\epsilon}$ change orientation. In accordance with structure, the period of C^* dielectric properties along Z coincides with the helical pitch, while in a CLC this period is $p/2$.

The optical properties of a C^* are determined by its tensor $\hat{\epsilon}(z)$ [40, 41]

$$\hat{\epsilon}(\mathbf{r}) = \begin{pmatrix} \epsilon_{11}^0 + \epsilon_a \cos 2\varphi & \epsilon_a \sin 2\varphi & \epsilon_a' \cos\varphi \\ \epsilon_a \sin 2\varphi & \epsilon_{22}^0 - \epsilon_a \cos 2\varphi & \epsilon_a' \sin\varphi \\ \epsilon_a' \cos\varphi & \epsilon_a' \sin\varphi & \epsilon_{33}^0 \end{pmatrix} \tag{3.50}$$

where

$$\begin{aligned} \epsilon_{11}^0 &= \epsilon_{22}^0 = (\epsilon_1 + \epsilon_2 \cos^2\Theta + \epsilon_3 \sin^2\Theta)/2 \\ \epsilon_{33}^0 &= \epsilon_2 \sin^2\Theta + \epsilon_3 \cos^2\Theta, \\ \epsilon_a &= (\epsilon_1 - \epsilon_2 \cos^2\Theta - \epsilon_3 \sin^2\Theta)/2 \\ E_a' &= (\epsilon_3 - \epsilon_2) \sin\Theta \cos\Theta \end{aligned} \tag{3.51}$$

In these formulae, Θ and φ are, respectively, the polar and azimuthal angles of one of the principal axes, namely axis "3" of the dielectric tensor of an individual smectic layer with respect to the fixed spatial axes X, Y, Z, the X- and Y-axes lying in the plane of the layer and Z perpendicular to the layer. The direction of axis "3" is determined by the long axis of molecules in a layer, while axis "1" lies in the plane of the layer: ϵ_1, ϵ_2, and ϵ_3 are the principal values of the $\hat{\epsilon}$ tensor corresponding to the "1", "2", and "3" axes. The angle Θ is constant over C^* volume, and φ changes with layer in agreement with helical structure

$$\varphi = 2\pi z/p \tag{3.52}$$

where p is helical pitch. Since φ depends on z, tensor $\hat{\epsilon}$ is a periodic function of z and is expanded into a Fourier series

$$\hat{\epsilon}(z) = \sum_{s=-\infty}^{\infty} \hat{\epsilon}_s \exp[is\tau z] \tag{3.53}$$

where $\tau = \frac{2\pi}{p}$. It follows from (3.50) and (3.52) that in a C^*, $\hat{\epsilon}_0$, $\hat{\epsilon}_{\pm 1}$ and $\hat{\epsilon}_{\pm 2}$ in (3.53) are nonzero and determined by the formulae

$$\hat{\epsilon}_1 = \hat{\epsilon}^*_{-1} = 1/4(\epsilon_2 - \epsilon_3)\sin 2\Theta \begin{pmatrix} 0 & 0 & \pm i \\ 0 & 0 & 1 \\ \pm i & 1 & 0 \end{pmatrix}$$

$$\hat{\epsilon}_{-2} = \hat{\epsilon}^*_2 = 1/4(\epsilon_1 - \epsilon_2 \cos^2\Theta - \epsilon_3 \sin^2\Theta) \begin{pmatrix} 1 & \pm i & 0 \\ \mp i & -1 & 0 \\ 0 & 0 & 0 \end{pmatrix} \tag{3.54}$$

$$\hat{\epsilon}_s \equiv 0, \ at\ |s| \geq 3.$$

The upper and lower signs in (3.54) correspond to the right- and left-handed helices, respectively. In the $\hat{\epsilon}_0$ tensor, only diagonal components are nonzero [see (3.50) and (3.51)]:

$$(\hat{\epsilon}_0)_{11} = (\hat{\epsilon}_0)_{22} = \epsilon^0_{11}, \quad (\hat{\epsilon}_0)_{33} = \epsilon^0_{33}. \tag{3.55}$$

Note that because of different periodicities in CLCs and C^*s, we must indicate that $\tau = 2\pi/p$ in (3.53) rather than $4\pi/p$ as for CLCs. Note also that the Fourier expansion (3.54) does not include C^* periodicity related to layered structure because the corresponding period is small and appears only in X-ray scattering.

3.3.2 *Second-Order Diffraction Reflection*

As in (3.54), the second harmonic in the Fourier expansion of $\hat{\epsilon}(z)$ for a C^* coincides with, to within a factor, the fundamental harmonic of expansion (2.15) for a CLC. This means that second-order diffraction scattering in a C^* is qualitatively similar to that of first order in a CLC. This case can be analyzed quantitatively using the equations of Sect. 3.1 with the Fourier components $\hat{\epsilon}_0$ and $\hat{\epsilon}_{\pm 2}$ from (3.54) and (3.55).

If light propagates along the C^* optical axis, according to the exact solution for this case, only the second-order diffraction reflection occurs. Then, the optical properties of a C^* are similar to those of a CLC and are described by the expressions given in Sect. 1.2 in which $\bar{\epsilon}$ is substituted by

$$\bar{\epsilon}' = \frac{1}{2}(\epsilon_1 + \epsilon_2 \cos^2\Theta + \epsilon_3 \sin^2\Theta) - \frac{1}{8\epsilon_3}(\epsilon_2 - \epsilon_3)^2 \sin^2\Theta$$

and δ is substituted by

$$\delta' = \frac{1}{2\bar{\epsilon}'}(\epsilon_1 - \epsilon_2 \cos^2\Theta - \epsilon_3 \sin^2\Theta) + \frac{1}{8\epsilon_3\bar{\epsilon}'}(\epsilon_2 - \epsilon_3)^2 \sin^2\Theta. \quad (3.56)$$

3.3.3 First-Order Reflection

Since the optical properties of a C^* differ from those of CLCs only for light propagation at an angle to the optical axis in first-order reflection, we restrict ourselves to first-order diffraction reflection in a C^*. If we substitute $\hat{\epsilon}$ in the form of (3.54) and (3.55) into (2.26), the equation system degrades into two independent systems:

$$\begin{cases} (\epsilon_\sigma - k_0^2/q_0^2)E_0^\sigma + fE_1^\pi = 0 \\ fE_0^\sigma + (\epsilon_\pi - k_1^2/q_0^2)E_1^\pi = 0 \end{cases}$$
$$\begin{cases} (\epsilon_\pi - k_0^2/q_0^2)E_0^\pi + fE_1^\sigma = 0 \\ fE_0^\pi + (\epsilon_\sigma - k_1^2/q_0^2)E_1^\sigma = 0 \end{cases} . \quad (3.57)$$

where E^σ and E^π are amplitudes of the σ- and π-polarized waves, $\epsilon_\sigma = (\hat{\epsilon}_0)_{11}, \epsilon_\pi = (\hat{\epsilon}_0)_{11} \sin^2\theta + (\hat{\epsilon}_0)_{33} \cos^2\theta, f = (\epsilon_2 - \epsilon_3) \sin 2\theta \cos\theta, q_0^2 = (\omega/c)^2$, and $\frac{\pi}{2} - \theta$ is the angle between the optical axis and direction of light propagation. It follows from (3.57) that a σ-polarization yields a π-polarized wave after diffraction and vice versa.

The solution for (3.57) and the corresponding boundary problem is more simple than in the case of a CLC and similar to the analogous solution for X-ray diffraction. Hence, we give the final results for a plane-parallel sample of C^*.

3.3.4 Boundary Problem

When the helical axis is perpendicular to a C^* surface (Bragg's case), the angular (frequency) reflection bands for σ- and π-polarizations given by (3.57) coincide. The reflection is then polarization and angle (frequency) dependent as

$$R = |\mathbf{e}_1^* \hat{\epsilon}_1 \mathbf{e}_0|^2 \frac{\sin^2\left[\frac{q_0 L(\Delta_B^2 - f^2)^{1/2}}{2\bar{\epsilon}^{1/2}\sin\theta}\right]}{\Delta_B^2 - f^2\cos^2\left[\frac{q_0 L(\Delta_B^2 - f^2)^{1/2}}{2\bar{\epsilon}^{1/2}\sin\theta}\right]} \quad (3.58)$$

where $\mathbf{e}_0$ and $\mathbf{e}_1$ are the polarization vectors of the incident and diffracted waves, $\bar{\epsilon} = (\epsilon_\sigma + \epsilon_\pi)/2$, L is crystal thickness, and parameter Δ_B characterizes the deviation from the Bragg condition as above:

$$\Delta_{\mathrm{B}} = (\tau/2q_0^2)\left(2q_0\sqrt{\bar{\epsilon} - \bar{\epsilon}\cos^2\theta} - \tau\right).$$

Note that the result of summation of (3.58) over final polarization $\mathbf{e}_1$ does not depend on $\mathbf{e}_0$; i.e., only scattered wave polarization, rather than intensity, depends on initial wave polarization. The reflection band is limited by values $\Delta_{\mathrm{B}} = \pm f$, and its center lies at $\Delta_{\mathrm{B}} = 0$.

Polarization parameters are easily revealed by analyzing factor $|\mathbf{e}_1^*\hat{\epsilon}\mathbf{e}_0|^2$ in (3.58). That is, in first-order diffraction reflection from a C^*, the right-handed circularly polarized wave becomes left-handed polarized and vice versa (at normal incidence in a CLC however, one of the circularly polarized waves is not reflected, and the other is reflected and retains its polarization). Linearly polarized light remains linearly polarized, and angles φ_0 and φ_1 between the scattering plane and incident planes and scattered light polarization, respectively, are related as $\varphi_1 = \pi/2 - \varphi_0$.

When the helical axis is parallel to the sample surface (Laue case), the reflection bands for σ- and π-polarizations generally do not coincide although they may partially overlap. When $L \gg q_0 p^2 \cos\theta$, the reflection and its frequency (angle) dependence for the σ- and π-polarizations are presented as

$$R = f^2[(\Delta_L \pm m)^2 + f^2]^{-1} \sin^2\left[\frac{q\sqrt{(\Delta_L \pm m)^2 + f^2}}{2\sqrt{\bar{\epsilon}}\cos\theta}\right] \tag{3.59}$$

where $m = (\epsilon_\sigma - \epsilon_\pi)/2 = \cos^2\theta[\epsilon_1 + \epsilon_2(1 - 3\sin^2\Theta) + \epsilon_3(1 - 3\cos^2\Theta)]/4\Delta_L = (\tau/2q_0^2)(\tau - 2q_0\sin\theta)$ and the "plus" and "minus' signs correspond to σ- and π-polarizations, respectively. Note that at $m = 0$ reflection bands coincide and polarization properties are similar to those of the Bragg case. It follows from (3.59) that reflection coefficient R in the Laue case is non-monotonically dependent on thickness L; these are the pendulum beats known in the theory of X-ray diffraction. These beats occur particularly in smectic liquid crystals with long period for diffraction in the forward direction [42].

These analytical results (exact and from dynamic diffraction theory) agree with the numerical calculations of smectic crystal optical properties [27, 40]. Note that, for simplicity, we used a physically justified constraint for the $\hat{\epsilon}(z)$ tensor of a C^*, viz. one of its principal axes lies in the plane layer. If that were not the case, no new effect and no principal difficulties in solving the problem would appear but the corresponding expressions would be cumbersome; thus, we omit them.

In conclusion, we note that C^*s may be characterized by ferroelectric properties [43] which make their structure and optical properties sensitive to electric fields. Thus, C^*s with ferroelectric properties show promise for applications in electro-optical imaging devices. Hence, the electro-optics of C^*s is studied extensively [1, 44, 45] but this topic is out of the scope of this book.

3.4 Blue Phase of Liquid Crystals

The blue phase of CLCs exhibits the most sophisticated optical properties of all liquid crystals [1, 2, 46]. However, the nature and structure of the blue phases are not determined with confidence at the present time; but, it is clear that optical methods are the most useful in the study of blue phases in general and spatial structures of their dielectric properties in particular. Hence, we do not assume a concrete form for the dielectric tensor of blue phases but place the emphasis at specific optical features which may help to reveal blue phase structure in the future experiments.

3.4.1 Observed Properties of Blue Phases

The blue phase was observed for some CLCs in a narrow temperature range of less than or about one degree in the transition between isotropic liquid and usual cholesteric phase. The studies used several physical methods and showed that the blue phase is a thermodynamically stable state. However, in that narrow temperature range, some compounds exhibit up to three different blue phases labeled BPI, BPII, and BPIII in the order they appear as temperature increases. Phase BPIII is the least investigated; hence, we restrict ourselves to the two former cases.

Without going into details of blue phase properties (see [1, 2, 46]), we discuss the most unusual optical properties. First, the blue phase selectively reflects visible light—sometimes, it appears as blue (Figs. 3.16 and 3.17). In contrast to CLCs where

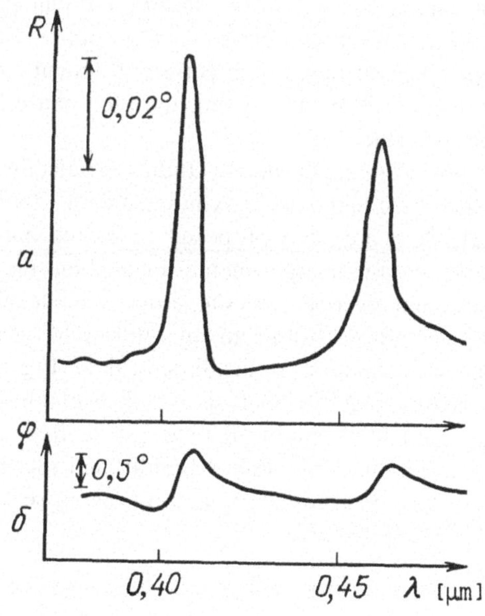

Fig. 3.16 Dependences of selective reflection (upper curve) and optical rotation (lower curve) on wavelength in cholesteryl pelargonate at 91.35 °C [47] corresponding to the transition from BPI to BPII. Sample thickness is 25 and 12 μm, respectively

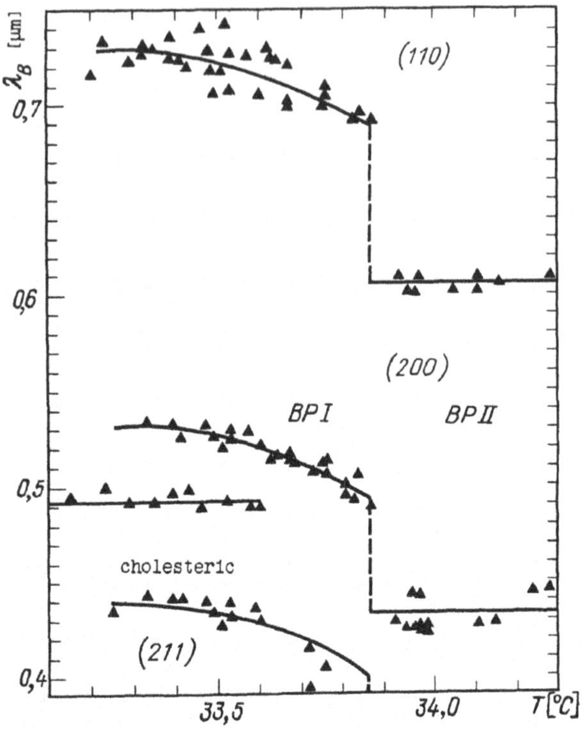

Fig. 3.17 Selective reflection dependence on the temperature of a CLC and blue phases BPI and BPII for the 50–50% mixture of chiral and non-chiral biphenyls (BPI is supercooled below 33.6 °C) [52]

only first-order selective reflection is strong, the blue phase yields several reflexes of comparable intensity. Definite circular polarization, as in CLCs, is selectively reflected backward from the blue phase. Second, the blue phase exhibits strong optical gyrotropy, the direction of optical rotation that inverts sign at the selective reflection frequency. Third, the blue phase is optically isotropic—it has no linear birefringence.

In recent experimental studies of selective reflection, the consensus was that the blue phase had a cubic symmetry with either a primitive or body-centered unit cell [2]. There are many proposed structures that agree with the experimental data or, at least, the main experimental facts about the blue phase. Specifically, these are the three-dimensional periodic lattice of the locally biaxial dielectric permittivity tensor which follows from Landau's theory [48–50] and the three-dimensional lattice of linear declinations in the director field (Fig. 3.18 [51]). Determination of blue phase structure requires analysis of all available experimental data and the preferential assignment of one of the model structures. If this is not possible from the available data, we must recommend additional measurements which are necessary for the structure determination. The following sections of this chapter describe a way of achieving this goal.

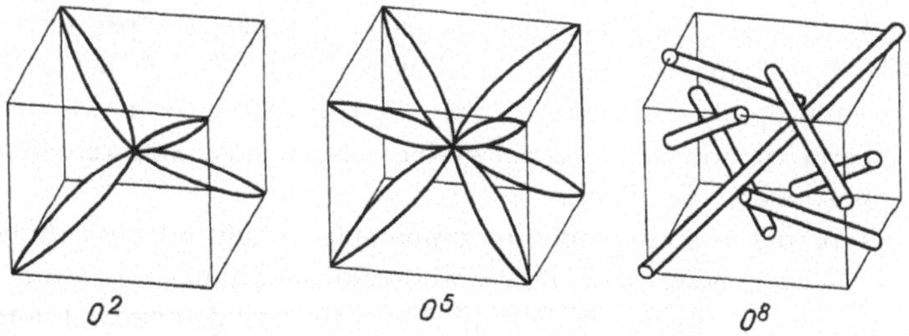

Fig. 3.18 Structure of declinations in cubic cells of various space groups [53]

3.4.2 Symmetry Restrictions for the Dielectric Permittivity Tensor

Since the dielectric permittivity tensor $\hat{\epsilon}(\mathbf{r})$ contains all the information about the optical properties of a blue phase, it is important to find its explicit form. However, blue phase structure is not known; therefore, we do not assume a specific form of $\hat{\epsilon}(\mathbf{r})$ but account for its properties based on known experimental facts. This is done by expanding $\hat{\epsilon}(\mathbf{r})$ in a Fourier series

$$\hat{\epsilon}(\mathbf{r}) = \sum_{\tau} \hat{\epsilon}_{\tau} e^{i\tau\mathbf{r}} \tag{3.60}$$

where $\boldsymbol{\tau}$ are reciprocal lattice vectors of the blue phase.

We set constraints for coefficients of (3.60) from known optical properties. First, the presence of selective scattering with several reflexes means that, in contrast to expansion (2.15) for a CLC, expansion (3.60) contains more than three nonzero terms. Also, the known polarization properties of selective scattering for these reflexes determine certain constraints for the corresponding Fourier components $\hat{\epsilon}_{\tau}$ of the tensor. Second, the optical activity and the sign inversion at the selective reflection frequency support the conclusion that blue phase optical activity is due to structural properties. Hence, the molecular gyrotropy in $\hat{\epsilon}$ may be neglected at least to a first approximation. Third, the absence of birefringence means that the zero harmonic $\hat{\epsilon}_0$ is proportional to the unit tensor.

In view of these considerations and the experimental data, we assume that the blue phase has cubic structure with the tensor

$$\hat{\epsilon}(\mathbf{r}) = \hat{\epsilon}_0 + \hat{\epsilon}^{a}(x,\ y,\ z) \tag{3.61}$$

where $\hat{\epsilon}_0$ is the average dielectric permittivity and $\epsilon^{a}_{ik}(x,\ y,\ z)$ is the three-dimensional periodic part of $\hat{\epsilon}(\mathbf{r})$. We neglect the weak effects of possible spatial periodicity of blue phase density, light absorption, and molecular gyrotropy. The $\hat{\epsilon}^{a}(\mathbf{r})$ tensor is spurless, real, and symmetric under these conditions. In conventional crystal optics, crystals are assumed homogeneous since the light wavelength is longer than the lattice period; here, only the homogeneous part of the dielectric permittivity tensor is essential and its symmetry is determined by the point group of the crystal and is well known from optics. For cubic structures (analyzed here), in accordance with the adopted approximations, the part of $\hat{\epsilon}_0$ that corresponds to the zero harmonic of (3.60) is proportional to the identity tensor.

The lattice period of a blue phase is comparable to light wavelength; hence, we include the inhomogeneous part, $\hat{\epsilon}^{\mathrm{a}}(\mathbf{r})$, whose local symmetry is not necessarily identical with cubic symmetry, which varies over the unit cell and is limited by the crystal space group. We use these symmetry restrictions to establish the most general form of $\hat{\epsilon}^{\mathrm{a}}(\mathbf{r})$ for specific space groups describing blue phase structure. Since the blue phase is chiral, we restrict ourselves to space groups without inversion centers, i.e., enantiomorphic groups.

The symmetry properties permit us to determine the most general form of the blue phase tensor $\hat{\epsilon}(\mathbf{r})$ [54], i.e., the form which is invariant with respect to all transformations of the space group describing blue phase symmetry properties. According to experiment [2], the observed blue phase symmetry is described by cubic space groups.

In the study of blue phase $\hat{\epsilon}(\mathbf{r})$ symmetry properties, we account for that fact that all cubic space groups have a third-order symmetry axis directed along the cube diagonal. As a result, all components of the $\hat{\epsilon}(\mathbf{r})$ tensor depend on two arbitrary periodic functions of coordinates $f_1(\mathbf{r})$ and $f_2(\mathbf{r})$:

$$\hat{\epsilon}(x, y, z) = \begin{pmatrix} f_1(x, y, z) & f_2(z, x, y) & f_2(y, z, x) \\ f_2(z, x, y) & f_1(y, z, x) & f_2(x, y, z) \\ f_2(y, z, x) & f_2(x, y, z) & f_1(z, x, y) \end{pmatrix}. \tag{3.62}$$

We find directly that the tensor of (3.62) is invariant under rotations of 120° and 240° around the third-order axis; this is equivalent to the cyclic permutation of coordinates X, Y, and Z with a corresponding cyclic permutation of $\hat{\epsilon}(\mathbf{r})$ tensor components, e.g., $x \to y \to z \to x$, $\epsilon_{xx} \to \epsilon_{yy} \to \epsilon_{zz} \to \epsilon_{xx}$, etc.

In addition to third-order axes, a cubic group necessarily includes the other elements of symmetry, namely the second- and fourth-order axes, screw axes, and translations. These elements yield additional restrictions for functions $f_i(\mathbf{r})$ [54, 55]; specifically, groups $T^1 - P23$, $T^2 - F23$, $T^3 - I23$, yield the following relationships for $f_i(\mathbf{r})$ because they have second-order rotation axes:

$$\begin{aligned} f_1(x, y, z) &= f_1(x, \bar{y}, \bar{z}) = f_1(\bar{x}, \bar{y}, z) = f_1(\bar{x}, y, \bar{z}) \\ f_2(x, y, z) &= f_2(x, \bar{y}, \bar{z}) = -f_2(\bar{x}, \bar{y}, z) = -f_2(\bar{x}, y, \bar{z}) \end{aligned} \tag{3.63}$$

where $\bar{x}_i = -x_i$. Since groups $T^4 - P2_13$ and $T^5 - I2_13$ have the screw 2_1 axes, they yield

$$f_1(x,y,z)=f_1\left(\frac{1}{2}+x,\frac{1}{2}-y,\bar{z}\right)=f_1\left(\frac{1}{2}-x,\bar{y},\frac{1}{2}+z\right)$$
$$=f_1\left(\bar{x},\frac{1}{2}+y,\frac{1}{2}-z\right)$$
$$f_2(x,y,z)=f_2\left(\frac{1}{2}+x,\frac{1}{2}-y,\bar{z}\right)=-f_2\left(\frac{1}{2}-x,\bar{y},\frac{1}{2}+z\right)$$
$$=-f_2\left(\bar{x},\frac{1}{2}+y,\frac{1}{2}-z\right). \tag{3.64}$$

All other cubic groups contain subgroups T^1 or T^4; hence, relationships (3.63) and (3.64) are valid for them, plus an additional relationship. Namely, those relationships are: (3.63) and

$$f_i(x,y,z)=f_i(\bar{x},\bar{y},\bar{z}) \tag{3.65}$$

for $O^1 - P432$, $O^3 - F432$ and $O^5 - I432$; (3.63) and

$$f_i(x,y,z)=f_i\left(\frac{1}{2}-x,\frac{1}{2}-z,\frac{1}{2}-y\right) \tag{3.66}$$

for $O^2 - P4_243$; (3.64) and

$$f_i(x,y,z)=f_i\left(\frac{1}{4}-x,\frac{1}{4}-z,\frac{1}{4}-y\right) \tag{3.67}$$

for $O^4 - F4_132$, $O^6 - P4_332$ and $O^8 - I4_732$; and (3.64) and

$$f_i(x,y,z)=f_i\left(\frac{3}{4}-x,\frac{3}{4}-z,\frac{3}{4}-y\right) \tag{3.68}$$

for $O^7 - P4_132$. Beyond this, for body-centered groups ($I23$, $I2_13$, $I432$, $I4_132$) relationship$f_i(x,y,z)=f_i(\frac{1}{2}+x,\frac{1}{2}+y,\frac{1}{2}+z)$ holds and for face-centered groups ($F23$, $F432$, $F4_132$) the additional relationship $f_i(x,y,z)=f_i(\frac{1}{2}+x,\frac{1}{2}+y,z)=f_i(\frac{1}{2}+z,y,z+\frac{1}{2})=f_i(x,\frac{1}{2}+y,\frac{1}{2}+z)$ applies. We note that relationships (3.65)–(3.68) are given for a coordinate system with axes directed along the edges of the unit cubic cell and origin at the point of maximum point symmetry [56]. No other relationship stems from the requirement that $\hat{\epsilon}(\mathbf{r})$ be invariant under transformations of enantiomorphic space groups. Relationships (3.53)–(3.68) and the relationships connected with the centering properties of unit cells are not sufficient to determine functions $f_i(\mathbf{r})$, although they impose stringent constraints on the $\hat{\epsilon}(\mathbf{r})$ tensor, its Fourier harmonics, and phase relationships between different Fourier harmonics of (3.60).

3.4.3 Fourier Harmonics of $\hat{\epsilon}(\mathbf{r})$

Consider the vector of the reciprocal lattice in expansion (3.60) in the form $\boldsymbol{\tau} = (\frac{2\pi}{d})(h\mathbf{x} + k\mathbf{y} + \ell\mathbf{z})$ where $\mathbf{x}, \mathbf{y}, \mathbf{z}$ are the unit vectors of the coordinate axis, h, k, and ℓ are arbitrary integers, the so-called Miller indices, and d is the dimension of the crystal unit cell. Then, the Fourier harmonics are

$$\hat{\epsilon}_{\tau} = \frac{1}{d^3}\int \hat{\epsilon}(\mathbf{r}) \exp(-i\boldsymbol{\tau}\mathbf{r}) \mathrm{d}r \tag{3.69}$$

and integration is over the volume of a unit cell. First of all, formulae (3.62) and (3.69) yield that in cubic crystals

$$\hat{\epsilon}_{hk\ell} = \begin{pmatrix} f_1^{hk\ell} & f_2^{\ell hk} & f_2^{k\ell h} \\ f_2^{\ell hk} & f_1^{k\ell h} & f_2^{hk\ell} \\ f_2^{k\ell h} & f_2^{hk\ell} & f_1^{\ell hk} \end{pmatrix} \tag{3.70}$$

where

$$f_i^{hk\ell} = \frac{1}{d^3}\int f_i(x,\ y,\ z) \exp[2\pi i(hx + ky + \ell z)] \mathrm{d}r. \tag{3.71}$$

Further, as follows from (3.70) and from the fact that the $\hat{\epsilon}^{\mathrm{a}}(\mathbf{r})$ tensor has zero spur, the Fourier harmonics $\epsilon_{hk\ell}$, $\epsilon_{k\ell h}$ and $\epsilon_{\ell hk}$ are expressed by a single set of five numbers which generally are complex. Using the properties of $f_i(\mathbf{r})$ given by (3.63)–(3.68), we obtain restrictions on $f_i^{hk\ell}$ for reflexes of a special type presented in Table 3.1. These relationships simplify the study of blue phase optical properties and the phase transitions in it. R, I, and C are real, imaginary, and complex quantities, respectively; the letter c denotes chiral reflections, while the asterisk denotes reflections that become allowed as a result of the local anisotropy of $\hat{\epsilon}$. The condition that reflections for body-centered lattices ($T^3, T^5,\ O^5,\ O^8$) is $h + k + \ell = 2n$ while, for face-centered lattices ($T^2,\ O^3,\ O^4$), the condition is that h, k and ℓ have the same parity.

The expressions for $\hat{\epsilon}(\mathbf{r})$ and its Fourier harmonics were obtained within the above specified single coordinate system. It is often convenient to consider each Fourier harmonic $\hat{\epsilon}_{\tau}$ in its own coordinate system in which one of the axes (say Z) is directed along $\boldsymbol{\tau}$ and the two others perpendicular to $\boldsymbol{\tau}$. In such a system, the ϵ_{τ} tensor is expanded with five basic tensors $\hat{\sigma}_m (m = 0,\ \pm 1,\ \pm 2)$, each with the property that transformation of rotation around τ by an angle φ reduces to multiplication by $\exp(im\varphi)$. That is, the $\hat{\epsilon}_{\tau}$ tensor is expanded over the irreducible representations of the group of rotations around τ, namely

$$\hat{\epsilon}_{\tau} = \sum_{m=-2}^{2} \epsilon(\boldsymbol{\tau}; m)\hat{\sigma}_m \tag{3.72}$$

Table 3.1 Reflections in cubic groups and restrictions on the components $\hat{\epsilon}^{\tau}$

Indices of reflections	Types	ϵ^{τ}_{xx}	ϵ^{τ}_{yy}	ϵ^{τ}_{xy}	ϵ^{τ}_{xz}	ϵ^{τ}_{yz}	m
$T^1,\ T^2,\ T^3$							
$h00$	c	R	R_1	0	0	I	$0,\ \pm 2$
$hk0$	c	R	R_1	R_2	I	I_1	$0,\ \pm 1_i \pm 2$
hhh		0	0	C	C	C	0
hkl	c	C	C_1	C_2	C_3	C_4	$0, \pm 1, \pm 2$
$T^4,\ T^5$							
$h = 2n + 1$	*	0	0	I	R	0	± 1
$h00\ h = 2n$	c	R	R_1	0	0	I	$0, \pm 2$
$h = 2n + 1$	c	I	I_1	I_2	R	R_1	$0, \pm 1, \pm 2$
$hk0\ h = 2n$	c	R	R_1	R_2	I	I_1	$0, \pm 1, \pm 2$
hhh		0	0	C	C	C	0
hkl	c	C	C_1	C_2	C_3	C_4	$0, \pm 1, \pm 2$
$O^6,\ O^8$							
$h = 4n \pm 1$	*	0	0	$\mp iR$	R	0	1
$h00\ h = 4n + 2$	$*, c$	0	R	0	0	I	± 2
$h = 4$		$-2R$	R	0	0	0	0
$hh0\ h = 2n + 1$	c	I	I	I_1	R	$-R$	$0, \pm 2$
$h = 2n$	c	R	R	R_1	I	$-I$	$0, \pm 2$
$h = 4n \pm 1$		0	0	$(1 \mp i)R$	$(1 \mp i)R$	$(I \mp i)R$	0
$hhh\ h = 4n + 2$		0	0	I	I	I	0
$h = 4n$		0	0	R	R	R	0
hkl	c	C	C_1	C_2	C_3	C_4	$0, \pm 1, \pm 2$
$O^1,\ O^3,\ O^5$							
$h00$		$-2R$	R	0	0	0	0
$hh0$	c	R	R	R_1	I	$-I$	$0, \pm 2$
hhh		0	0	R	R	R	0
hkl	c	C	C_1	C_2	C_3	C_4	$0, \pm 1, \pm 2$
O^2							
$h = 2n + 1$	$*, c$	0	R	0	0	I	± 2
$h00\ h = 2n$		$-2R$	R	0	0	0	0
$hh0$	c	R	R	R_1	I	$-I$	$0, \pm 2$
$hk0$	c	R	R_1	R_2	I	I_1	$0, \pm 1, \pm 2$
$h = 2n + 1$		0	0	I	I	I	0
$hhh\ h = 2n$		0	0	R	R	R	0
hkl	c	C	C_1	C_2	C_3	C_4	$0, \pm, \pm 2$
$O^4,\ O^7$							
$h = 4n \pm 1$	*	0	0	$\pm iR$	R	0	-1
$h00\ h = 4n + 2$	$*, c$	0	R	0	0	I	± 2
$h = 4n$		$-2R$	R	0	0	0	0
$h = 2n + 1$	c	I	I	I_1	R	$-R$	$0, \pm 2$
$hh00\ h = 2n$	c	R	R	R_1	I	$-I$	$0, \pm 2$
$h = 4n \pm 1$		0	0	$(1 \pm i)R$	$(1 \pm i)R$	$(1 \pm i)R$	0
$hhh\ h = 4n + 2$		0	0	I	I	I	0
$h = 4n$		0	0	R	R	R	0
hkl	c	C	C_1	C_2	C_3	C_4	$0, \pm 1, \pm 2$

where $\epsilon(\tau; m)$ are the expansion coefficients and

$$\hat{\sigma}_0 = \frac{1}{\sqrt{6}} \begin{pmatrix} -1 & 0 & 0 \\ 0 & -1 & 0 \\ 0 & 0 & 2 \end{pmatrix}, \quad \hat{\sigma}_{\pm 1} = \frac{1}{2} \begin{pmatrix} 0 & 0 & \pm i \\ 0 & 0 & 1 \\ \pm i & 1 & 0 \end{pmatrix},$$

$$\hat{\sigma}_{\pm 2} = \frac{1}{2} \begin{pmatrix} 1 & \mp i & 0 \\ \mp i & -1 & 0 \\ 0 & 0 & 0 \end{pmatrix}$$

(below we omit the index a at ϵ).

Expansion (3.72) is especially useful if $\boldsymbol{\tau}$ is parallel to one of the crystal symmetry axes. In that case, there are selection rules [57] that characterize the space group for corresponding Fourier harmonics. These rules determine, for each group and every $\boldsymbol{\tau}$, the value of m such that $\epsilon(\boldsymbol{\tau}; m \not\equiv 0)$ in (3.72). A specific example of this is one in which we suppose that vector $\boldsymbol{\tau}$ is parallel to the second-order axis. Neither $\hat{\epsilon}_{\boldsymbol{\tau}}$ tensor nor applicable Fourier component changes with rotation by angle π about that axis. However, basis tensors change by a factor of $\exp(im\pi)$ after each rotation. For $m = \pm 1$, these factors are -1 and the terms with $m = \pm 1$ must be absent in (3.72). If τ is parallel to the third- or fourth-order axes, then (3.72) retains only the term with $m = 0$. The rules for the case when $\boldsymbol{\tau}$ is parallel to the screw axis n_j are determined similarly. In this case, terms in (3.72) whose m's are such that $(m - jh)/n$ are not integers vanish. (Here, we assumed for simplicity that the screw axis is parallel to X and $\boldsymbol{\tau} = 2\pi h\mathbf{x}/d$.) The selection rules so determined are used to compile Table 3.1 (see last column).

3.4.4 The Explicit Form of $\hat{\epsilon}(\mathbf{r})$

Thus, symmetry restrictions are the most essential for Fourier harmonics of $\hat{\epsilon}_{\boldsymbol{\tau}}$ with $\boldsymbol{\tau}$ vectors parallel to symmetry axes. For cubic crystals, these vectors are (h, o, o), (h, h, o) and (h, h, h). In addition, using (3.63)–(3.68), (3.70), and (3.71) we find the relationships between all equivalent harmonics and express them in terms of one of them. We call equivalent such Fourier harmonics that correspond to the τ vectors which transform into one other under the symmetry operations belonging to the crystal space group. Moreover, components (modes) of (3.72) with fixed m that correspond to equivalent Fourier harmonics are expressed via one of these components. We denote a set of such equivalent modes by $[hk\ell, m]$. However, symmetry restrictions reveal no relationship between non-equivalent Fourier harmonics, e.g., between Fourier harmonics with different $|\boldsymbol{\tau}|$. Therefore, the number of parameters needed to determine the structure of the blue phase is generally infinite.

According to the experimental data, however, a few parameters are actually important. For example, *BP* (probably with group O^2) exhibits two reflections (100) and (110) (and equivalent to them) and *BP* (group O^8) exhibits three reflections (110),

(200), and (211) (equivalent). Polarization measurements [52, 58, 59] show that the dominant contribution to observed reflections originates from modes with $m = 2$. If we consider only these modes, we obtain, from (3.60) and (3.72), the following expressions for $f_i(r)$ which determine the spatial structure of the $\hat{\epsilon}(\mathbf{r})$ tensor for O^2 in accordance with (3.62):

$$\begin{aligned} f_1(x, y, z) &= \epsilon(100; 2)(C_z - C_y) + \epsilon(110; 2)[2C_yC_z - C_x(C_y + C_z)] \\ f_2(x, y, z) &= -\epsilon(100; 2)S_x + \epsilon(110; 2)\left[\sqrt{2}S_x(C_y - C_z) - S_yS_z\right], \end{aligned} \tag{3.73}$$

for O^8

$$\begin{aligned} f_1(x, y, z) &= \epsilon(110; 2)(2S_yC_z - S_xC_y - S_zC_x) + \epsilon(200; 2)(C_{2z} - C_{2y}) \\ &\quad + \frac{2}{3}\epsilon(211; 2)\Big[2\sqrt{6}(C_{2z}S_xC_y - C_{2y}S_zC_x) - 2S_{2x}C_yS_z + S_{2y}C_zS_x \\ &\quad + S_{2x}C_xS_y\Big], \\ f_2(x, y, z) &= \epsilon(110; 2)\left[S_zC_y - \sqrt{2}(C_xC_y + S_zS_x)\right] - \epsilon(200; 2)S_{2x} \\ &\quad + \frac{2}{3}\epsilon(211; 2)\Big[2C_{2y}S_zS_x - \sqrt{6}(S_{2z}S_xS_y + S_{2y}C_zC_x) \\ &\quad - 5S_{2x}S_yC_z - 2C_{2z}C_xC_y + 2C_{2y}S_zS_x\Big] \end{aligned} \tag{3.74}$$

where we denote $C_x = \cos(2\pi x/d), S_x = \sin(2\pi x/d)$, etc. The coefficients $\epsilon(hk\ell; 2)$ in (3.73) and (3.74) are real for symmetry reasons, and their absolute values and signs are determined either experimentally or theoretically. If the theory of phase transition is invoked, in addition to symmetry restrictions, the number of nonzero parameters in (3.60) and (3.72) may decrease thus refining blue phase structure (see [2]).

3.4.5 Optical Properties

The experimentally observed specific optical features of the blue phase and CLCs are connected with light diffraction at periodic structures. Since the blue phase has no birefringence, a theoretical description of its optical properties is more simple than that of a CLC and similar to X-ray diffraction theory in usual crystals except that polarization properties are different [54, 57, 60].

The experimentally observed reflections are identified using the Bragg condition (2.4), and possible structures of the blue phase particularly its possible space group are extremely limited [52, 58, 61, 62]. This situation is similar to X-ray diffraction structural analysis in which an observed set of reflexes is used to determine the symmetry of a crystal and the orientation of crystallographic axes.

As shown below, more detailed information about blue phase structure is contained in the intensity and polarization properties of the reflections and their angular and frequency widths.

Kinematic Theory. Let us start our consideration of the optical properties of the blue phase from the simplest kinematic approximation.

From the standpoint of the simplest kinematic approximation [see (2.5) and (2.16)], the intensity I_τ of a reflex and reflection coefficient R is proportional to the square of the corresponding Fourier harmonic

$$R(\mathbf{e}_0; \mathbf{e}_1) = \frac{I_\tau}{I_0} |\mathbf{e}_1^* \hat{\epsilon}_\tau \mathbf{e}_0|^2 R_\tau \tag{3.75}$$

where I_0 is the incident intensity, $|\mathbf{R}_1^* \hat{\epsilon} \mathbf{e}_0|^2$ a polarization-structural factor (see below), and factor R_τ depends on crystal dimensions and shape and determines the reflex angular and frequency width.

This formula reveals an essential property of light scattering by the blue phase—scattering shows a strong and complicated dependence on polarization that distinguishes light scattering in the blue phase from X-ray diffraction. The polarization dependence of R and the information on crystal structure are contained in the polarization-structural factor $|\mathbf{e}_1^* \hat{\epsilon}_\tau \mathbf{e}_0|^2$ which describes diffraction scattering of the $\mathbf{e}_0$-polarized wave into an $\mathbf{e}_1$-polarized wave. If we are not interested in the polarization of the diffracted wave, the factor $|\mathbf{e}_1^* \hat{\epsilon}_\tau \mathbf{e}_0|^2$ is summed over all $\mathbf{e}_1$ and the reflection coefficient becomes

$$R(\mathbf{e}_0) = (|\mathbf{A}_\tau|^2 - \mathbf{q}_\tau^{-2} |\mathbf{q}_\tau \mathbf{A}_\tau|^2) R_\tau \tag{3.76}$$

where $\mathbf{A}_\tau = \hat{\epsilon}_\tau \mathbf{e}_0$, $\mathbf{q}_\tau = \mathbf{q} + \boldsymbol{\tau}$ and $\mathbf{q}$ is the incident wave vector.

If the incident wave is polarized, the diffracted wave is polarized with polarization vector

$$\mathbf{e}_\tau = [\mathbf{A}_\tau - \mathbf{q}_\tau (\mathbf{q}_\tau \mathbf{A}_\tau) \mathbf{q}_\tau^{-2}] |\mathbf{A}_\tau - \mathbf{q}_\tau (\mathbf{q}_\tau \mathbf{A}_\tau) \mathbf{q}_\tau^{-2}|^{-1}. \tag{3.77}$$

Note that at $\mathbf{e}_1 = \mathbf{e}_\tau$ the reflection in (3.75) is at a maximum.

Let us consider polarization properties of scattering in the blue phase in more detail [54]. If the Bragg condition is satisfied approximately for a certain reflex corresponding to the reciprocal lattice vector $\boldsymbol{\tau}$, the polarization properties of scattering are determined by the form of the Fourier harmonic $\hat{\epsilon}_\tau$ in (3.75)–(3.77).

Note that, as in the case of Mössbauer diffraction with polarization-independent forward scattering amplitude (Sect. 3.4.5), polarization properties of diffraction scattering are conveniently described by introducing vectors of eigenpolarizations $\mathbf{n}_i$ and $\mathbf{n}_i'(i = 1, 2)$ [see (3.38)]. For the blue phase, the operators $\hat{S}$ and $\hat{S}'$ in (3.38) are given in the form

$$\begin{aligned} S_{ik} &= (\epsilon_{\tau ie})^* \epsilon_{\tau mk} (\delta_{m\ell} - q_\ell q_m / q^2) \\ S_{ik}' &= (\epsilon_{\tau i\ell})^* \epsilon_{\tau mk} (\delta_{m\ell} - q_{\tau\ell} q_{\tau m} / q_\tau^2). \end{aligned} \tag{3.78}$$

Expression (3.76) reaches its maximum and minimum values for the eigenpolarizations, and the diffractive scattering of each eigenpolarization $\mathbf{n}_i$ results in the corresponding eigenpolarization of the diffracted wave $\mathbf{n}'_i$ because $(\mathbf{n}'^*_2\hat{\epsilon}_\tau\mathbf{n}_1) = (\mathbf{n}'^*_1\hat{\epsilon}_\tau\mathbf{n}_2) = 0$. The polarization-structural factor of an eigenpolarization is $|F^\tau_\sigma|^2$, where

$$F^\tau_\sigma = (\mathbf{n}'^*_\sigma\hat{\epsilon}_\tau\mathbf{n}_\sigma). \tag{3.79}$$

The kinematic approach permits the description and qualitative understanding of many specific features of blue phase optics. In the general case of an arbitrary relationship between components with different m values in (3.72), eigenpolarizations given by polarization vectors n_i and $\mathbf{n}'_i$ are elliptical and depend on the Bragg angle θ_B. We say that reflections with elliptical eigenpolarizations are chiral since they have different reflection coefficients for right- and left-handed polarized light; the chiral reflections are labeled "*C*" in Table 3.1. The eigenpolarizations of non-chiral reflections are linear.

When only one term with $m = 2$ or $m = -2$ is present in (3.72), maximum diffractive scattering is experienced by light whose eigenpolarization is the same as in the case of a CLC; within the kinematic approximation, we denote this polarization by $\hat{\mathbf{n}}_d$ [see (2.21)] and the polarization-structural factor in (3.76) is $|\epsilon(\boldsymbol{\tau}_i \pm 2)|^2(1 + \sin^2\theta_B)|\mathbf{n}^*_d\mathbf{e}_i|^2/4$. For $\theta_B = \pi/2$ (backward scattering) in our case, only right-handed or only left-handed circular polarization is subjected to diffraction. The experimentally observed polarization properties of scattering indicate that the amplitudes of modes with $m = 2$ or -2 are greater than those of all other modes.

Generally, if non-polarized light is diffractively scattered by the blue phase, the scattered light is partially polarized. Using (3.75)–(3.79), we find the degree of polarization of scattered light P_r as

$$P_r = ||F_1|^2 - |F_2|^2|(|F_1|^2 + |F_2|^2)^{-1} \tag{3.80}$$

where incident light is non-polarized. The scattered light polarization is described by vector $\mathbf{n}'_1$ if $|F_1| > |F_2|$, and $\mathbf{n}'_2$ if $|F_2| > |F_1|$. In a special case, the scattered light may be completely polarized, even for unpolarized incident light was not polarized. That situation occurs when one of the eigenpolarizations is not subjected to diffraction scattering; i.e., one of the amplitudes F_1 or F_2 is zero [e.g., a single component with $m = 2$ or -2 is presented in (3.72)].

Circular and Linear Dichroism. Up till now we have considered selective reflection from the blue phase. When light reflection is at a maximum, the intensity of light transmitted through the sample is at a minimum; this fact was exploited in experimental studies of the blue phase [58, 61, 62]. Specifically, it is informative to measure the circular dichroism that depends on the difference between transmission coefficients T for the left-handed $(-)$ and right-handed $(+)$ polarized light as described in Sect. 3.1.5 [see (3.35) and Fig. 3.19].

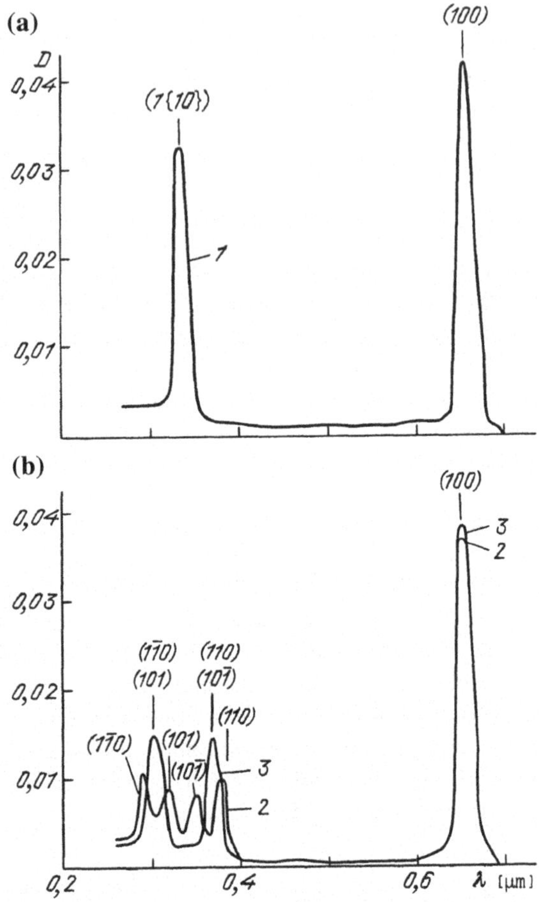

Fig. 3.19 Circular dichroism spectra of BPII obtained from the mixture of cholesteryl nonanoate and cholesteryl chloride [61]: **a** normal incidence and **b** oblique incidence; curves 2 and 3 were obtained with different incidence directions

When the sample does not absorb light, the values of D are easily expressed within the kinematic approximation in terms of the reflection coefficient $R_{\pm}$ for right (left)-handed polarization:

$$D = \frac{1}{2}(R_{-} - R_{+}). \qquad (3.81)$$

Assuming that only the mode with $m = \pm 2$ is present in (3.72), (3.81) and (3.76) yield an expression for the circular dichroism averaged over wavelength for each individual reflex:

$$\int D d\lambda \sim |\epsilon(\boldsymbol{\tau}_j \pm 2)|^2 (1 + \sin^2 \theta_{\mathrm{B}})(\sin \theta_{\mathrm{B}})^{-1}. \qquad (3.82)$$

In addition to the circular dichroism, the blue phase exhibits linear dichroism D_{e} of diffraction origin. Linear dichroism is connected with the difference between reflection coefficients of two orthogonal linear polarizations, one of which, for example, is parallel to the plane of vectors $\mathbf{q}$ and τ, and the other orthogonal to this polarization (3.11). It is essential that circular and linear dichroisms have different dependences on components of the $\hat{\epsilon}_\tau$ tensor and Bragg angle. Hence, if measured simultaneously these two effects provide additional information about $\hat{\epsilon}_\tau$. If only the components with $m = 2$ or -2 contribute to $\hat{\epsilon}_\tau$, $D_e = \cos^2\theta_B D/2\sin\theta_B$, while in backward diffraction reflection ($\theta_B = \pi/2$) and linear dichroism vanish.

We emphasize that the existence of linear dichroism does not contradict the assignment of cubic crystal symmetry. When light propagates exactly along the third- or fourth-order axis, it follows that linear dichroism must vanish for symmetry reasons. The physical cause for this is that the linear dichroisms due to different reflections compensate each other. However, if the propagation direction deviates slightly from a third- or fourth-order axis by an angle of the order of the reflection width, this compensation is violated and considerable linear dichroism occurs.

Optics of Perfect Samples. We assumed everywhere above that the scattering is small within an individual perfect region of the blue phase. However, now samples are available with large perfect regions [61–65], and for these regions the kinematic approximation may become invalid since it does not account for the processes of multiple scattering that are essential in thick ($L \gg (\mathbf{qs})/(q^2|F_\sigma|)$) samples. Since the form of the dielectric permittivity tensor is complicated, there is neither an exact analytical solution of Maxwell equations therefore nor an exact description of multiple scattering. The restrictions on $\hat{\epsilon}$, however, permit a general description of blue phase optical properties within the framework of the two-wave approximation of dynamic diffraction theory.

Since the blue phase exhibits no birefringence, the solutions of the dynamic system (2.26) have less complex polarization properties than the corresponding solutions for a CLC; in addition, they are analogous to those of higher-order reflections in a CLC and first-order reflexes in chiral smectics. Thus, the eigenpolarizations in the blue phase are always separated; i.e., the system of four equations (2.26) for vector amplitudes $\mathbf{E}_0$ and $\mathbf{E}_1$ splits into two uncoupled systems of two equations for eigenpolarization wave amplitudes; these polarizations are not π- and σ-polarizations—they are determined by (3.38) where $\hat{S}$ and $\hat{S}'$ are given by (3.38). The systems are:

$$\begin{cases} (1 - k_0^2/q^2)E_0^\sigma + \tilde{F}_\sigma^{-\tau} E_1^\sigma = 0 \\ \tilde{F}_\sigma^\tau E_0^\sigma + (1 - k_1^2/q^2)E_1^\sigma = 0 \end{cases} \tag{3.83}$$

$$\tilde{F}_\sigma^\tau = (\mathbf{n}_\sigma'^* \hat{\epsilon}_\tau \mathbf{n}_\sigma)/\epsilon_0.$$

Note that the dynamic system (3.83) is completely equivalent to the corresponding system in X-ray diffraction theory. Thus, we use the results of X- and γ-ray theory to describe light diffraction in the blue phases as in first-order reflexes in chiral smectics and high-order reflexes in CLC. The only specifics of the diffraction in the blue phase are in its polarization properties.

The polarization properties of diffraction in the blue phase differ from those of a CLC. In a CLC, the eigenpolarizations of light propagating at an angle to the cholesteric axis vary in the angular (or frequency) selective reflection band, while in the blue phase, polarization properties of eigenwaves remain essentially constant over the entire selective reflection band since the reflections are relatively narrow.

Reflection and Transmission of Light. Using the analogous relationship between X-ray diffraction and blue phase optics, we give the final results for the optical parameters of a plane-parallel layer of a blue phase illuminated by a plane monochromatic wave (Fig. 2.5). We expand the amplitudes of incident $\mathbf{E}^{\mathrm{e}}$, diffracted $\mathbf{E}^{\mathrm{r}}$, and transmitted $\mathbf{E}^{\mathrm{t}}$ waves into eigenpolarizations

$$\mathbf{E}^{\mathrm{e}} = \sum_{\sigma=1}^{2} E_{\sigma}^{\mathrm{e}} \mathbf{n}_{\sigma}, \; \mathbf{E}^{\mathrm{r}} = \sum_{\sigma=1}^{2} E_{\sigma}^{\mathrm{r}} \mathbf{n}_{\sigma}^{1}, \; \mathbf{E}^{\mathrm{t}} = \sum_{\sigma=1}^{2} E_{\sigma}^{\mathrm{t}} \mathbf{n}_{\sigma}.$$

For Bragg geometry ($b < 0$, see Fig. 2.5),

$$\begin{aligned} E_{\sigma}^{\mathrm{r}} &= E_{\sigma}^{\mathrm{e}} \tilde{F}_{\sigma}^{\tau} (\alpha + i\Delta ctg\ell)^{-1} \\ E_{\sigma}^{\mathrm{t}} &= E_{\sigma}^{\mathrm{e}} \Delta(\Delta \cos \ell - i\alpha \sin \ell)^{-1} \end{aligned} \tag{3.84}$$

and Laue geometry ($b > 0$, in Fig. 2.5)

$$\begin{aligned} E_{\sigma}^{\mathrm{r}} &= E_{\sigma}^{\mathrm{e}} \tilde{F}_{\sigma}^{\tau} \Delta^{-1} \sin \ell \\ E_{\sigma}^{\mathrm{t}} &= E_{\sigma}^{\mathrm{e}} (\cos \ell + i\alpha\Delta^{-1} \sin \ell) \end{aligned} \tag{3.85}$$

where $\alpha = (\tau^2 + 2q_0\boldsymbol{\tau})/2q_0^2$; $\Delta = [\alpha^2 + \tilde{F}_{\sigma}^{\tau}\tilde{F}_{\sigma}^{-\tau}/b]^{1/2}$; $\ell = \Delta q_0^2 L/2(q\mathbf{s})$; $b = (q_0\mathbf{s})/(q_{\tau}\mathbf{s})$, and $\mathbf{s}$ is the inner normal to the sample surface; parameter α characterizes the deviation of wavelength (or incidence angle) from that given by the Bragg condition (2.4). Assuming that deviations are small, ($|\alpha| \ll 1$), we obtain that

$$\alpha = (\theta_{\mathrm{B}} - \theta) \sin 2\theta_{\mathrm{B}} \tag{3.86}$$

when incident angle θ changes and

$$\alpha = (\lambda - \lambda_{\mathrm{B}}) \sin^2 \theta \lambda_{\mathrm{B}}^{-1} \tag{3.87}$$

when wavelength λ changes where $\lambda_{\mathrm{B}} = 4\pi\sqrt{\epsilon_0} \sin\theta/|\boldsymbol{\tau}|$. Quantities ℓ in (3.74) are the dimensionless thickness of the crystal, while quantity Δ is proportional to the diffraction correction to the wave vector. The relationships (3.84) and (3.85) completely solve the problem of light diffraction in the blue phase within the two-wave approximation. For example, the reflection coefficients for the waves with eigenpolarizations are

$$R_{\sigma} = |\tilde{F}_{\sigma}^{\tau}|b^{-1} \sin^2 \ell(\alpha^2 + \Delta^2 \cos^2 \ell)^{-1} \tag{3.88}$$

for Bragg geometry and

$$R_\sigma = |\tilde{F}_\sigma^\tau|^2 b^{-1} \Delta^{-2} \sin^2 \ell \tag{3.89}$$

for Laue geometry.

Polarization Parameters. If the incident wave has an arbitrary polarization vector e and polarization degree P, the reflection coefficient is

$$R(P,\ \mathbf{e}) = \frac{1}{2}(1-P)(R_1+R_2) + P(R_1|\mathbf{n}_1^*\mathbf{e}|^2 + R_2|\mathbf{n}_2^*\mathbf{e}|^2). \tag{3.90}$$

If $P = 0$ (non-polarized incident light), the degree of polarization of reflected and transmitted light becomes

$$\begin{aligned} P^r &= |R_1 - R_2|(R_1+R_2)^{-1} \\ P^t &= |R_1 - R_2|(2 - R_1 - R_2)^{-1}. \end{aligned} \tag{3.91}$$

The reflected light polarization vector coincides with $\mathbf{n}_1'$, if $R_1 > R_2$, and with $\mathbf{n}_2'$, if $R_2 > R_1$. The transmitted light partial polarization is described by $\mathbf{n}_1$ at $R_2 > R_1$ and $\mathbf{n}_2$ at $R_1 > R_2$. If only one component with $m = 2$ or -2 is present in $\hat{\epsilon}_\tau$, polarization parameters are identical with those obtained within the kinematic approximation. Note that these expressions were obtained by neglecting light reflection at the dielectric boundary, i.e., when the blue phase and surrounding medium have the same average dielectric constant. When this assumption is not valid, the polarization parameters may be changed. There is no principal difficulty in solving the problem while accounting for dielectric boundaries (see also [1]).

Effects of Multiple Scattering. The formulae of dynamic theory (3.48)–(3.91) describe a series of effects related to multiple scattering which do not appear in the kinematic consideration. For example, in thick crystals, the frequency (or angular) width of the diffraction scattering region, i.e., the range of frequencies or angles in which the eigenpolarization waves experience selective reflection, is proportional to $|F_\sigma|$ (as seen in (3.84) and (3.85)) and hence proportional to the corresponding harmonic in the Fourier expansion of $\hat{\epsilon}(\mathbf{r})$.

The dynamical approach also describes experimentally observed rotation of polarization plane in the blue phases and, particularly, the inversion of rotation sign (see Fig. 3.17). From (3.84) and (3.85), we find, for the case of circular polarizations $\mathbf{n}_i$ and $\mathbf{n}_i'$, an expression for the rotation angle of polarization plane φ_t for a wave transmitted through a layer of the blue phase

$$\varphi_t = \frac{1}{2}\left\{\left[arctg\left(\frac{\alpha}{\Delta}tg\ell\right)\right]_{\sigma=1} - \left[arctg\left(\frac{\alpha}{\Delta}tg\ell\right)\right]_{\sigma=2}\right\}. \tag{3.92}$$

The point at which the rotation changes sign corresponds to $\alpha = 0$. Note that in perfect samples of the blue phase, just as in CLCs, the rotation angle nonlinearly depends on sample thickness as follows from (3.22) (see [42]).

The other effect of multiple scattering is the pendulum beats which were considered in Chap. 3 of this book first edition. These beats are a manifestation of light frequency dependence (a deviation of the angle from Bragg's value) of reflection and transmission coefficients at a fixed thickness, when all other parameters are fixed. The period of these dynamical beats is directly expressed by $|F_{\sigma}^{\tau}|$. These values are measurable directly without involving absolute reflection intensities.

Another result of the dynamic consideration is that the blue phase birefringence and linear dichroism due to multiple scattering do not contradict the fact that the blue phase structure belongs to the cubic space group. We are referring to diffraction birefringence with generally elliptical but circular eigenpolarizations for particular propagation directions (as in the case of light propagation along the CLC axis). The difference between eigenwave effective refraction indices Δn is proportional to the phase difference ψ between eigenpolarization waves transmitted through the crystal: $\Delta n = (q_0 \mathbf{s})\psi / q_0^2 L$, where ψ is given by (3.92) without factor 1/2. The predominant presence of a linear component in the elliptical eigenpolarizations produces the linear birefringence. This birefringence exhibits strong frequency (angular) dependence and changes sign in the selective reflection band when the sign of α in (3.92) changes sign. In the special case when only components with $m = 2$ or -2 are present in ϵ_{τ}, the linear dichroism and linear birefringence vanish at $\theta_B = \pi/2$ but increase when the Bragg angle decreases. Diffraction birefringence for non-chiral reflection is linear as described in [54]. Generally, diffractive birefringence is due to many reflexes; however, that case needs a special study [66, 67].

Hence, we conclude that the experimentally observed linear birefringence [58, 68] may have diffraction origin which does not deny blue phase cubic structure.

3.4.6 Structure Studies of the Blue Phase

In this section, we consider possible ways of determining the structure of blue phases and the practical results obtained in this field. Structure studies of a blue phase (and any other crystal) include three stages: (i) determination of translational periodicity (or dimensions of a unit cell), (ii) determination of lattice symmetry properties, i.e., its space group, and (iii) exact determination of order parameter $\hat{\epsilon}(\mathbf{r})$ dependence on coordinates over the unit cell (or electron density if ordinary crystals are involved). These stages are listed in order of their complexity; when (iii) is achieved, the structure is determined completely.

Optical structural analysis. Let us consider the most informative methods, namely the optical methods. As shown above, every Fourier harmonic of the $\hat{\epsilon}(\mathbf{r})$ expansion in (3.60) results in reflective light scattering by the blue phase. Hence, the structure of the blue phase order parameter is determined completely if we experimentally delineate all Fourier harmonics and reconstruct $\hat{\epsilon}(\mathbf{r})$ from (3.60). Formally, expansion (3.60) may include an infinite number of harmonics. However, only a few reflections have experimentally noticeable intensities (see [2]) and optical structure analysis reduces, in practice, to determination of a small number of tensor Fourier harmonics. Let us now discuss this issue in more detail. Specifically, in the most general case, every Fourier component is determined by twelve parameters because there are six independent components of the symmetric tensor $\hat{\epsilon}(\mathbf{r})$ which generally are complex quantities. We do not assume that $S_p(\hat{\epsilon}_\tau) = 0$; i.e., we account for the spatial modulation of the isotropic part of the dielectric permittivity. The condition that an isotropic Fourier component exists for every space group is the same as for the component with $m = 0$ (for X-rays [69]). Note that the allowed reflections for cubic space groups given in Table 3.1 differ from the corresponding X-ray reflections [69]. Some reflections are allowed because the dielectric permittivity of the blue phase is anisotropic while that for X-rays is scalar. Note also that there may be density modulation of the isotropic part of $\hat{\epsilon}_\tau$; however, the practice is to neglect this effect.

If we use some additional factors or suggestions in studying the blue phase, we reduce the number of parameters that require experimental determination. For example, if we use the symmetry properties of $\hat{\epsilon}_\tau$ in cubic crystals, we reduce the number of independent parameters in $\hat{\epsilon}_\tau$ for reflexes of a particular type (see Table 3.1). Also, if the results of phase transition theory are involved in the analysis of the blue phase, we reduce the number of free parameters. That is, within the framework of Landau's theory, for every $\hat{\epsilon}_\tau$ in (3.72) there is one nonzero term with $m = 2$ or -2. Consequently, the number of free parameters in each harmonic reduces to two.

If we do not use additional non-optical information, one parameter, $\hat{\epsilon}_\tau$ phase, cannot be determined from light intensity and polarization measurements because they depend on quadratic combinations of $\hat{\epsilon}_\tau$. In principle, it is possible to determine all other parameters. Measurements of scattering polarization parameters in a fixed geometry permit determination of seven parameters of $\hat{\epsilon}_\tau$ [70]. This limitation originates from the fact that light waves are transverse; e.g., the components with $m = 1$ cannot be measured in the case of backward scattering ($\theta_B = \pi/2$). Therefore, the scattered light intensity and polarization for two arbitrary directions are sufficient for determining $\hat{\epsilon}_\tau$, apart from its phase.

The structural and symmetry properties of blue phases and their correlations with optical parameters discussed above result in the following conclusions: (i) The dimensions of a unit cell are defined by the frequency (angular) positions of selective reflection band [see (2.4)], and (ii) the space group of the structure is given by the set of $\boldsymbol{\tau}$ (Miller's indices of observed reflexes), i.e., by the set of selective reflection band. Note that in some cases one set of observed reflections may correspond to several space groups, as in X-ray structural analysis. The situation is more complicated in practice because a few reflexes of the longest wavelengths are observed while, for example, the simple and body-centered groups are distinguishable from the seventh reflection only [71] in the order of increasing $|\boldsymbol{\tau}|$. In that case, unambiguous space group determination requires additional information.

Polarization Measurement and the Phase Problem. Optical structure analysis of blue phases includes the possibility of obtaining more information than is possible from X-ray structure analysis. This is related to the fact that the routine polarization measurement for visible light is nearly impossible in X-rays. Measuring polarization properties of reflexes permits either complete re-establishment of space group or dramatic reduction in the number of competing options. Hence, it is informative to measure the chiral properties of certain reflections, i.e., the difference in scattering of opposite circular polarizations. For example, in groups O^5, O^2, and O^8 the second, third, and sixth reflections are not chiral (see Table 3.1). In principle, all groups, except T^3 and T^5 [54, 60], are distinguished by polarization measurements (the polarization properties of reflections in T^3 and T^5 coincide, and only relative phases of certain reflections differ). The complete delineation of a structure requires determination of $\hat{\epsilon}_{\tau}$ for the observed reflections and accounting for relative phase. According to (3.60), this is equivalent to unambiguously establishing the coordinate dependence of order parameter $\hat{\epsilon}(\mathbf{r})$. We emphasize again that usual diffraction measurements including polarization in both X-ray and visible regions, give no information about $\hat{\epsilon}_{\tau}$ phases which is necessary to determine $\hat{\epsilon}(\mathbf{r})$ in accordance with (3.60). There are sources of coherent light in the visible range; hence, practical phase measurements in that range appear to be a less complex problem. Specifically, we can use the practical technique, often used in recording holograms, of detecting the interference between a subject beam (in our case, the diffracted beam) and a coherent reference beam of known phase. We note that for determining $\hat{\epsilon}(\mathbf{r})$, only relative phases of Fourier harmonics with different $\boldsymbol{\tau}$ are needed. To find those phases, one can use the multiwave diffraction, as in the case of X-rays [72, 73]. When the Bragg condition (2.4) is satisfied for several reciprocal lattice vectors $\boldsymbol{\tau}$ simultaneously, the intensity and polarization of diffracted waves carry information about relative phases corresponding to $\hat{\epsilon}_{\tau}$ (no coherent source needed in this case).

In blue phases, information is obtained from the selective scattering dependence on azimuthal angle, i.e., the dependence on the angle of sample rotation round the reciprocal lattice vector τ present in (2.4). Components $\epsilon(\boldsymbol{\tau}; m)$ with $m = 0, \pm 1, \pm 2$ and the interference between them are responsible for the modulation of measured values with azimuthal period 2π, π, $2\pi/3$, and $\pi/2$. Hence, by measuring scattering parameter (e.g., the scattering intensity of two circular and two linear polarizations) dependence on the azimuth we can determine the amplitude and the relative phase of components $\epsilon(\boldsymbol{\tau}; m)$ in (3.72).

The Results of Structure Investigations. As mentioned above, the procedure for determination of blue phase structure may be executed on the basis of optical measurements alone without auxiliary information or assumptions. Experimental studies of pure substances and mixtures yield valuable information about blue phase structure, but they do not permit unambiguous structure determination. The consensus regarding the period of blue phases is that blue phases occur in substances (mixtures with short pitches of the cholesteric phases ($p \lesssim 5000$ Å). In all substances studied, the ratio of maximum selective scattering wavelength λ_{Bp} (in medium) to cholesteric pitch is approximately the same, viz. $\lambda_{Bp_1} \approx 1.4p$, $\lambda_{Bp_2} \approx 1.2p$. The period of blue phases increases at lower temperatures. This effect is strong in *BRI* and is weaker [58] or completely vanishes [52] in *BPII*. According to theory (see [2]), period is almost independent of temperature and $\lambda_{BPI} \approx 1.3p$ and $\Lambda_{BPII} \approx 1..13p$. This slight discrepancy with experiment is probably related to the harmonics of $\hat{\epsilon}_{\tau}$ neglected in the theory.

Blue phase space groups are less clear, and there are several space groups proposed as possible candidates for *BPI* and *BPII*. The only unanimous conclusion is that *BPI* and *BPII* have cubic structure. Many researchers (see [2]) think that *BPI* has a cubic body-centered lattice (group O^8) and *BPII* has the primitive cubic cell (group O^2). The results of polarization measurements on oriented samples are the most convincing in this connection [61, 62] because the reciprocal lattice vector orientation is established with respect to crystallographic axes. However, we cannot completely exclude the other possibilities because of the finite accuracy of measurements. Specifically, groups O^2 and O^8 have subgroups T^1 and T^5, respectively. Two sets of reflexes (100) and (110) observed in *BPII* and attributed to the O^2 group may also be identified as (200) and (220) reflexes of any cubic group except O^1, O^2, O^3, and O^5 (see Table 3.1). A more reliable determination of group requires a larger number of reflection measurements.

Results which differ from those above were obtained when studying selective light scattering in cholesteric *n*-alkanoates ($3 \leq n \leq 18$) [71], one of the phases of which was assigned to the O^5 group. In [74], the *BPII* lattice was assigned as body-centered since its third reflection is chiral. Discrepancies in space group designation for blue phases are apparently associated with the fact that a small number of reflections were considered, although it is possible that the symmetries of blue phases in different compounds or mixtures differ. Moreover, space group identification may be hampered by multiwave effects due to which reflection–polarization properties differ from those predicted by theory; e.g., non-chiral reflexes may exhibit chirality. Some effects which make identification difficult may appear in mixtures with decomposed phases [75].

Until now, the optical methods did not reveal the complete structure of compounds; however, some auxiliary concepts may change this. If we assume, according to theory, that the main structural contribution comes from modes $\epsilon(\boldsymbol{\tau}; 2)$, we can calculate the ratio of the amplitude moduli of different harmonics from the data on the circular dichroism [61, 62]. Taking the values of $\epsilon(\tau; 2)$ with the signs which follow from theory, we obtain $\hat{\epsilon}(\mathbf{r})$ from (3.73) and (3.74) to within a factor. We should note that for a series of mixtures of cholesteryl nanoate and cholesteryl chloride these ratios are, on average, $|\epsilon(110; 2)| : |\epsilon(200; 2)| : |\epsilon(211; 2)| \approx 1 : 1 : 0.33$ for *BPI* and $|\epsilon(100; 2)| : |\epsilon(110; 2)| \approx 2.5$ for *BPII* [61, 62, 67]; that is, they are close to values of Landau's theory. We can determine the signs of $\epsilon(\boldsymbol{\tau}; 2)$, i.e., solve the phase problem experimentally using NMR, an effective and useful instrument for studying blue phase structure [76, 77]. NMR spectra may contain information about the dielectric permittivity tensor $\hat{\epsilon}(\mathbf{r})$ and its spatial distribution in the unit cell. Using the values of $\hat{\epsilon}_\tau$ obtained by optical methods for calculation of NMR spectra and retaining $\hat{\epsilon}_\tau$ phases as free parameters, we can then fit the measured and calculated spectra, thus find the phases, and recover the distributions of $\hat{\epsilon}(\mathbf{r})$ in the structure involved.

To conclude this chapter, we note that one of the main problems associated with optical studies of blue phases is the structure of *BPIII* which has no long-range translational order [78]. One of the possible explanations for this is that *BPIII* is the phase of strongly developed fluctuations in an isotropic liquid [48, 60], since the estimates made [60] indicate that common fluctuations are not sufficiently large to cause the circular dichroism observed in this phase. It is also possible that *BPIII* consists of an emulsion of *BPI* or *BPII* in an isotropic liquid [79].

The optics of chiral liquid crystals, especially the optics of blue phases, remains the subject of intense investigation [80–111] because of the increasing number of possible new applications for this class of liquid crystals [82–85]; hence, the physics of this type of liquid crystal is considered interesting by many researchers [81–83]. The chiral liquid crystal, particularly the chiral smectics, is regarded as prospective material in LC display systems [84, 85]. New applications of this material may be connected with surface-guided optical waves [90, 99] whose parameters are changed easily by weak external forces. The most significant point concerning the physics of chiral liquid crystals and blue phases is that primary methods of investigation are optical in nature.

There was recent progress in blue phase structure determination due to application of the Kossel diagram technique [89]. New *BP* phases induced by electric field were observed [88, 90] by the electro-optics methods [86] which also enable us to observe [102] unusual electrostriction phenomena which along with the structure phase transitions in the electric field [88, 101] found their theoretical explanation [87, 92, 102]. The previously unexplained reflections [104] which contradicted to the attributed to *BP* space groups found their explanation in the framework of many wave optics [3, 44, 105]. The main attention in these investigations is paid to the *BPIII*.

Recently, it has been suggested that *BIII* has a quasicrystal structure [112–114], but there was no experimental confirmation of this suggestion [86]. Very interesting possibilities for its structure investigations are opened by the method of fast freezing of blue phase which permits to transform the blue phase into solid state maintaining its liquid crystal structure properties [115]. However, the nature of *BPIII* as well as its structure is not understood. It is why the fog phase (*BPIII*) remains to be a mystery which is more than a hundred years old and is an intriguing object of investigations.

References

1. V.A. Belyakov, A.S. Sonin, *Optics of Cholesteric Crystals* (Nauka, Moscow, 1982) (in Russian)
2. V.A. Belyakov, V.E. Dmitrienko, Uspakhi Fiz. Nauk **146**, 369 (1985) [Sov. Phys. - Usp. **28**, 535 (1985)]
3. P.G. de Gennes, *The Principles of Liquid Crystals* (Charendon Press, Oxford, 1974)
4. S. Chandrasekhar, *Liquid Crystals* (Cambridge University Press, Cambridge-New York-Melburn, 1977)
5. M.A. Osipov, Fiz. Tv. Tela **27**, 1651 (1985)
6. S.M. Osadchii, V.A. Belyakov, Kristallografiya **28**, 123 (1983)
7. R. Dreher, G. Meier, Solid State Commun. **13**, 607 (1973). Phys. Rev. **8A**, 1616 (1973)
8. D. Taupin, J. de Phys. Coll. C-4 **30**, 32 (1969)
9. E. Elachi, O. Yeh, J. Opt. Soc. Am. **53**, 840 (1973)
10. V.A. Belyakov, V.E. Dmitrienko: Fiz. Tv. Tela **15**, 2724 and 3540 (1973) [Sov. Phys. - Solid State, **15**, 1811, 2364 (1974)]

11. D.W. Berreman, T.J. Sheffer, Phys. Rev. Lett. **25**, 577 (1970), Mol. Cryst. Liq. Cryst. **11**, 395 (1970), Phys. Rev. Ser. A. **5**, 1397 (1971)
12. R.W. James, *The Optical Principles of the Diffraction of X-rays* (Bell and Sons, London, 1967)
13. B.W. Batterman, H. Cole, Rev. Mod. Phys. **36**, 681 (1964)
14. Z.G. Pinsker, *Dynamical Scattering of X-rays in Crystals* (Springer, Berlin, 1978)
15. G.S. Ranganath, S. Chandraskekhar, U.D. Kini, K.A. Suresh, S. Ramaseshan, Chem. Phys. Lett. **19**, 556 (1973)
16. M. Born, E. Wofl, *Principles of Optics* (Pergamon Press, Oxford, 1965)
17. H. DeVries, Acta Cryst. **4**, 219 (1951)
18. S. Chandrasekhar, G.S. Ranganath, Mol. Cryst. Liq. Cryst. **25**, 195 (1974)
19. H. Takezoe, Y. Ouchi, M. Hara, A. Fukuda, E. Kuze, Jpn. J. Appl. Phys. **22**, 1080 (1983)
20. H. Takezoe, Y. Ouchi, A. Sugita, M. Hara, A. Fukuda, E. Kuze, N. Goto, Jpn. J. Appl. Phys. **21**, L390 (1982)
21. H. Takezoe, Y. Ouchi, M. Hara, A. Fukuda, E. Kuze, N. Goto, Jpn J. appl. Phys. **22**, L185 (1983)
22. H. Takezoe, K. Hashimoto, Y. Ouchi, M. Hara, A. Fukuda, E. Kuze, Mol. Cryst. Liq. Cryst. **101**, 329 (1983)
23. R.S. Akopyan, B.Ya. Zeldovich, N.V. Tabiryan, ZhETF **83**, 1770 (1982)
24. M.A. Paterson, Phys. Rev. A **27**, 520 (1983)
25. C. Oldano, Phys. Rev. A **31**, 1014 (1985)
26. C. Oldano, P. Allia, L. Trossi, J. Phys. **46**, 573 (1985)
27. S. Garoff, R.S. Meyer, R. Barakat, J. Opt. Soc. Am. **68**, 1217 (1978)
28. D.G. Khoshtariya, S.M. Osadchii, G.S. Chilaya, Kristallografiya **30**, 755 (1985)
29. V.V. Zheleznyakov, B.V.V. Kocharovskii, V.V. Kocharovskii, Uspekhi Fiz. Nauk. **141**, 257 (1983)
30. V.A. Belyakov, V.E. Dmitrienko, V.P. Orlov, Uspekhi Fiz. Nauk **127**, 221 (1979) [Sov. Phys. - Usp. **22**, 63 (1979)]
31. M.J. Stephen, J.P. Straley, Rev. Mod. Phys. **46**, 617 (1974)
32. S. Chandrasekhar, Rep. Progr. Phys. **39**, 613 (1976)
33. G. Holzwarth, N.A.W. Holzwarth, J. Opt. Soc. Am. **63**, 324 (1973)
34. R. Nitiananda, U.D. Kini, S. Chandrasekhar, K.A. Suresh, in *Proceedings of International Conference on Liquid Crystals Banglore-Pramana* vol. 1, Suppl. (1975), p. 325
35. V.A. Belyakov, V.E. Dmitrienko, FIz. Tv. Tela **18**, 2880 (1976) [Sov. Phys. - Solid State, **18**, 1681 (1976)]
36. Y. Kagan, A.M. Afanas'ev, ZheTF **48**, 327 (1965)
37. K.A. Suresh, Mol. Crys. Liquid. Crys. **35**, 267 (1976)
38. S.N. Andronishidze, V.E. Dmitrienko, D.G. Khoshtariya, G.S. Chilaya, Pis'ma ZhETF **32**, 19 (1980)
39. S. Endo, T. Kuribara, T. Akahane, Jpn. J. Appl. Phys. **22**, 499 (1983)
40. B.W. Berreman, Mol. Cryst. Liq. Cryst. **22**, 175 (1973)
41. V.A. Belyakov, V. E. Dmitrienko, in *Light Scattering in Solids*, ed. by J.L. Birman, H.Z. Cummins, K.K. Rebane (Plenum Press, 1979, New York, London), p. 377
42. S.M. Osadchii, Kristallografiya **28**, 758 (1983)
43. S.A. Plkin, *Structural Transformations in Liquid Crystals* (Nauka, Moscow, 1981) (in Russian)
44. L.M. Blinov, *Electro- and Magnetooptics of Liquid Crystals* (Nauka, Moscow, 1978). (in Russian)
45. V.E. Dmitrienko, V.A. Belyakov, ZhETF **78**, 1568 (1980) [Sov. Phys. -JETP **51**, 787 (1980)]
46. H. Stegemeyer, K. Bergman, in *Liquid Crystals of One and Two-Dimensional Order*, ed. by W. Helfrich, G. Heppke (Springer, Berlin, Heidelberg, 1980) p. 1611, H. Stegemeyer, Liq. Cryst. **1**, 3 (1986)
47. K. Bergman, R. Pollmann, G. Scherer, H. Stegemeyer, Z. Naturefor., **34a**, 253 (1979)
48. S.A. Brazovskii, S.G. Dmitriev, ZhETF **69**, 979 (1975)
49. S.A. Brazovskii, V.M. Filev, ZhETF **75**, 1140 (1978) [Sov. Phys. - JETP **48**, 573 (1978)]

50. H. Grebel, R.M. Hornreich, S. Shtrikman, Phys. Rev. A **28**, 1114 (1983)
51. S. Meiboom, J.P. Sethna, P.S. Andersen, W.F. Brinkman, Phys. Rev. Lett. **46**, 1216 (1981)
52. D.L. Johnson, J.H. Flack, P.P. Crooker, Phys. Rev. Lett. **45**, 641 (1980), Phys. Lett. **82A**, 247 (1981)
53. S. Meiboom, M. Sammon, D.W. Berreman, Phys. Rev. A **28**, 3553 (1984)
54. V.A. Belyakov, V.E. Dmitrienko, S.M. Osadchii, ZhETF **83**, 585 (1982) [Sov. Phys. - JETP **56**, 322 (1982)]
55. V.E. Dmitrienko, Acts. Cryst. **A40**, 89 (1984)
56. *Atlas of Space Groups of the Cubic System* (Moscow, Nauka, 1980) (in Russian)
57. R.M. Honreich, S. Shtrikman, Phys. Lett. **82a**, 345 (1981)
58. S. Meiboom, M. Sammon, Phys. Rev. Lett. **44**, 882 (1980), Phys. Rev. **24A**, 468 (1981)
59. J.H. Flack, P.P. Crooker, H.C. Svoboda, Phys. Rev. A. **26**, 723 (1982)
60. R.M. Hornreich, S. Shtrikman, Phys. Rev. A **28**, 1791 (1983)
61. V.A. Kizel, V.V. Prokhorov, Pis'ma ZhETF **38**, 283 (1983)
62. V.A. Kizel, V.V. Prokhorov, ZhETF **87**, 450 (1984) [Sov. Phys. - JETP **60**, 257 (1984)], V.A. Kizel, V.V. Prokhorov, D.G. Khoshtariya, G.S. Chilaya, Kristallografiya **31**, 130 (1986)
63. M. Marcus, Phys. Rev. A **25**, 2276 (1982)
64. R. Barbett-Massin, P. Cladis, P. Pieranski, Phys. Rev. A **31**, 3912 (1985)
65. H. Stegemeyer, F. Prosch, Phys. Rev. A **30**, 3369 (1984)
66. V.A. Belyakov, V.E. Dmitrienko, Kristallografiya **27**, 14 (1982)
67. V.A. Belyakov, E.I. Demiknov, V.E. Dmitrienko, V.K. Dolganov, ZhETF **89**, 2035 (1985) [Sov. Phys. - JETP **62**, 1173 (1985)]
68. P.H. Keyes, A.J. Nicastro, E.M. McKinnon, Nol. Crys. Liq. Crys. **67**, 715 (1981)
69. *International Tables for X-ray Crystallography*, vol. 1 (Birmingham, Kynoch Press, 1952) p. 306
70. R.M.A. Azzam, N.M. Bashara, *Ellipsometry and Polarized Light* (North-Holland, Amsterdam, 1977)
71. A.J. Nicastro, P.H. Keyes, PHys. Rev. A **27**, 431 (1983)
72. B. Post, Acta Cryst. **A35**, 17 (1979)
73. L.D. Chapman, D.R. Yoder, R. Colella, Phys. Rev. Lett. **46**, 1578 (1981)
74. K. Tanimoto, P.P. Crooker, Phys. Rev. A **29**, 1596 (1984)
75. M. Marcus, Molec. Cryst. Liq. Cryst. **82**, 33 (1982)
76. E.T. Samulski, A. Luz, J. Chem. Phys. **73**(1), 142 (1980)
77. Z. Yanis, G. Chidichimo, J.W. Doane, Phys. Rev. A. **28**, 3012 (1983)
78. E.I. Demikhov, V.K. Dolganovv, Pis'ma ZhETF **42**, 15 (1985)
79. R.M. Hornreich, M. Kugler, S. Shtrikman, Phys. Rev. Lett. **48**, 1404 (1982)
80. V.A. Belyakov, V.E. Dmitrienko, Optics of chiral liquid crystals, ed. by I.M. Khalatnikov, in *Soviet Scientific Reviews/Section A* (Harwood Academic Publishers GmbH, London, 1989)
81. D.C. Wright, N.D. Mermin, Rev. Mod. Phys. **61**, 385 (1989)
82. G.S. Chilaya, V.G. Cigrinov, Kristallografiya **33**, 260 (1988)
83. G.S. Chilaya, L.N. Lisetski, Mol. Cryst. Liq. Cryst. **140**, 243 (1980)
84. L.A. Bersenev, V.G. Chigrinov, D.I. Dergachov et al., Liq. Cryst. **5**, 1171 (1989)
85. A.V. Parfenov, V.G. Chigrinov, LIq. Cryst. **7**, 131 (1990)
86. P.P. Crooker, Liq. Cryst. **5**, 751 (1989)
87. R.M. Hornreich, S. Shtrikman, Liq. Crust. **5**, 777 (1989)
88. F. Porsch, H. Stegemeyer, Liq. Cryst. **5**, 791 (1989)
89. B. Jerome, P. Pieranski, Liq. Cryst. **5**, 799 (1989)
90. G. Heppke, B. Jerome, H.-S. Kitzerow, P. Pieranski, Liq. Cryst. **5**, 813 (1989)
91. P. Pieranski, P.E. Cladis, R. Barbet-Massin, Liq. Cryst. **5**, 829 (1989)
92. V.A. Belyakov, V.E. Dmitrienko, Liq. Cryst. **5**, 839 (1989)
93. V.E. Dmitrienko, Liq. Cryst. **5**, 847 (1989)
94. G. Voets, H. Martin, W. van Dael, Liq. Cryst. **5**, 871 (1989)
95. H. Zink, W. van Daeil, Liq. Cryst. **5**, 899 (1989)
96. H.F. Glesson, H. Cole, Liq. Cryst. **5**, 917 (1989)

97. G. Heppke, H.-S. Ktzerow, D. Ltzsh, H. Papenfuss, Liq. Cryst. **8**, 407 (1990)
98. H.-S. Kitzerow, P.P. Crooker, S.L. Kwok, Phys. Rev. A **42**, 3442 (1990)
99. V.A. Belyakov, V.P. Orlov, Poverkhnost **5**, 13 (1990)
100. V.A. Belyakov, V.P. Orlov, Mol. Cryst. Liq. Cryst. Lett. **8**, 1 (1991)
101. P. Pieranski, P.E. Cladis, Phys. Rev. A **35**, 3552 (1987)
102. H.-S. Kitzerow. P.P. Crooker, S.L. Kwok, G. Heppke, J. Phys. France **51**, 1303 (1990)
103. D. Lubin, R.M. Hornreich, Phys. Rev. A **36**, 849 (1987)
104. W. Kuczinski, Phys. Lett. A **110**, 405 (1987)
105. J.L. Birman, R.M. Hornreich, S. Shtrikman, Phys. Rev. A **39**, 1591 (1989)
106. P.L. Finn, P.E. Cladis, Mol. Crys. Liq. Cryst. **84**, 159 (1982)
107. R. Barbet-Massin, P.E. Cladis, Pieranksi: Phys. Rev. A **30**, 1161 (1984)
108. P. Pieranski, P.E. Cladis, R. Barbet-Massin, J. Physique Lett. **46**, L-973 (1985)
109. W. Kuczynski, Phys. Stat. Sol. (a) **96**, K127 (1986)
110. R. Barbet-Massin, P. Pieranski, J. Physique Coll. C3 Suppl. to No. 3 **46**, C3-61 (1985)
111. V.A. Belyakov, S.M. Osadchii, V.A. Korotkov, Kristallografiya **31**, 522 (1986)
112. V.M. Filev, Pis'ma ZhETF **43**, 523 (1986)
113. R.M. Hornreich, S. Shtrikman, Phys. Rev. Lett. **56**, 1723 (1986)
114. D.S. Rokhsar, J.P. Sethna, Phys. Rev. Lett. **56**, 1727 (1986)
115. J.A. Zasadzinski, S. Meiboom, D.W Berreman, Phys. Rev. Lett. **57**, 364 (1986)

Chapter 4
Nonlinear Optics of Periodic Media

This chapter is devoted to nonlinear optical phenomenon in periodic media, mainly the case in which radiation wavelength is close to the spatial period of the nonlinear medium. Under this condition, effects can appear which are not observed in the usual crystals where the spatial period is much smaller than optical wavelength so that periodicity does not affect nonlinear optical phenomena [1, 2].

To be more specific, we will consider as an example the generation of high optical harmonics. Here, the periodicity of a medium leads not only to the well-known modification of the phase-matching conditions in harmonic generation [3, 4] (reciprocal lattice vector of the periodic medium can now be included in the phase-matching conditions), but also to the theoretically predicted enhancement of the efficiency of the nonlinear transformation of optical frequency when the harmonic generation in a nonlinear periodic medium undergoes diffraction [5]. For the sake of simplicity, we will consider the generation of high harmonics in periodic media within the approximation of a nondepleted pumping wave.

Before considering concrete examples of the generation of high harmonics, we note that the descriptions of charged particle radiation discussed in Chap. 5 of [19] and of nonlinear harmonic generation (in the case of a given pumping wave) are somewhat analogous. In both cases, the problem is reduced to the solution of inhomogeneous Maxwell equations, the inhomogeneity being due to either particle current or nonlinear polarization induced by the pumping wave [see (4.2)]. Therefore, the radiation of charged particles and the nonlinear generation of high harmonics in periodic media possess qualitatively similar features. One of the peculiarities they have in common, namely the enhancement of efficiency of nonlinear generation under diffraction conditions, was already mentioned. Analogous to this effect is the enhancement of the charged particle radiation intensity under diffraction conditions described in Chap. 5 of [19].

V. Belyakov, *Diffraction Optics of Complex-Structured Periodic Media*, Springer Series in Optical Sciences 203,
https://doi.org/10.1007/978-3-319-43482-7_4

4.1 Enhancement of Efficiency of the Nonlinear Transformation of Optical Frequencies in Periodic Medium

Considerable attention has recently been paid to the nonlinear optics of periodic media [3, 6–8]. These media are of interest because they provide wider potentialities in phase-matched frequency transformation than uniform media. In this section, it will be shown that a periodic medium has a greater number of phase matchings than a uniform medium and provides more efficient nonlinear frequency transformation. We will consider these phenomena taking second-harmonic generation (SHG) in one-dimensional periodic media as an example. As we will see, maximum enhancement of the SHG for collinear geometry is reached when the frequency (2ω) of the wave generated under phase matching coincides with that of the frequency boundaries ω_e of the stop band (see Chap. 1). If the absorption can be neglected, the maximum of SHG phase-matched intensity is proportional to the fourth power of the sample thickness L and its ratio to the SHG intensity in a homogeneous medium is proportional to $(\delta L\omega/c)^2$, where δ is the amplitude of the nonlinear medium dielectric constant spatial modulation.

As was mentioned above, an analogous increase in Cherenkov radiation intensity also occurs at the frequencies coinciding with the boundary frequencies of a forbidden band [9, 10]. Since the specific features of the periodic medium are not important here, we, as was already mentioned, will choose for simplicity's sake a one-dimensional periodic nonlinear medium in which second-harmonic generation occurs [5]. Let a pumping wave with frequency ω and wave vector $\mathbf{k}$ propagate in a plane-parallel sample of thickness L (Fig. 4.1), the dielectric properties of the medium outside the sample being described by the tensor $\hat{\epsilon}$, and those of the sample being harmonically modulated in space with small amplitude δ and described by the tensor

$$\hat{\epsilon}(\mathbf{r}) = \hat{\epsilon}(1 + 2\delta \cos \boldsymbol{\tau}\mathbf{r}) \tag{4.1}$$

where $\boldsymbol{\tau}$ is the reciprocal lattice vector of the periodic structure ($\tau = 2\pi/d$, d being the period). We will assume that the sample's properties allow for nonlinear generation of the second harmonic and the direction of $\mathbf{k}$ is close to the second-harmonic Bragg's direction. Then, we can apply the dynamic diffraction theory discussed in

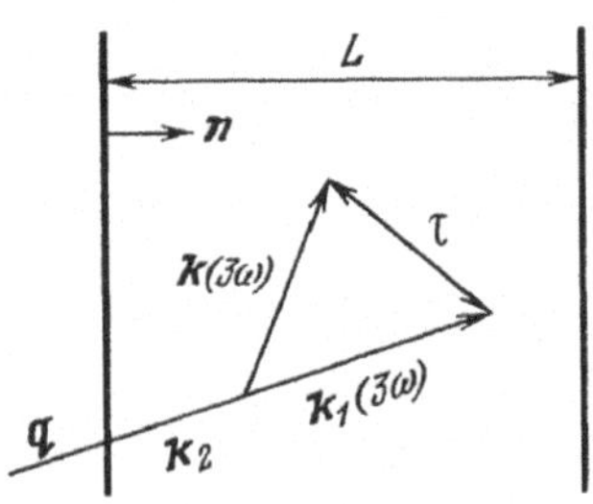

Fig. 4.1 Geometry of nonlinear frequency transformation in a periodic medium

Chap. 2. In a periodic medium, the second-harmonic wave is a sum of two plane waves $\mathbf{E} = \mathbf{E}_0 e^{i\mathbf{k}_0\mathbf{r}} + \mathbf{E}_1 e^{i\mathbf{k}_1\mathbf{r}}$ where $\mathbf{k}_1 = \mathbf{k}_0 + \boldsymbol{\tau}$. Assume also that one of the principal axes of the tensor $\hat{\epsilon}$ is perpendicular to the plane $\mathbf{k}_0, \mathbf{k}_1$. Then, the system of two equations can be obtained from Maxwell equations [see (2.23)–(2.26)] for $\mathbf{E}_0, \mathbf{E}_1$ within the approximation of a nondepleted pumping wave (see [7]) by replacing the inhomogeneity related to particle current by $4\pi \left(\frac{2\omega}{c}\right)^2 P^{(2)}(r, 2\omega)$ where $P^{(2)}$ is the nonlinear polarization:

$$
\begin{aligned}
(1 - k_0^2/q_q^2)E_0^q + \delta^q E_1^q &= -\frac{4\pi}{\epsilon} P_0^{qmn} \delta(\mathbf{k}_0 - \mathbf{k}_m - \mathbf{k}_n) \\
\delta^q E_0^q + (1 - k_1^2/q_q^2)E_1^q &= -\frac{4\pi}{\epsilon} P_1^{qmn} \delta(\mathbf{k}_0 - \mathbf{k}_m - \mathbf{k}_n).
\end{aligned} \tag{4.2}
$$

Here, $\mathbf{q}_q$ is the wave vector in the homogeneous medium with the dielectric tensor $\hat{\epsilon}$, and P_0 and P_1 are the corresponding zero and first components of the Fourier expansion of the nonlinear polarization $P^{(2)}_{(r,2\omega)} = \sum_j \mathbf{P}_j e^{i\tau_j \mathbf{r}}$. All quantities in (4.2) are marked by polarization indices q, m, n which are either π or σ, denoting, respectively, the linear polarization in the plane $\mathbf{k}_0, \mathbf{k}_1$ or perpendicular to this plane:

$$
\delta^\pi = \delta \cos \mathbf{k}_0 \hat{\mathbf{k}}_1, \quad \delta^\sigma = \delta.
$$

The solution of system (4.2) essentially depends on the ratio of δ to the value of frequency dispersion $\eta = (1 - k_0^2/q_q^2)/\delta$ (see the next section for more details), on the geometry of the experiment (see Fig. 4.1), in particular, on the parameter $b = \cos(\mathbf{k}_1 \hat{,} \mathbf{n}) / \cos(\mathbf{k}\hat{\mathbf{n}})$ where $\mathbf{n}$ is the normal to the sample's surface, and on the closeness of the harmonic wave to the diffraction conditions as defined by the parameter

$$
\alpha = \frac{1}{2}[\cos \mathbf{k}\hat{\mathbf{n}} \cos \mathbf{k}\hat{\mathbf{k}}_1 (\eta + 2\nu) + \eta \cos \mathbf{k}_1 \hat{\mathbf{n}}](\cos \mathbf{k}\hat{\mathbf{n}} \cos \mathbf{k}_1 \hat{\mathbf{n}})^{-1}
$$

where $\nu = -(2\boldsymbol{\tau}\mathbf{q} + \boldsymbol{\tau}^2)/(2\mathbf{q}^2 \delta \cos \mathbf{k}\hat{\mathbf{k}}_1)$, the vector $\mathbf{q}$ being directed along $\mathbf{k}$, $\mathbf{q}^2 = \left(\frac{2\omega}{c}\right)^2 \epsilon(2\omega)$. Solving the boundary problem for $b < 0$ (Bragg geometry) as was done in Chaps. 2 and 3, we obtain the electric field amplitudes for the harmonic wave at the sample's boundary:

$$
\begin{aligned}
E_0 &= \left[E_0^i + \frac{E_0^i(\xi_+ - \xi_-)e^{-i\alpha\ell} + 2iE_1^i \sin \Delta\ell}{\xi_- e^{-i\Delta\ell} - \xi_+ e^{i\Delta\ell}} \right] \cdot D^{-1} \\
E_1 &= \left[E_1^i + \frac{E_1^i(\xi_+ - \xi_-)e^{i\alpha\ell} + 2iE_0^i \xi_- \xi_+ \sin \Delta\ell}{\xi_- e^{-i\Delta\ell} - \xi_+ e^{i\Delta\ell}} \right] \cdot D^{-1}
\end{aligned} \tag{4.3}
$$

where we have put

$$D = \cos\widehat{\mathbf{k}\mathbf{k}_1}\,\eta(\eta + 2\nu) - 1$$
$$\Delta = [\alpha^2 - D/(\cos\widehat{\mathbf{k}_1\mathbf{n}}\cos\widehat{\mathbf{k}_1\mathbf{n}}]^{1/2}$$
$$\ell = \delta q L/2, \quad \xi_\pm = \eta + (\alpha \pm \Delta)\cos\widehat{\mathbf{k}\mathbf{n}},$$
$$E_0^i = [\cos\widehat{\mathbf{k}_1\mathbf{k}}(\eta + 2\nu)P_0 - P_1]/\delta, \quad E_1^i = (\eta P_1 - P_0)/\delta.$$

In (4.3) and everywhere below, the polarization indices are omitted for the sake of simplicity. So, the above formulas describe the generation of the σ-polarized harmonic only. When analyzing other physical situations, the same formulas are used, provided that the corresponding values are marked with the proper polarization indices.

It follows from (4.3) that phase matching occurs when the argument of the delta function (4.2) is zero; i.e., phase-matching conditions can be represented in the form

$$2\alpha = (\eta\cos\widehat{\mathbf{k}\mathbf{n}}\cos\widehat{\mathbf{k}_1\mathbf{n}})^{-1}(\cos\widehat{\mathbf{k}\mathbf{n}} + \eta^2\cos\widehat{\mathbf{k}_1\mathbf{n}}). \tag{4.4}$$

The matching conditions of (4.4) with $k_m(2\omega) = k_q(\omega) + k_\ell(\omega) + s\tau$ are more diverse than in the case of a homogeneous medium. For example, for fixed values of ω and δ, condition (4.4) can be satisfied by varying the direction of the pumping wave or the modulation period. Condition (4.4) being satisfied, the harmonic wave intensities in the directions k_0 and k_1 are described by

$$|E_q|^2 = \eta^2 E_q^{i2}(\xi_+ - \xi_-)\ell^2(\cos\widehat{\mathbf{k}\mathbf{n}} + \eta^2\cos\widehat{\mathbf{k}_1\mathbf{n}})^{-2}\left|\xi_- \mathrm{e}^{-\mathrm{i}\Delta\ell} - \xi_+ \mathrm{e}^{\mathrm{i}\Delta\ell}\right|^2. \tag{4.5}$$

It follows from (4.5) that if the ratio $\eta = \left(1 - \frac{\epsilon(\omega)}{\epsilon(2\omega)}\right)/\delta$ is close to the critical value $\eta = \eta_c = 1/\sqrt{|b|}$, i.e., if the harmonic frequency exactly equals the boundary frequency of the forbidden band, then the second harmonic experiences strong diffraction scattering and the generation efficiency increases sharply, compared with that in a homogeneous medium.

In this case, i.e., when $(\delta q L)^{-1} \ll |\eta - \eta_c| \leq 1$, (4.5) becomes

$$\begin{aligned} |E_0|^2 &= |\eta E_1|^2 \\ &= \left[\frac{2\pi(P_0 - P_1\eta)qL}{\epsilon\cos\widehat{\mathbf{k}\mathbf{n}}}\right]^2 \left[(b\eta^2 + 1) - 4b\eta\sin^2 a\left(\frac{b\eta^2 + 1}{q}\right)\right]^{-1}, \\ a &= \ell/\cos\widehat{\mathbf{k}_1\mathbf{n}}. \end{aligned} \tag{4.6}$$

When η tends to η_c, the radiation intensity oscillates and increases (see Fig. 4.2). The maximum values of the intensities $|E_q|^2$ are reached when η differs slightly from

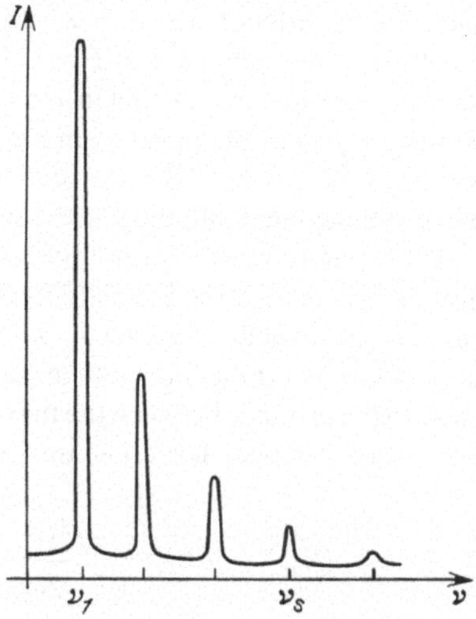

Fig. 4.2 Frequency dependence of the amplitude of nonlinear harmonic generation for the matching conditions satisfied near the boundary of the selective reflection band: $\nu = \frac{2\omega-\omega_e}{\omega_B}$, $\nu_s = \delta\left(\frac{\pi s}{a}\right)^2/(4\sqrt{|b|}\cdot\cos\hat{\mathbf{k}_1\tau})$

its critical value: $\eta - \eta_c = \pi s/(ab)$, $s = \pm 1, \pm 2, \pm 3$ and are proportional to the fourth power of the sample thickness:

$$|E_0|^2 = |bE_1|^2 = (qL)^4[\delta(P_1 - P_0\sqrt{|b|})/(2s\epsilon\cos\hat{\mathbf{kn}}\cos\hat{\mathbf{k}_1\mathbf{n}})]^2. \tag{4.7}$$

When the pumping wave direction is fixed, the maximum amplitudes can be expressed in terms of the difference between second-harmonic frequency 2ω and that of the forbidden band boundary ω_e:

$$|E_0|^2 = |bE_1|^2 = \left[\frac{\pi qL(P_1 - P_0\sqrt{|b|})}{2\epsilon\cos\hat{\mathbf{kn}}}\right]^2 \delta\omega_e/\left[\sqrt{|b|}(2\omega-\omega_e)\cos^2\hat{\mathbf{k\tau}}\right] \tag{4.8}$$

where

$$2\omega - \omega_e = \omega_B\delta\left(\frac{\pi s}{a}\right)^2/(4\sqrt{|b|}\cos^2\hat{\mathbf{k\tau}}), \quad \omega_e = \omega_B\left(1 + \frac{\delta\sqrt{|b|}}{2\cos^2\hat{\mathbf{k\tau}}}\right)$$

$$\omega_B = -c\tau\left(1 + \frac{\eta}{2}tg\hat{\mathbf{k\tau}}tg\hat{\mathbf{kn}}\right)/[2\sqrt{\epsilon(2\omega)}\cos\hat{\mathbf{k\tau}}].$$

We will now discuss the conditions under which the enhancement of efficiency of the nonlinear transformation described by (4.7), (4.8) can be at a maximum. In the simplest collinear geometry, for instance, a strong correlation of the frequency dispersion of ϵ and δ is required for the phase-matching conditions to be satisfied. Such a correlation is not required in noncollinear geometry. For the maximum efficiency of the nonlinear transformation to be achieved, the linewidth of the pumping wave must be narrow compared with the frequency width of the first maximum of the amplitude oscillations (see Fig. 4.2). This width depends on the sample thickness and is of the order of $\Delta\omega/\omega = \lambda^2/(\delta L^2)$. If the pumping wavelength $\lambda \sim 10^4$ Å and $\delta = 0.01$, $\Delta\omega/\omega = 10^{-4}$ at $L = 0.1$ cm, the maximum enhancement of harmonic intensity (compared with that in a homogeneous sample with constant dielectricbreak permittivity ϵ) must appear even for nanosecond pulses (more so for longer pulses) and be of the order of 10^2, provided the enhancement remains within the framework of the nondepleted pumping wave approximation.

If the pumping wave is not monochromatic and the harmonic linewidth exceeds those of the generation intensity maxima in (4.6), then SHG efficiency is low, compared with that at the maxima of (4.7), (4.8). Harmonic intensity can now be obtained by averaging over the linewidth in accordance with (4.3). If the harmonic linewidth is close to the distance between the maxima in (4.8), then averaging yields the following expression for mean harmonic intensity

$$|E_0|^2 = |bE_1|^2 = \frac{\pi^2(P_1 - P_0\sqrt{|b|})^2(kL)^2}{2\epsilon^2\cos^2\hat{\mathbf{kn}}\cos^2\hat{\mathbf{k\tau}}}\left(\frac{\delta\omega_e\sqrt{|b|}}{|2\omega - \omega_e|}\right)^{1/2}. \quad (4.9)$$

Note that the simplifications made here do not affect the above qualitative results and can be removed. For example, we can easily take account of the diffraction of the pumping wave, reflection at the sample's boundaries, etc. The smallness of δ is, however, essential in quantitative analysis, numerical calculations being necessary if this condition is not met. The requirement $b < 0$ is important for the enhancement of efficiency of the nonlinear transformation, which for $b > 0$ does not arise. As for media in which the effect considered can be observed, there are, in addition to artificial heterostructures, incommensurable crystal structures, crystals with helicoidal magnetic ordering, and chiral liquid crystals. The latter crystals seem the most promising because their parameters can be varied easily.

4.2 Third-Harmonic Generation in a Cholesteric Liquid Crystal

The peculiarities of nonlinear generation of the second harmonic in periodic media discussed in the previous section are very general, and they also manifest themselves in other nonlinear optical phenomena. Here, third-harmonic generation (THG) will be considered for cholesteric crystals in collinear geometry, both the pumping wave and the harmonic being directed along the cholesteric axis. Cholesteric crystals were chosen since they are used in experiments on harmonic generation [4, 8], and the exact solutions of Maxwell equations within linear optics are known for them, unlike the previous section. Hence, the analytical description of THG is not limited to the case of small dielectric anisotropy δ. Besides, the nonlinear optical phenomena in liquid crystals provide important information on the structures of those crystals, revealing their nonlinear susceptibilities. For the sake of brevity, we will not give here a detailed description of this topic that is completely outlined in [8]. We shall only note that in order to observe in liquid crystals the nonlinear optical effects related to the electron mechanism discussed below, pulse lasers (e.g., *Q*-switched Nd lasers) as a rule are used [8] in order to provide high intensity of light and to avoid overheating of liquid crystals and the distortion of their structures. The effects of orientational optical nonlinearity, which are not considered here, can be observed with low-power lasers, such as a He–Ne laser, in a continuous mode [11].

Within the approximation of a nondepleted pumping wave of frequency ω which propagates along the cholesteric axis, the generation of the third harmonic is described by an equation of the type of (1.8) which includes the nonlinear polarization vector P as the inhomogeneity:

$$\left[\frac{\partial^2}{\partial z^2} + \left(\frac{3\omega}{c}\right)^2 \hat{\epsilon}(z,\ 3\omega)\right] \boldsymbol{\epsilon}(z,\ 3\omega) = -4\pi \left(\frac{3\omega}{c}\right)^2 \mathbf{P}^{(3)}\mathbf{E}(z,\ 3\omega) \tag{4.10}$$

where the nonlinear polarization vector $\mathbf{P}^{(3)}$ is defined from the cubic tensor of nonlinear susceptibility [1, 2] as

$$\mathbf{P}^{(3)}(z,\ 3\omega) = \hat{\chi}(z, 3\omega)\mathbf{E}(z, \omega)\mathbf{E}(z, \omega)\mathbf{E}(z, \omega). \tag{4.11}$$

All quantities in (4.10) and (4.11) are functions of the coordinate Z and frequency ω, which emphasize their spatial inhomogeneity and the frequency dispersion of the dielectric properties of liquid crystals.

4.2.1 Conditions for Phase Matching

In order to determine the intensity of the third-harmonic radiation from a cholesteric sample, as well as its direction and polarization, we must solve (4.10) with regard to boundary conditions. However, before solving the boundary problem, let us find the conditions for the most effective generation of the third harmonic, i.e., phase-matching conditions [1].

We use the fact that the pumping wave in a CLC must satisfy (1.7); hence, we represent this wave $\mathbf{E}(z, \omega)$ as the sum, as in (1.10), over the eigensolutions of (1.13) and perform Fourier expansion of $\mathbf{P}^{(3)}$ in (4.10).

The particular solution of (4.10) will be looked for, as was done in Chap. 1, in the form:

$$\mathbf{E}(z, 3\omega) = \left[E^+\mathbf{n}_+ e^{i\tilde{K}^+(3\omega)z} + E^-\mathbf{n}_- e^{i\tilde{K}^-(3\omega, z)}\right] e^{-3i\omega t} \tag{4.12}$$

where $\tilde{K}^+ - \tilde{K}^- = \tau$.

After substituting the Fourier expansions of the nonlinear polarization of (4.11) and (4.12) into (4.10), we obtain the following system of equations for the amplitudes E^+ and E^-:

$$\left[1-\left(\frac{\tilde{K}^{+}}{q}\right)^{2}\right]E^{+}+\delta E^{-}=\sum_{s\ell mn}\mathcal{P}^{\ell mn}_{+s}(j_1j_2j_3)\delta(\tilde{K}^{+}-k^{m}_{j_2}-k^{\ell}_{j_1}+k^{n}_{j_3}+s\tau)$$
$$\delta E^{+}+\left[1-\left(\frac{\tilde{K}^{-}}{q}\right)^{2}\right]E^{-}=\sum_{s\ell mn}\mathcal{P}^{\ell mn}_{-s}(j_1j_2j_3)\delta(\tilde{K}^{-}-k^{\ell}_{j_1}-k^{m}_{j_2}-k^{n}_{j_3}+s\tau)$$
(4.13)

where the right-hand sides are defined by the Fourier coefficients of the nonlinear polarization vector obtained when substituting into (4.11) the pumping wave field in the form of expansion (1.10), where the indices ℓ, m, n take the values $+$ and $-$

$$\mathbf{P}(3\omega z)=-\left(\frac{4\pi}{\epsilon_3}\right)^{-1}\sum_{s,\ell,m,n=\pm}\mathbf{n}_{\pm}\mathcal{P}^{\ell mn}_{\pm s}(j_1j_2j_3)$$
$$\cdot\exp\{\mathrm{i}[(k^{\ell}_{j_1}+k^{m}_{j_2}+k^{n}_{j_3}+s\tau)-3\omega t]\}. \tag{4.14}$$

For the sake of generality, when deriving (4.14), we have assumed that the pumping wave is a superposition of all four eigenwaves of the CLC. Then expressing $K^{\pm}_{j}$ through $k^{\pm}$ and taking $\tilde{K}^{\pm}=\tilde{k}_3\pm\tau/2$ (see the notation of Chap. 1), we find that delta functions on the right-hand sides of (4.13) can be represented in the form $\delta(\tilde{k}_3\pm k^{\ell}\pm k^{m}\pm k^{n}+s_{+}\tau)$ with $s_{+}=3,2,1,0,-1,-2$ for the first equation, and in the form $\delta(\tilde{k}_3\pm k^{\ell}\pm k^{m}\pm k^{n}+s_{-}\tau)$ with $s_{-}=2,1,0,-1,-2,-3$ for the second one; all possible combinations of signs in the arguments of the delta functions are admissible. Formally, the value of $\tilde{k}_3(\tilde{K}^{\pm})$ in the inhomogeneous solution does not necessarily satisfy the dispersion equation (1.14) at the frequency 3ω. However, harmonic generation can actually be effective, provided that $\tilde{k}_3$ satisfies or almost satisfies the dispersion equation. Hence, the quantity $\tilde{k}_3$ is convenient to represent as $\tilde{k}_3=\Delta k^{\pm}+k^{\pm}_{3}$ where $k^{\pm}_{3}$ satisfies the dispersion equation; i.e., it describes the eigenwaves in the CLC at the frequency 3ω. Then, the condition that an individual addend of the right-hand side of (4.13) contributes to the solution is

$$\Delta k^{\pm}_{\ell mn}=k^{\ell}+k^{m}+k^{n}-k^{\pm}_{3}+s_{\pm}\tau. \tag{4.15}$$

The indices ℓ, m, n marking $\Delta k^{\pm}$ in (4.15) characterize the pumping field. Note that (4.15) was obtained using very general suppositions about the linear and nonlinear dielectric properties of CLC.

It is usually assumed (also below) that the local dielectric parameters of a CLC are similar to those of nematic crystals [4, 8], i.e., to those of uniaxial crystals. This assumption is a sufficiently good approximation; it means that in (4.15) $s_{+}=s_{-}=0$.

Without fixing our attention on the explicit dependence of the coefficients $\mathcal{P}^{\ell mn}_{\pm}$ on the tensor of nonlinear susceptibility $\hat{\chi}$, we can easily see that the maximum efficiency of THG is achieved when the right-hand side of (4.15) is zero.

The solution of the inhomogeneous system (4.13) (the particular solution) can be written as

$$E^+ = \frac{D^+}{D^0}, \quad E^- = \frac{D^-}{D^0} \tag{4.16}$$

where

$$D^0 = \left[1 - \left(\frac{\tilde{K}^+}{q}\right)^2\right]\left[1 - \left(\frac{\tilde{K}^-}{q}\right)^2\right] - \delta^2$$

$$D^+ = \mathcal{P}_+^{\ell mn}\left[1 - \left(\frac{\tilde{K}^-}{q}\right)^2\right] - \delta \mathcal{P}_-^{\ell mn}$$

$$D^- = \mathcal{P}_-^{\ell mn}\left[1 - \left(\frac{\tilde{K}^+}{q}\right)^2\right] - \delta \mathcal{P}_+^{\ell mn}.$$

Thus, expression (4.15) being zero determines the conditions for phase-matched generation, the number of such conditions being much greater than in the case of a homogeneous medium. As follows from (1.11) and (4.12), the conditions do not reduce now to the relationship between the wave vectors of the pumping wave and the harmonic; they may also include the reciprocal lattice vector τ. Formally speaking, relationship (4.15) determines 40 different matching conditions corresponding to various combinations of indices with both positive and negative values of $K^\pm$. However, not every such condition can be achieved in practice. This will be discussed below.

4.2.2 Solution of the Boundary Problem

In order to find THG intensity in a sample with finite dimensions, we have to determine the field **E** in the sample, this field being the superposition of the particular solution of (4.13), (4.16) and the eigensolutions of the homogeneous system obtained from (4.13) by canceling the right-hand sides

$$\mathbf{E}(z, 3\omega) = \mathrm{e}^{-3\mathrm{i}\omega t}\Bigg[E^+\mathbf{n}_+\mathrm{e}^{\mathrm{i}\tilde{K}^+ z} + E^-\mathbf{n}_-\mathrm{e}^{\mathrm{i}\tilde{K}^- z} + \sum_{j=1}^{4} C_j\left(E_j^+\mathbf{n}_+\mathrm{e}^{\mathrm{i}K_{3j}^+ z} + E_j^-\mathbf{n}_-\mathrm{e}^{\mathrm{i}K_{3j}^- z}\right)\Bigg]. \tag{4.17}$$

The coefficients C_j in this superposition are determined from the boundary conditions.

When the pumping wave propagates in a CLC along the optic axis, the harmonic propagation direction may be either the same or opposite, or the harmonic may propagate in both directions simultaneously. For the three cases, the boundary conditions reduce to the following:

1. The field of the third harmonic is zero at the front surface of the sample.
2. The field is zero at the exit (with respect to the pumping wave) surface.
3. The field component of the harmonic directed along the pumping wave is zero at the front surface, and the component corresponding to the backward propagation harmonic is zero at the exit surface.

Using one of the above conditions, we can find the coefficients C_j in (4.17) and then the field and flux of the third harmonic at the boundary that gives the final solution to the problem in question. Note that the boundary conditions formulated above correspond to the case where $\bar{\epsilon}$ coincides with the dielectric permittivity of the medium outside the sample, i.e., when there is no reflection at the boundaries. The reflection can be easily taken into account, but the corresponding formulae prove to be very cumbersome.

4.2.3 *Third-Harmonic Generation in the Band of Selective Reflection*

So far, we have dealt with the general analysis of the third harmonic in CLC using the exact solution of the corresponding linear problem. Nothing was said about specific values of the frequency of pumping wave ω and of 3ω and about how they are located with respect to the bands of selective reflection.

Let us now consider the situation in which there is a strong selective reflection in CLC of either the pumping wave or the harmonic (other situations are considered in [4, 7, 12]). The qualitative features of the nonlinear frequency transformation will be shown to appear due to light diffraction and to the specific features of the propagation of electromagnetic waves in the periodic structure, these features having been discussed in Sect. 4.1 for a simple periodic medium. Selective reflection, in particular, can change the efficiency of phase-matched harmonic generation so that the matching conditions which were not effective when there was no selective reflection may now prove otherwise.

Even before solving the boundary problem, we can conclude from the form of the particular solution of the inhomogeneous problem that harmonic generation efficiency increases if the phase-matching conditions are satisfied at the frequency 3ω corresponding to the multiple roots of the dispersion equation (1.14) or near this frequency. This is so because at those frequencies, the determinant in the denominator of the solutions of (4.16) tends to zero as $(\Delta k_{\ell mn})^2$ when approaching the phase-matching conditions, whereas in the absence of multiple roots, it tends to zero as $\Delta k_{\ell mn}$. Since the multiple roots of (1.14) correspond to the boundary frequencies of the selective reflection band, qualitatively new features in harmonic generation must be expected when the phase-matching conditions are satisfied at the frequency 3ω near the boundary of the selective reflection band.

We will now consider the quantitative description of THG under the conditions of selective scattering of the third harmonic. That is, we shall solve the boundary problem formulated above and find the following formulae for the amplitudes of the third harmonic at the front and exit surface of a CLC sample:

$$E^+(3\omega, L) = D^{-1}\left[D^+ + \frac{D^+(\xi_+ - \xi_-)e^{-i\tilde{k}_3 L} + 2iD^- \sin k_3^- L}{\xi_- e^{-ik_3^- L} - \xi_+ e^{ik_3^- L}}\right] e^{i(\tilde{k}_3 + \tau/2)L}$$

$$E^-(3\omega, L) = D^{-1}\left[D^- + \frac{D^-(\xi_+ - \xi_-)e^{-i\tilde{k}_3 L} + 2i\xi_+\xi_- \sin k_3^- L D^+}{\xi_- e^{-ik_3^- L} - \xi_+ e^{ik_3^- L}}\right] \tag{4.18}$$

where $D, D^\pm$ are defined in (4.16), $\xi_\pm = \delta^{-1}\{\kappa_3^{-2}(\tau/2 \pm k_3^-)^2 - 1\}$, and $\epsilon_1, \epsilon_3, \delta_1$, and δ_3 are, respectively, the dielectric permittivity and anisotropy at the frequencies ω and 3ω.

According to our assumption that the polarization of the harmonic corresponds to the diffracting polarization, we find that diffractive scattering of the harmonic may occur for the following phase-matching conditions:

$$\begin{aligned} 3k^+(\omega) &= k^-(3\omega) \\ 3k^-(\omega) &= k^-(3\omega) \\ 2k^+(\omega) + k^-(\omega) &= k^-(3\omega) \\ k^+(\omega) + 2k^-(\omega) &= k^-(3\omega) \end{aligned} \tag{4.19}$$

For any given frequency ω, the matching conditions can, in general, be provided by the proper choice of the cholesteric helical pitch whose phase-matched value depends on the dispersion of dielectric permittivity and anisotropy δ. Under this condition, the phase-matching conditions are satisfied, in general, outside the band of selective reflection of the harmonic.

As was shown before, it is of special interest to analyze the situations when phase matching occurs near the selective reflection band. However, in order for the phase-matching conditions (4.19) to be satisfied at the boundary of the selective reflection band or near it, certain relationships between the dispersion of dielectric permittivity and the anisotropy δ must be satisfied. The existence of frequency dispersion of the dielectric permittivity of a CLC means that the corresponding conditions are satisfied (if they can be satisfied at all) for a specific frequency of the pumping wave and for a specific helical pitch.

Assuming that the frequency of the third harmonic coincides with one of the edges of the selective reflection band, i.e., that $k^-(3\omega) = 0$ and

$$\tau = 2q(3\omega)(1 \pm \delta_3)^{1/2} \tag{4.20}$$

we see from (4.19) that only the last of the conditions of that equation can be met.

The condition that phase matching occurs at the boundary of the selective reflection band is

$$k^+(\omega) + 2k^-(\omega) = 0. \tag{4.21}$$

It follows from (4.19) and (4.21) that the pumping field must contain the waves with both diffracting and nondiffracting polarizations.

Substituting the expressions for $k^\pm$ from (1.15) into (4.21) and taking into account (4.20), we find the phase-matching conditions at the boundary of the selective reflection band:

$$d^2\left[1 - \left(\frac{25}{9}\right)\delta_1^2\right] - 82d + 81 = 0 \tag{4.22}$$

where

$$d = \bar{\epsilon}_1[\bar{\epsilon}_3(1 \pm \delta_3)]^{-1}.$$

Hence,

$$\frac{\epsilon_1}{\epsilon_3} = \frac{1 \pm \delta_3}{1 - 25\delta_1^2/9}\left[41 \pm \sqrt{41^2 - 81\left(1 - \frac{25}{9}\delta_1^2\right)}\right]. \tag{4.23}$$

Assuming the frequency dependence of ϵ and δ to be known, we can find from (4.22) the pumping wave frequency ω and from (4.20) the helical pitch which corresponds to the phase matching at the boundary of the selective reflection band. It follows from (4.22) that the phase-matching condition can be satisfied near or at the boundary of the selective reflection band, provided that the frequency dispersion of the dielectric permittivity $|1 - \epsilon_1/\epsilon_3|$ is of the order of the anisotropy δ_3 at the frequency 3ω. This condition requires that the type of CLC be properly chosen. Nematic–cholesteric mixtures seem to be the most suitable to obtain the needed relationship between the dispersion and the anisotropy.

Let us now consider in more detail the expressions (4.18) which yield the third-harmonic amplitude in the case when the phase-matching conditions are satisfied at the boundary of the selective reflection band. These expressions show that the third harmonic's amplitude oscillates as a function of the pumping wave frequency or of the difference between the boundary frequency and 3ω.

In order to prove this, we will find the phase-matched values of the amplitudes $E^\pm$, i.e., when the last of the conditions (4.19) is satisfied and the values $\tilde{K}^\pm$ satisfy the dispersion equation [the determinant D of (4.13) is zero]. For this purpose, it is sufficient to discard the indeterminacy 0/0 for the limit $D \to 0$ in (4.18). As a result, the phase-matched amplitudes of the harmonic exiting the crystal are

$$E^\pm = \pm\frac{\mathrm{i}q_3L(\xi_+ - \xi_-)\mathrm{e}^{\mathrm{i}k_3^- L}D^\pm}{4(k_3^-/q_3)[(\tau/q_3)^2 + \delta_3^2](\xi_-\mathrm{e}^{-\mathrm{i}k_3^- L} - \xi_+\mathrm{e}^{\mathrm{i}k_3^- L})}. \tag{4.24}$$

As follows from (4.24), the efficiency of the nonlinear frequency transformation can be much greater when phase matching is achieved near the frequency boundary of the selective reflection band: $|k_3^-/q_3| \leq \delta_3$. In fact, the amplitudes in (4.24) oscillate strongly as a function of the phase-matching values of k_3^-, these amplitudes being proportional to sample thickness at the minima and to squared thickness at the maxima.

4.2.4 Extreme Efficiency of Transformation

As follows from (4.24), maxima of amplitudes of (4.24) are reached at $k_3^- = s\pi/L$ and minima at $k_3^- = (2s+1)\pi/2L$ where s is an integer. These extreme amplitudes are:

$$
\begin{aligned}
E^{\pm}_{\max} &= -\frac{\mathrm{i}(q_3L)^2 D^{\pm}}{4s\pi} \sim \frac{\mathrm{i}\delta(q_3L)^2\mathcal{P}}{4\pi s} \\
E^{\pm}_{\min} &= \frac{\mathrm{i}\sin k_3^-(q_3L)(\tau+2k_3^-)q_3^{-1}D^{\pm}}{4\left[\frac{1}{2}(\tau/q_3)^2-\sqrt{(\tau/q_3)^2+\delta_3^2}\right]\sqrt{(\delta/q_3)^2+\delta_3^2}}T \sim \mathrm{i}(q_3L)\mathcal{P}
\end{aligned}
\tag{4.25}
$$

where $\mathcal{P}$ is the characteristic value of nonlinear polarization. It is seen in (4.24), (4.25) that the minimum of at least one of the amplitudes E^+ or E^- is of the order of the amplitude in the absence of diffraction. As for the maximum value, it corresponds to the enhanced efficiency of nonlinear transformation in the range of noticeable diffraction scattering of the harmonics, but outside this region ($|k_3^-/q_3| > \delta_3$), this maximum reduces to the known expressions proportional to q_3L [1, 2]. Also, from (4.24), (4.25), it follows that the generation maxima occur when the parameter $\nu = (\omega_3 - \omega_e)/\omega_e$ describing the location of the harmonic frequency outside the boundaries of the selective reflection band $\omega_e = \tau c/[2\sqrt{\epsilon_e}(1 \pm \delta_3)]$ is

$$
\nu = \frac{3\omega - \omega_e}{\omega_e} = \frac{1}{\delta_3\epsilon_3}\left(\frac{c\pi s}{\omega_e L}\right)^2 \approx \frac{1}{\delta_3}\left(\frac{2\pi s}{\tau L}\right)^2. \tag{4.26}
$$

It is assumed that when the harmonic frequency 3ω satisfies (4.26), the phase-matching condition is also satisfied. Comparing (4.25) and (4.26), we can see that the maximum intensity of THG decreases as

$$
I^{\max}_{(3\omega)} = I_0 \frac{\delta_3\omega_3}{|3\omega - \omega_e|} \tag{4.27}
$$

as the phase-matching frequency 3ω goes from the boundary of the selective reflection band, where the frequency 3ω takes discrete values given by (4.26);

I_0 is approximately equal to the THG intensity in the same sample but far from the selective reflection band.

The latter formula can be represented in another form if we assume that the frequency of the light is fixed and the helical pitch is variable so that phase matching is achieved at the boundary of the selective reflection band by varying the pitch, i.e.,

$$I_{(\tau)}^{\max} = I_0 \frac{\delta\tau_e(3\omega)}{|\tau - \tau_e(3\omega)|} \tag{4.28}$$

where $4\pi/\tau_e$ is the helical pitch corresponding to the coincidence of the frequency 3ω with the boundary of the selective reflection band, and the discrete values of τ_e are given by (4.26) if 3ω is replaced by τ and ω_e by τ_e. Note that harmonic amplitude turns out to be proportional to L^2 near the selective reflection band not only when the determinant D of the system (4.13) is exactly zero, in particular, when $\tilde{k}_3$ coincides with k_3^-, but also near these conditions at $\tilde{k}_3 = \pi n/L$ and $k_3^- = \pi s/L$. Hence, the formulae (4.19), (4.21) prove to be the conditions of proportionality of $E^\pm$ to L^2, provided that $\pm\pi n/L$ is added to the right-hand side of (4.19), (4.21) and k_3^- is set equal to $\pi s/L$. In this case, the maximum amplitudes are

$$E^\pm = \frac{(q_3L)^2 D^\pm}{\pi^2(n^2 - s^2)} \left[1 - \mathrm{e}^{\mathrm{i}(s\pm n)\pi}\right] \left(\frac{q_3}{k_3^+}\right)^2 \tag{4.29}$$

where s is an integer and n is a number with the opposite parity or a half-integer. It is seen from (4.25), (4.26), and (4.29) that the absolute maximum of harmonic intensity is, in fact, at the boundary of the selective reflection band. When the phase-matching condition (4.21) is exactly satisfied, the maximum occurs at $s = 1$ in (4.26), which means that the frequency slightly deviates from the boundary of the selective reflection band $[\nu = \delta^1(\pi s/q_3L)^2]$. When the deviation of the dispersion $\Delta d = n\pi/q_3L$ from the boundary value given by (4.22) is small, an absolute maximum occurs at $n = 1$ and $s = 0$ in (4.29), i.e., when the harmonic frequency exactly coincides with the boundary of the selective reflection band. The absolute value of the maximum amplitude in (4.29) is close to that of (4.25); they differ only by factor.

4.2.5 *The Conditions for the Enhancement Effect*

It is then natural to ask what the conditions are for maximum amplitude of the harmonic, if the dispersion differs from the boundary value of (4.22) by more than $\Delta d = \pi/q_3L$. Since the cholesteric pitch can be easily varied in an experiment, we will take this parameter τ as the free parameter of the problem while the other parameters, and the frequency ω also, are fixed. By varying τ, we can, for example, satisfy the phase-matching conditions (4.19) which do not correspond to the boundary of the selective reflection band ω_e, or we can superimpose the harmonic frequency with ω_e, the phase-matching conditions not being satisfied in this case. The former may mean that harmonic amplitude is described by (4.25) with $s > 1$, and the latter that it is described by (4.29) at $s = 0$ and $n > 1$. An analysis of (4.22)–(4.29) shows that when the value $\nu = |\tau_e - \tau_p|/\tau_e$ is greater than $\delta^{-1}(\pi/q_3L)$ where τ_p corresponds to the phase-matched pitch and $\tau_e^{-1} = 2(3\omega/c)\sqrt{\epsilon_3(1 \pm \delta_3)}$, the absolute maximum of the harmonic amplitude proportional to $(q_3L)^2$ is reached under phase-matching conditions and is given by (4.25). The condition of the absolute maximum is the requirement that $q_3L\sqrt{\delta_3\nu} = s$ should be an integer. If s is not an integer, then a greater value of amplitude may correspond to a small deviation of τ from τ_p. Whether or not the amplitude is described by (4.29) depends on the compatibility of the considered variations of τ with the conditions for n and s in (4.29).

The maximum value of the phase-matched generation intensity determined by (4.25), (4.27), and (4.29) is related to a strictly monochromatic wave. An effect of the same order can be achieved for a wave with a finite width $\Delta\omega/\omega$ if the linewidth is less than that of the harmonic's maximum, which is about $(\pi s/\delta)^2/(q_3/L)^3$, and this determines the sharpness of phase matching under diffraction conditions.

If the linewidth is of the order of the frequency width of oscillations in (4.24), i.e., $(s\pi/\delta L)^2/\delta_3$, then the generation efficiency proves to be less than the maximum value and is described by averaging the square of the modulus of (4.24) over the linewidth. This efficiency is

$$|E^{\pm}|^2_{ev} \sim I_0\sqrt{\frac{\delta_3\tau_e}{|\tau - \tau_e|}} = I_0\sqrt{\frac{3\omega\delta}{|3\omega - \omega_e|}} \tag{4.30}$$

where 3ω now stands for the frequency position of the line's center and can vary continuously. It is seen in (4.30) that the enhancement of the nonlinear frequency transformation occurs when the detuning from ω_e and the width of the pumping line are not greater than δ_3.

Thus, when the phase-matching conditions are satisfied at the boundary of the selective reflection band, the nature of nonlinear generation greatly changes compared with that apart from the selective reflection band [4, 7] and the angular distribution of the generated harmonic also changes. (The generations with comparable intensities run both in the direction of the pumping wave and in the opposite direction, and generation efficiency is enhanced.)

The possibility of achieving phase matching at the boundary of the selective reflection band depends on certain relationships between the frequency dispersions of dielectric permittivity $\bar{\epsilon}$ and anisotropy δ [see (4.22) and (4.23)]. As mentioned above, these relationships could be satisfied in properly chosen cholesteric crystals, since the dielectric and structure parameters of available cholesteric crystals and cholesteric compositions indicate that such a choice is indeed possible [7]. Up until now, we have completely neglected CLC absorption, which results in a lessening of the effects in question, and this must be taken into account when describing a real experiment (see [13]). So far, we have considered only third-harmonic generation in a CLC. However, all the results hold, with obvious modifications, for second-harmonic generation in chiral smectics also [13].

The specific features of the nonlinear frequency transformation in CLC discussed above are not inherent in the propagation of waves along the cholesteric axis only. They also appear when light propagates at an angle to the axis. We will not consider here the case of oblique incidence and refer the reader to the original papers on the subject [4, 14]. We only note that the polarization parameters of generation prove to be very complicated and a theoretical description is more difficult because exact solutions to Maxwell equations in linear optics are not known for this case.

4.3 Phase-Matching Conditions Independent of Frequency Dispersion

In Sects. 4.1 and 4.2, we analyzed the increase in efficiency of the phase-matched nonlinear transformation of frequency under the conditions of diffraction of the generated wave, those conditions requiring certain relationships between the frequency dispersion of dielectric permittivity $\hat{\epsilon}(\mathbf{r})$ and the amplitude of its modulation δ to be fulfilled in the periodic medium. Here, we will study the enhancement of efficiency of the nonlinear frequency transformation when the phase-matching conditions do not depend on the frequency dispersion of ϵ [15] but, to be satisfied, require that the pumping field has a special configuration, for example, two oppositely directed pumping waves.

Let us consider second-harmonic generation in a periodic nonlinear medium having the shape of a plane-parallel sample with thickness L. For the sake of simplicity, we assume this medium is periodic in one dimension and the direction Z along which the dielectric and nonlinear properties are periodic is perpendicular to the sample surfaces (see Fig. 4.3). The harmonic field is described by the solutions of

$$\text{grad div}\,\mathbf{E} - \Delta\mathbf{E} - \frac{4\omega^2}{c^2}\epsilon(z, 2\omega)\mathbf{E} = \mathbf{P}(z, 2\omega) \tag{4.31}$$

where ϵ is the dielectric permittivity and $\mathcal{P}$ is the quadratic nonlinear polarization.

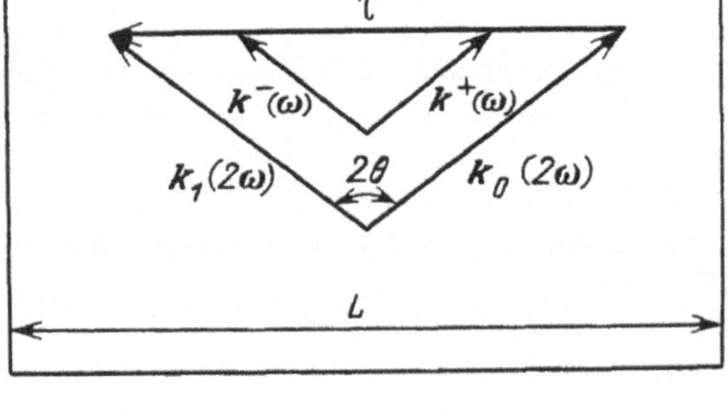

Fig. 4.3 Pumping geometry which provides the fulfillment of the phase-matching conditions, independent of the dispersion of dielectric permittivity. $\mathbf{k}^{\pm}(\omega)$ and $\mathbf{k}_{0,1}(2\omega)$ are the wave vectors of the pumping wave and harmonic, respectively

Keeping in mind that the efficiency of nonlinear frequency transformation increases under the diffraction of harmonic waves, we assume that the frequency 2ω coincides with or is close to the boundary of the frequency band forbidden for propagating solutions, ω_e. The analysis of (4.31), similar to that given above, indicates that along with the phase-matching conditions depending on the frequency dispersion ϵ and on the amplitude of its modulation δ, there is one more condition which does not depend on ϵ. This condition is realized if the frequency 2ω coincides, practically speaking, with the forbidden band boundary and the nonlinear polarization in (4.31) contains a component modulated with a standing wave, with a period determined by the medium's periodicity.

In order to describe harmonic generation under the above conditions, we consider on the right-hand side of (4.31) only the components with a nonlinear polarization of the form

$$\mathcal{P} \sim \mathrm{e}^{-2i\omega t}\mathrm{e}^{2i\mathbf{k}_\perp(\omega)\mathbf{r}}\cos(\boldsymbol{\tau}\mathbf{r}+\varphi)$$

where φ is the phase determining the positions of nodes of the standing wave modulating the nonlinear susceptibility, τ is the reciprocal lattice vector of the periodic structure, and $\mathbf{k}_\perp$ is the vector perpendicular to the periodicity direction. It is evident that such components do exist in the nonlinear polarization, provided that the pumping field (see Fig. 4.3) is the coherent superposition of two plane waves of the type $A_+\mathrm{e}^{i\mathbf{k}_1(\omega)\mathbf{r}} + A_-\mathrm{e}^{\mathbf{k}_2(\omega)\mathbf{r}}$, in which the projections of the wave vectors on the periodicity direction have opposite signs (in one particular case, it may be two oppositely directed plane waves).

Let us assume for the sake of simplicity that the modulation amplitude in the sample is small and that the mean value of $\epsilon(r)$ coincides with the dielectric permittivity of the external medium. Then, the solution of (4.31) found within the two-wave approximation of dynamic diffraction theory, neglecting the depletion of the pumping wave (see the sections above), shows that harmonic amplitude at the exit sample surface oscillates as a function of frequency and sharply decreases when 2ω moves away from the forbidden band. This amplitude outside the forbidden band is described by

$$E = \frac{8\pi\Delta\chi A_+A_-}{\epsilon\delta}\frac{(\alpha\mathrm{e}^{i\varphi}-\mathrm{e}^{-i\varphi})(1-\cos\ell\sqrt{\nu}) - i\sqrt{\nu}\mathrm{e}^{i\varphi}\sin(\ell\sqrt{\nu})}{(\sqrt{\nu}\cos\ell\sqrt{\nu}+i\alpha\sin\ell\sqrt{\nu})\sqrt{\nu}} \tag{4.32}$$

where $\ell = \delta\omega\sqrt{\epsilon}L(c\sin\theta)^{-1}$; $\nu = 4\sin\theta|2\omega-\omega_e|(\delta\omega_\mathrm{B})^{-1}$ is the parameter describing the shift of harmonic frequency from the forbidden band whose boundaries are $\omega_e = \omega_\mathrm{B}[1\pm\delta(2\sin^2\theta)^{-1}]$ and $\omega_\mathrm{B} = c\tau[2\sqrt{\epsilon(2\omega)}\sin\theta]^{-1}$, $\alpha = \pm(1+\nu/2)$ and $\alpha > 0$, $\alpha < 0$, respectively, for the high- and low-frequency edges of the forbidden band for light polarized in the plane perpendicular to the plane of Fig. 4.3. $\Delta\chi$ is the modulation amplitude of nonlinear susceptibility with the period $4\pi/\tau$, and $\delta\epsilon$ is that of ϵ with the period $2\pi/\tau$.

At the oscillation maxima, which are reached at $\nu = (\pi s/\ell)^2$ with odd s, E takes the values

$$E_{\max} = 8\pi\Delta\chi\delta A_+A_-(\alpha\mathrm{e}^{i\varphi}-\mathrm{e}^{-i\varphi})\left(\frac{\omega L}{c\sin\theta}\right)^2(\pi s)^{-2}. \tag{4.33}$$

The frequency width of maxima in (4.33) is of the order of $\omega\delta^{-2}(c/\omega L)^3$, the amplitude of these maxima being proportional to L^2 and the distance between the maxima being of the order of $\omega\delta^{-1}(c/\omega L)$. The intensity of the harmonic at the maxima, as a function of frequency, behaves as $I \sim I_0\left(\frac{c}{\omega L}\right)^2 \omega_e^2/(2\omega-\omega_e)^2$ when 2ω tends to ω_e where I_0 is the phase-matched generation intensity if the diffraction conditions do not hold. The greatest value of intensity defined by (4.33) is reached at $s = 1$, this value being $(\delta L\omega/c)^2$ times greater than I_0 just at the boundary of the forbidden band. This value can be reached only for a sufficiently monochromatic pumping wave $[\Delta\omega/\omega \leq \delta^{-2}(c/\omega L)^3]$. If, on the contrary, the pumping linewidth is wide, SGH enhancement can occur only if $\Delta\omega/\omega < c/\omega L$. If the width of the pumping line is of the order of the distance between maxima, then intensity depends on the distance between the harmonic frequency and the forbidden band's boundary

$$I(2\omega) \sim \frac{I_0}{\sqrt{\delta}}\left(\frac{c}{\omega L}\right)^2\left(\frac{\omega_e}{|2\omega-\omega_e|}\right)^{3/2}$$

where ω now denotes the center of the line and can assume continuous values. The physical reasons for the possibility of the phase matching described are quite clear. For example, in collinear geometry ($\theta = \pi/2$), the eigensolutions of (4.31) at the boundary of the forbidden band are standing waves, with the period determined by medium periodicity. The standing waves of nonlinear polarization with the same period (the pumping field as two oppositely directed waves) are provided by the spatial modulation of the nonlinear susceptibility caused by the pumping field in the form of a standing wave. This is just the basis for the phase-matching conditions independent of the frequency dispersion of ϵ.

In addition to the frequency oscillations of harmonic intensity, there are oscillations connected to small changes of the pumping wave direction, the latter oscillations also being described by the above formulae if the value of ν is set equal to $2(|\Delta\theta|/\delta)\sin 2\theta$ where $\Delta\theta$ describes the deviation of harmonic direction from those defining the angular boundaries of the forbidden band. Since the phase matching under discussion can be achieved in both collinear and noncollinear geometries, it is convenient to use these conditions for experimental studies of the effect of the enhancement of efficiency of a nonlinear frequency transformation in periodic media.

To conclude this section, we note that the phase-matching conditions independent of the frequency dispersion of and the enhancement of efficiency of the nonlinear process in a periodic medium are of a very general nature and their analysis for a simple periodic medium permitting SGH was carried out only for the sake of simplicity. The same effects appear, for example, in a generation of the third harmonic if the pumping field is a coherent superposition of three plane waves, with a zero-sum for their wave vectors. Similar effects may surely occur in more complicated periodic media, for example, in three-dimensional periodic media and chiral liquid crystals [7], the latter crystals being a convenient subject for experimental studies.

4.4 Nonlinear Generation of Harmonics in Blue Phase of Liquid Crystals

In [16], we considered the optical properties of the blue phases and discussed the possibility of revealing their structure by means of linear optics. In this section, we use as an example of the nonlinear generation of the second harmonic in the blue phases [16] to analyze the possibility of nonlinear optics in studying blue phase structure. Since nonlinear harmonic generation in the blue phase is described in a similar way to that outlined above for CLCs and simple periodic media, we will focus our attention on the symmetry analysis of the nonlinear susceptibility tensor of the blue phases and finding restrictions on its components for specific space groups which may describe the symmetry properties of blue phases.

4.4.1 Phase-Matching Condition

As is known from [1, 2], nonlinear generation is most effective under phase-matching conditions which have the following well-known form for a homogeneous medium:

$$\mathbf{k}^{\sigma_2}(2\omega) = \mathbf{k}^{\sigma_1}(\omega) + \mathbf{k}^{\sigma_3}(\omega) \tag{4.34}$$

where $\mathbf{k}(m\omega)$ are the wave vectors of the harmonic and pumping waves, and σ_i are the indices of the eigenpolarizations. Because of the optical isotropy of the blue phases, the matching conditions (4.34) cannot be satisfied when there is a frequency dispersion of dielectric permittivity if the waves of harmonic and pumping propagate in the sample as plane waves. In this case, the frequency dispersion in (4.34) cannot be compensated by birefringence.

However, the phase-matching conditions (4.34) may be met in the blue phases, provided that either the pumping wave or the harmonic is subjected to diffraction. In this case, the renormalization of the wave vectors due to diffraction [6, 14] can compensate for the frequency dispersion of dielectric permittivity if the latter is not too large. Hence, under phase-matching conditions (4.34), the blue phase can generate the second harmonic if the light in this phase is subject to diffraction. Below, we will assume that only the harmonic, but not the pumping wave, is subjected or almost subjected to diffraction. Note that the phase-matching condition (4.34) corresponds to SGH due to homogeneous component of nonlinear susceptibility. Under diffraction conditions, these phase-matching conditions become [14, 15]:

$$\mathbf{k}^{\sigma_1}(2\omega) = \mathbf{k}^{\sigma_2}(\omega) + \mathbf{k}^{\sigma_3}(\omega) + s\boldsymbol{\tau} \tag{4.35}$$

where $s = \pm 1, \pm 2, \ldots$, and $\boldsymbol{\tau}$ is the reciprocal lattice vector of the blue phase. The phase-matching conditions (4.34) are easily satisfied, provided that the light within the blue phase experiences diffraction or the diffraction conditions are almost satisfied. Formally speaking, the frequency dispersion in a blue phase with a large period might be compensated in (4.34) by the term $s\boldsymbol{\tau}$.

In order to describe second-harmonic generation in a blue phase, it is necessary to solve equations of the type of (4.13) within the two-wave approximation, as was done in [14].

4.4.2 Symmetry Restrictions for the Tensor of Nonlinear Susceptibility

Leaving it to the reader to use the equations presented above to describe SGH in the blue phases, we will instead place our emphasis on an analysis of symmetry restrictions on the quadratic nonlinear susceptibility for BP.

It was mentioned previously that possible spatial structures of the order parameter of BPI and BPII are surely described by enantiomorphic cubic space groups, there being 13 such groups. Hence, the study of nonlinear optical properties of the blue phases, which is conducted next using the example of SHG, must include, as a stage, an analysis of nonlinear susceptibilities for corresponding groups.

The specifics of nonlinear optics of blue phases require that the structure period of the order parameter, and hence of other physical parameters, lies within the optical wavelength band. Unlike ordinary crystals in which spatially inhomogeneous components of the tensor of nonlinear susceptibility are described by the value averaged over the crystal unit cell (in fact, over a much larger volume), in the case of blue phases an essential role is also played by the tensor component varying in space. Since the local symmetry of general locations in a unit cell is, in general, lower than that of the crystal class, the symmetry of inhomogeneous components of the nonlinear susceptibility tensor is, in general, lower. Hence, this symmetry can permit nonlinear processes prohibited by the point group, i.e., by the crystal class of the space group determining the symmetry of the averaged nonlinear susceptibility tensor.

In this connection, the tensor of quadratic nonlinear susceptibility of the blue phase which determines SHG is conveniently expanded into a Fourier series:

$$\hat{\chi}(\mathbf{r}) = \sum_{\tau} \hat{\chi}_{\tau} e^{i\boldsymbol{\tau}\mathbf{r}} \tag{4.36}$$

where $\boldsymbol{\tau}$ is the blue phase reciprocal lattice vector. The zero harmonic of this expansion corresponds to the nonlinear susceptibility tensor averaged over a unit cell. Just this quantity is essential to the optics of usual crystals, and its symmetry properties have been well studied [1, 2].

Using the results of symmetry analysis [16], we conclude that the averaged component of quadratic nonlinear susceptibility is nonzero only in groups $T^i (i = 1, \ldots, 5)$ and is exactly zero in groups $O^i (i = 1, \ldots, 8)$. For groups with point symmetry 23: $\chi_{123} = \chi_{132} = \chi_{213} \neq \chi_{231} = \chi_{312} = \chi_{321} \neq 0$, and for groups with point symmetry

$43m : \chi_{123} = \chi_{132} = \chi_{213} = \chi_{231} = \chi_{312} = \chi_{321} \neq 0$. These relationships enable us to distinguish the groups T^1, T^2, T^3, T^4, T^5 and $O^1, O^2, O^3, O^4, O^5, O^6, O^7, O^8$ on the basis of their nonlinear optical properties, such as SHG, by considering whether they have a homogeneous component.

Let us now consider the symmetry restrictions for the spatially inhomogeneous part of the second-order nonlinear susceptibility tensor. An arbitrary tensor $T(\mathbf{r})$ must remain unchanged under symmetry transformations which belong to the space symmetry group of the crystal. Hence, for any space group, the general form of the tensor satisfying all symmetry restrictions for this group can, as mentioned in [16], be written as

$$T(\mathbf{r}) = \frac{1}{N} \sum_a \hat{T}^a(\mathbf{r}) \tag{4.37}$$

where summation is performed over all transformations a included in the crystal symmetry group [17], N is the number of elements in the group, and T^a is the result of the $\hat{a}$ transformation of T

$$T^a_{ijk}(\mathbf{r}) = a_{ii'}a_{jj'}a_{kk'}T_{i'j'k'}(\mathbf{r}) \tag{4.38}$$

where $\hat{a}$ is the coordinate-transforming matrix $\mathbf{r}' = \hat{a}(\mathbf{r} - \mathbf{b}_a)$, and $\mathbf{b}_a$ is the translation vector.

The presence of the third-order axis in all cubic groups results in the following restrictions for the tensor $\hat{\chi}(\mathbf{r})$. If we denote

$$\begin{aligned} \chi_{111}(\mathbf{r}) &= d_1(x, y, z), & \chi_{123}(\mathbf{r}) &= d_4(x, y, z) \\ \chi_{112}(\mathbf{r}) &= d_2(x, y, z), & \chi_{122}(\mathbf{r}) &= d_5(x, y, z) \\ \chi_{113}(\mathbf{r}) &= d_3(x, y, z), & \chi_{133}(\mathbf{r}) &= d_6(x, y, z) \end{aligned} \tag{4.39}$$

we can easily see that all components of $\chi(\mathbf{r})$ are obtained by a cyclic permutation of the coordinates x, y, z in the functions $d_i(x, y, z)$.

Thus, the second-order nonlinear susceptibility tensor is determined from six arbitrary periodic functions

$$\begin{vmatrix} d_1(x,y,z) & d_5(x,y,z) & d_6(x,y,z) & d_4(x,y,z) & d_3(x,y,z) & d_2(x,y,z) \\ d_6(y,z,x) & d_1(y,z,x) & d_5(y,z,x) & d_2(y,z,x) & d_4(y,z,x) & d_3(y,z,x) \\ d_5(z,x,y) & d_6(z,x,y) & d_1(z,x,y) & d_3(z,x,y) & d_2(z,x,y) & d_4(z,y,x) \end{vmatrix}. \tag{4.40}$$

Here, the third-rank tensor is written as a matrix, the first index of the tensor component χ_{ijk} corresponds to the first index of the matrix elements d, and the other two indices are determined from the second and third indices χ_{ijk} as 1, 1 = 1; 2, 2 = 2; 3, 3 = 3; 2, 3 = 4; 1, 3 = 5; 1, 2 = 6. In (4.40), we introduced additional symmetry restrictions which are not related to the third-order axis, but which assume the tensor is symmetric with respect to the two latter indices (in what follows, we will be interested in SHG when the pumping waves have the same frequency [1]).

The rest of the symmetry elements for the space groups T result in additional restrictions on the functions $d_i(\mathbf{r})$, but no relationships between these functions occur (and for the dielectric permittivity tensor as well, see [16]).

For example, for the space groups T^4 and T^5, the functions d_i are related by

$$\begin{aligned}
d_1(x,y,z) &= d_1\left(\frac{1}{2}+x, \frac{1}{2}-y, \overline{vz}\right) = -d_1\left(\bar{x}, \frac{1}{2}+y, \frac{1}{2}-z\right) \\
&= -d_1\left(\frac{1}{2}-x, \overline{v}y, \frac{1}{2}+z\right) \\
d_2(x,y,z) &= -d_2\left(\frac{1}{2}+x, \frac{1}{2}-y, \overline{vz}\right) = d_2\left(\bar{x}, \frac{1}{2}+y, \frac{1}{2}-z\right) \\
&= -d_2\left(\frac{1}{2}-x, \overline{v}y, \frac{1}{2}+z\right) \\
d_3(x,y,z) &= -d_3\left(\frac{1}{2}+x, \frac{1}{2}-y, \overline{vz}\right) = -d_3\left(\bar{x}, \frac{1}{2}+y, \frac{1}{2}-z\right) \\
&= d_3\left(\frac{1}{2}-x, \overline{v}y, \frac{1}{2}+z\right) \qquad (4.41) \\
d_4(x,y,z) &= d_4\left(\frac{1}{2}+x, \frac{1}{2}-y, \overline{vz}\right) = d_4\left(\bar{x}, \frac{1}{2}+y, \frac{1}{2}-z\right) \\
&= d_4\left(\frac{1}{2}-x, \overline{v}y, \frac{1}{2}+z\right) \\
d_5(x,y,z) &= d_5\left(\frac{1}{2}+x, \frac{1}{2}-y, \overline{vz}\right) = -d_5\left(\bar{x}, \frac{1}{2}+y, \frac{1}{2}-z\right) \\
&= -d_5\left(\frac{1}{2}-x, \overline{v}y, \frac{1}{2}+z\right) \\
d_6(x,y,z) &= d_6\left(\frac{1}{2}+x, \frac{1}{2}-y, \overline{vz}\right) = -d_6\left(\bar{x}, \frac{1}{2}+y, \frac{1}{z}-z\right) \\
&= -d_6\left(\frac{1}{2}-x, \overline{v}y, \frac{1}{2}+z\right)
\end{aligned}$$

where $\bar{x} = -x$, etc. In contrast to the T group, the O groups possess a fourth-order axis that results in the following relationships between the functions d_i themselves. For the groups O^6, O^8, and O^4

$$\begin{aligned}
d_3(x,y,z) &= -d_2\left(\frac{1}{4}-x, \frac{1}{4}-z, \frac{1}{4}-y\right) = -d_2\left(\frac{3}{4}+x, \frac{3}{4}-z, \frac{1}{4}+y\right) \\
&= d_2\left(\frac{3}{4}-x, \frac{1}{4}+z, \frac{3}{4}+y\right) = d_2\left(\frac{1}{4}+x, \frac{3}{4}+z, \frac{3}{4}-y\right)
\end{aligned}$$

$$\begin{aligned} d_5(x, y, z) &= -d_6\left(\frac{1}{4} - x, \frac{1}{4} - z, \frac{1}{4} - y\right) = d_6\left(\frac{3}{4} + x, \frac{3}{4} - z, \frac{1}{4} + y\right) \\ &= -d_6\left(\frac{3}{4} - x, \frac{1}{4} + z, \frac{3}{4} - y\right) = d_6\left(\frac{1}{4} + x, \frac{3}{4} + z, \frac{3}{4} - y\right) \end{aligned} \quad (4.42)$$

and for the groups O^1, O^2, and O^3,

$$\begin{aligned} d_3(x, y, z) &= -d_2(\overline{x}, \overline{z}, \overline{y}) = d_2(x, \overline{z}, y) \\ &= d_2(\overline{x}, z, y) = -d_2(x, z, \overline{y}) \\ d_6(x, y, z) &= -d_5(\overline{x}, \overline{z}, \overline{y}) = d_5(x, \overline{z}, y) \\ &= -d_5(\overline{x}, z, y) = d_5(x, z, \overline{y}). \end{aligned}$$

As a result, of the 18 components of the O groups (any third-rank tensor symmetric in its two indices has 18 components), 12 components (with two coinciding indices) are similar to those of the corresponding T groups: T^1 and O^1, O^2; T^2 and O^3, O^4; T^3 and O^5; T^4 and O^6; and T^5 and O^8. Components χ_{xxx} and χ_{xyz} with all possible cyclic permutations of the indices in the O groups differ from those in the T groups. Thus, we obtained the restrictions for the third-rank tensor of the blue phase. As was mentioned in our discussion of the nonlinear optics of blue phases, the Fourier components of the nonlinear susceptibility are important. The above relationships for the components of $\hat{\chi}(\mathbf{r})$ can lead to the restrictions for the Fourier components χ_τ.

4.4.3 Fourier Components of the $\hat{\chi}(\mathbf{r})$ Tensor

The Fourier components of the tensor $\hat{\chi}(r)$ are determined as follows:

$$\hat{\chi}_\tau = \frac{1}{V}\int \hat{\chi}(\mathbf{r})e^{i\tau\mathbf{r}}d^3\mathbf{r} \quad (4.43)$$

where integration is carried out over the unit cell volume V.

Calculating the Fourier components of $\chi(\mathbf{r})$ using (4.43) and the relationships presented in the previous section, we can find the most general form of $\hat{\chi}_\tau$ which follows from the symmetry restrictions of the space groups considered above (see Table 4.1).

In Table 4.1, R_i are arbitrary real numbers, I_i are imaginary numbers, and C_i are complex numbers, these numbers being different for different reciprocal lattice vectors. For any reciprocal lattice vector, the conditions for a Fourier component to be nonzero are $h + k + l = 2n$ for the body-centered lattices T^3, T^5, O^5, O^8 and the same parity of h, k, l for the face-centered lattices T^2, O^3, O^4. The components of a general type ($h \neq k \neq l$) are determined in the general case by 18 complex numbers, without any relationship between χ_{pqs} with different subscripts for both the T and the O groups.

Table 4.1 Fourier components of quadratic nonlinear susceptibility tensor χ in cubic groups

T groups

	T^4, T^5	χ_{111}	χ_{222}	χ_{333}	χ_{122}	χ_{233}	χ_{311}	χ_{133}	χ_{211}	χ_{322}	χ_{123}	χ_{231}	χ_{312}	χ_{131}	χ_{212}	χ_{323}	χ_{312}	χ_{223}	χ_{331}	T^1, T^2, T^3
$h00$	$h=2n$	I_1	0	0	I_2	0	0	I_3	0	0	R_1	R_2	R_3	0	I_4	0	0	0	I_5	$h00$
	$h=2n+1$	0	R_1	I_1	0	R_2	I_2	0	R_3	I_3	0	0	0	I_4	0	R_4	R_5	I_5	0	
$hk0$	$h=2n$	I_1	I_2	R_1	I_3	I_4	R_2	I_5	I_6	R_3	R_4	R_5	R_6	R_7	I_7	I_8	I_9	R_8	I_{10}	$hk0$
	$h=2n+1$	R_1	R_2	I_1	R_3	R_4	I_2	R_5	R_6	I_3	I_4	I_5	I_6	I_7	R_7	R_8	R_9	I_8	R_{10}	
hhh	—	C_1	C_1	C_1	C_2	C_2	C_2	C_3	C_3	C_3	C_4	C_4	C_4	C_5	C_5	C_5	C_6	C_6	C_6	hhh

O groups

	O^2	O^4, O^6, O^8	χ_{111}	χ_{222}	χ_{333}	χ_{122}	χ_{233}	χ_{311}	χ_{133}	χ_{211}	χ_{322}	χ_{123}	χ_{231}	χ_{312}	χ_{131}	χ_{212}	χ_{323}	χ_{312}	χ_{223}	χ_{331}	O^1, O^3, O^5	O^7
$h00$	$h=2n$	$h=4n$	I_1	0	0	I_2	0	0	I_2	0	0	0	R_1	R_1	0	I_3	0	0	0	I_4	$h00$	$h=4n$
	$h=2n+1$	$h=4n+2$	0	0	0	I_1	0	0	I_1	0	0	R_1	R_2	R_2	0	I_2	0	0	0	$-I_2$		$h=4n+2$
		$h=4n\pm1$	0	R_1	$\pm iR_1$	0	R_2	$\pm iR_3$	0	R_3	$\pm iR_2$	0	0	0	$\pm iR_4$	0	R_5	R_4	$\pm iR_5$	0		$h=4n+1$
$hk0$	$h=n$	$h=2n$	I_1	I_2	R_1	I_3	I_4	R_2	I_5	I_6	R_3	R_4	R_5	R_6	R_7	I_7	I_8	I_9	R_8	I_{10}	$hk0$	$h=2n$
		$h=2n+1$	R_1	R_2	I_1	R_3	R_4	I_2	R_5	R_6	I_3	I_4	I_5	I_6	I_7	R_7	R_8	R_9	I_8	R_{10}		$h=2n+1$
$hh0$	$h=n$	$h=2n$	I_1	I_1	0	I_2	I_3	R_1	I_3	I_2	$-R_1$	R_2	$-R_2$	0	R_3	I_4	I_5	I_4	$-R_3$	I_5	$hh0$	$h=2n$
		$h=2n+1$	R_1	R_1	0	R_2	R_3	I_1	R_3	R_2	$-I_1$	I_2	$-I_2$	0	I_3	R_4	R_5	R_4	$-I_3$	R_5		$h=2n+1$
hhh	$h=2n$	$h=4n$	I_1	I_1	I_1	C_1	C_1	C_1	$-C_1^*$	$-C_1^*$	$-C_1^*$	I_2	I_2	I_2	C_2	C_2	C_2	$-C_2^*$	$-C_2^*$	$-C_2^*$	hhh	$h=4n$
	$h=2n+1$	$h=4n+2$	R_1	R_1	R_1	C_1	C_1	C_1	C_1^*	C_1^*	C_1^*	R_2	R_2	R_2	C_2	C_2	C_2	C_2^*	C_2^*	C_2^*		$h=4n+2$
		$h=4n\pm1$	$(1\mp i)R$	$(1\mp i)R$	$(1\mp i)R$	C_1	C_1	C_1	$\mp iC_1^*$	$\mp iC_1^*$	$\mp iC_1^*$	$(1\mp i)R_1$	$(1\mp i)R_1$	$(1\mp i)R_1$	C_2	C_2	C_2	$\mp iC_2^*$	$\mp iC_2^*$	$\mp iC_2^*$		$h=4n\mp1$

Here, we do not touch upon the calculations of intensity and polarization of the harmonics because they are similar to those given in previous sections of this chapter (see also [16, 18, 19]). We only illustrate how the above symmetry restrictions can be used in a study of blue phases. The symmetry analysis of quadratic nonlinear susceptibility shows that the phase-matched SHG due to the homogeneous component $\hat{\chi}_0$ (T groups) is rigorously forbidden if the pumping wave is directed along [100] and [110]. Since (hhh) reflections were not observed experimentally in the blue phases, there should not also be phase-matched SHG due to $\hat{\chi}_0$ for the [111] direction because the dielectric tensor frequency dispersion cannot be compensated by the diffraction effects. A corresponding symmetry analysis of the phase-matched SHG due to the spatially inhomogeneous components of $\hat{\chi}$ showed that it is not forbidden for the [100] direction only for groups T^4, O^6, and O^7. Therefore, for example, if SHG in colinear geometry for the [100] direction was observed, this same fact would allow us to distinguish these groups from the above-mentioned 13 enantiomorphic groups. An example of experimental studying of a nematic ordering by SHG is presented in [20].

References

1. S.A. Akhmanov, R.V. Khokhlov, *Problems of Nonlinear Optics* (VINITI, Moscow, 1964) (in Russian)
2. N. Bloembergen, *Nonlinear Optics* (W.A. Benjamin Inc., New York, 1965)
3. N. Bloembergen, A. Sielvers, Appl. Phys. Lett. **17**, 483 (1970)
4. J.W. Shelton, Y.R. Shen, Phys. Rev. A **5A**, 1867 (1972)
5. V.A. Belyakov, V.N. Shipov, Phys. Lett. **86**, 94 (1981)
6. A.A. Maier, A.P. Sukhorukov, R.N. Kuz'min, Zh. Eksp. Teor. Fiz. **77**, 1282 (1979)
7. V.A. Belyakov, A.S. Sonin, *Optics of Cholesteric Liquid Crystals* (Nauka, Moscow, 1982) (in Russian)
8. S.M. Arakelyan, Yu.S. Chiligaryan, *Nonlinear Optics of Liquid Crystals* (Nauka, Moscow, 1984) (in Russian)
9. V.A. Belyakov, V.E. Dmitrienko, V.P. Orlov, Pis'ma Zh. Eksp. Teor. Fiz. **1**, 978 (1975)
10. N.V. Shipov, V.A. Belyakov, Zh. Eksp. Teor. Fiz. **75**, 1589 (1978) [Sov. Phys. JETP **48**, 802 (1978)]
11. B.Ya. Zel'dovich, N.V. Tabiryan, Usp. Fiziol. Nauk **147**, 633 (1985)
12. N.V. Shipov, V.A. Belyakov, Zh. Eksp. Teor. Fiz. **50**, 205 (1980)
13. S.V. Shiyanovskii, Ukr. Fiz. Zh. **27**, 361 (1982)
14. V.A. Belyakov, N.V. Shipov, Zh. Eksp. Teor. Fiz. **82**, 1159 (1982) [Sov. Phys. JETP **55**, 674 (1982)]
15. V.A. Belyakov, N.V. Shipov, Pis'ma Zh. Eksp. Teor. Fiz. **9**, 22 (1983)
16. V.A. Belyakov, V.E. Dmitrienko, *Optics of Chiral Liquid Crystals*, ed. by I.M. Khalatnikov (Soviet Scientific Reviews, Section A, London, 1989)
17. *Atlas of Space Groups of the Cubic System* (Moscow, Nauka, 1980)
18. V.A. Belyakov, V.E. Dmitrienko, S.M. Osadchii, Zh. Eksp. Teor. Fiz. **83**, 585 (1982) [Sov. Phys. JETP **56**, 322 (1982)]
19. V.A. Belyakov, *Diffraction Optics of Complex Structured Periodic Media* (Springer, New York, 1992)
20. S.K. Saha, G.K. Wong, Appl. Phys. Lett. **34**, 423 (1979)

Chapter 5
Optical Edge Modes in Photonic Liquid Crystals

5.1 Introduction

Recently, there has been an explosion of interest in mirrorless distributed feedback (DFB) lasing in chiral liquid crystals (CLCs) [1]. The reason for this interest is related to the observed low-threshold lasing [2, 3], unusual polarization properties of lasing, and frequency tunability of this lasing by application of an external field [4, 5], by temperature pitch variations [6, 7], by application of a mechanical stress [8], etc. Some results have recently been summarized in review publications [9–11].

The DFB low-threshold lasing in CLCs occurs at frequencies close to the frequencies of the stop-band edges [2–8]. The corresponding frequencies were associated with so-called edge modes [1]. It also happens that, at the same edge mode frequencies, an anomalously strong absorption of the pumping wave occurs [12–16].

In general, the theory of edge modes in CLCs (and the more general DFB lasing in spiral media) is very similar to the corresponding theory for conventional periodic solid media that was initially developed by Kogelnik and Shank [17] in the coupled-wave approximation and was later treated similarly in many papers (see [18] and the references therein). But the theory of edge lasing modes in CLCs deserves a separate study due to the unusual optical properties of the CLCs, and also because, in contrast to all other periodic media, an exact analytic solution of the Maxwell equations is known for CLCs (and more generally for spiral media). Many related results, usually obtained in a numerical approach, may therefore be obtained analytically for CLCs. For example, the anomalously strong absorption effect existing in CLCs when the light frequency is close to the stop band has been treated analytically [12, 13]. (Later, the anomalously strong absorption effect for conventional periodic media was studied in [19]).

General analytic expressions for the solution of the boundary problem for the non-absorbing, absorbing, and amplifying CLC layer, i.e., for the "edge modes" (EM), are presented in this chapter when the light propagation direction coincides with the

V. Belyakov, *Diffraction Optics of Complex-Structured Periodic Media*, Springer Series in Optical Sciences 203,
https://doi.org/10.1007/978-3-319-43482-7_5

spiral axis [20]. In particular, the dispersion equation for the EMs determining their frequencies is presented; the lasing threshold gain is found, and an expression for the threshold in a specific limiting case is also presented, together with numerical solutions of the dispersion equation for typical values of the CLC parameters. The properties of the EM (coordinate intensity distribution, frequency width of the EM, etc.) are analyzed. It is also discussed how the revealed properties of the EM allow one to decrease the DFB lasing threshold thanks to a low gain for the lasing and a strong absorption for the pumping wave.

5.2 Eigenwaves in CLC

To solve the boundary problem related to EMs, we need to identify the eigenwaves in a CLC. As is known [13, 21–25], the eigenwaves corresponding to propagation of light along a spiral axis z in a CLC, i.e., the solutions of the Maxwell equation (1.8), are given by a superposition of two plane waves of the form given by the (1.10).

The dielectric tensor is given in (1.6), and the wave vectors $K^{\pm}$ in (1.10) satisfy the condition given in (1.11). The wave vectors $K^{\pm}$ in four eigensolutions of (1.8) are determined by (1.15) and with a slight modification of the notation in (1.15) are given by

$$K_j^+ = \tau/2 \pm \kappa\left\{1 + (\tau/2\kappa)^2 \pm \left[(\tau/\kappa)^2 + \delta^2\right]^{1/2}\right\}^{1/2}, \tag{5.1}$$

where j labels the eigensolutions with the ratio of amplitudes E^-/E^+ given, as follows from (1.16), by the expression

$$\left(E^-/E^+\right)_j = \delta \Big/ \left[\left(K_j^+ - \tau\right)^2/\kappa^2 - 1\right], \tag{5.2}$$

with $\kappa = \omega\varepsilon_0^{1/2}/c$. We do not specify the kind of CLCs under investigation here (chiral smectics or cholesterics) because the optics of light propagating along the spiral axis are identical for both LC types [13, 21–25]. For definiteness, we give the expressions for cholesterics below. The corresponding expressions for chiral smectics can be obtained by a simple redefinition of the relevant parameters [see [13], Chap. 2 and (3.50)].

Two of the eigenwaves corresponding to circular polarization in the sense of chirality coinciding with one of the CLC spirals experience strong diffraction scattering at frequencies in the region of the stop band. The other two eigenwaves corresponding to the opposite circular polarizations are almost completely unaffected by diffraction scattering even at the frequencies of the stop band for the former circular polarization.

Because, as we shall see in what follows, the specific features of the EMs in a CLC are related to eigenwaves with the diffracting polarization, we consider only the propagation of light with the diffracting polarization in the CLC.

where $\kappa = \omega\varepsilon_0^{1/2}/c$. We do not specify the kind of CLCs under investigation here (chiral smectics or cholesterics) because the optics of light propagating along the spiral axis is identical for both LC types [13, 21–25]. For definiteness, we give the expressions for cholesterics below. The corresponding expressions for chiral smectics can be obtained by a simple redefinition of the relevant parameters [see [13], Chap. 2 and (3.50)].

Two of the eigenwaves corresponding to the circular polarization with the sense of chirality coinciding with one of the CLC spiral experience strong diffraction scattering at the frequencies in the region of the stop band. The other two eigenwaves corresponding to the opposite circular polarizations are almost not unaffected by the diffraction scattering even at the frequencies of the stop band for the former circular polarization.

Because, as we see in what follows, the specific features of EM in the CLC are related to eigenwaves of diffracting polarization, we limit ourselves by considering the propagation of light of the diffracting polarization only in the CLC.

5.3 Boundary Problem

To investigate EMs in a CLC, we have to consider a boundary problem. We assume that the CLC is represented by a planar layer with the spiral axis perpendicular to the layer surfaces (Fig. 5.1). To justify our intention to consider only propagation of light with the diffracting polarization, we also assume that the average CLC dielectric constant ε_0 coincides with the dielectric constant of the ambient medium. This assumption effectively prevents the conversion of one circular polarization into

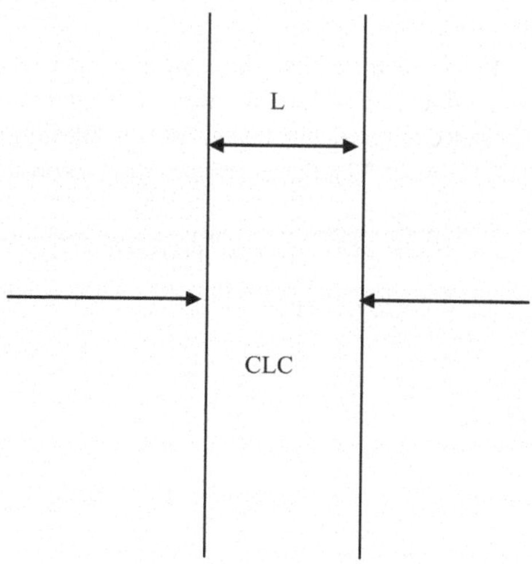

Fig. 5.1 Schematic of the boundary problem for edge modes

another at layer surfaces [13, 25] and allows one to take into account only two eigenwaves with diffracting circular polarization.

We begin with the linear boundary problem in the formulation where two plane waves with the diffracting polarization and the same frequency are incident along the spiral axis at the layer from opposite sides (see Fig. 5.1) and the dielectric tensor can have a nonzero imaginary part of any sign (which means that the CLC layer may be either absorbing or amplifying). The amplitudes E_j^+ of the two diffracting eigenwaves excited in the layer by the incident waves (they are denoted by $E_+^+$and E_-^+) are determined by the equations

$$E_+^+ + E_-^+ = E_{\mathrm{ir}} \tag{5.3}$$

$$\begin{aligned} &\exp\left[\mathrm{i}K^+{}_+L\right]\left\{\delta/\left[\left(K_+^+ - \tau\right)^2/\kappa^2 - 1\right]\right\}E_+^+ + \\ &\exp\left[\mathrm{i}K^+{}_-L\right]\left\{\delta/\left[\left(K_-^+ - \tau\right)^2/\kappa^2 - 1\right]\right\}E_-^+ = E_{\mathrm{il}} \end{aligned}$$

where E_{ir} and E_{il} are the amplitudes of the waves incident at the layer from the right and from the left, L is the layer thickness, and

$$K_\pm^+ = \tau/2 \pm \kappa\left\{1 + (\tau/2\kappa)^2 - \left[(\tau/\kappa)^2 + \delta^2\right]^{1/2}\right\}^{1/2}. \tag{5.4}$$

The amplitudes of waves exiting the layer on the right and left sides E_{er} and E_{el} are given by

$$\begin{aligned} E_{\mathrm{er}} &= \left\{\delta/\left[\left(\mathrm{i}K_+^+ - \tau\right)^2/\kappa^2 - 1\right]\right\}E_+^+ + \left\{\delta/\left[\left(K_-^+ - \tau\right)^2/\kappa^2 - 1\right]\right\}E_-^+ \\ E_{\mathrm{el}} &= \exp\left[\mathrm{i}\left(\mathrm{i}K_+^+ - \kappa\right)L\right]E_+^+ + \exp\left[\mathrm{i}\left(K_-^+ - \kappa\right)L\right]E_-^+. \end{aligned} \tag{5.5}$$

If we assume that the amplitude of only one incident wave is nonzero, (5.5) determines the reflected and transmitted waves (the reflection, R, and transmission, T, coefficients of the layer) and, in particular, their frequency dependence [13, 23, 25]. From (3.33), the corresponding expressions for R and T take the form

$$\begin{aligned} R &= \delta^2|\sin qL|^2/\left|\left(q\tau/\kappa^2\right)\cos qL + \mathrm{i}\left[(\tau/2\kappa)^2 + (q/\kappa)^2 - 1\right]\sin qL\right|^2 \\ T &= \left|\exp[\mathrm{i}\kappa L]\left(q\tau/\kappa^2\right)\right|^2/\left|\left(q\tau/\kappa^2\right)\cos qL + \mathrm{i}\left[(\tau/2\kappa)^2 + (q/\kappa)^2 - 1\right]\sin qL\right|^2, \end{aligned} \tag{5.6}$$

where

$$q = \kappa\left\{1 + (\tau/2\kappa)^2 - \left[(\tau/\kappa)^2 + \delta^2\right]^{1/2}\right\}^{1/2}. \tag{5.7}$$

If both amplitudes of the incident waves are equal to zero, and if the dielectric tensor has a positive (or a very small negative) imaginary part, no waves emerge from the layer.

The solution of the system (5.3) for the amplitudes E_+^+ and E_-^+ of the eigenwaves in the CLC layer is given by the following expressions (in the case of a wave incident only at one surface of the layer):

$$\begin{aligned} E_+^+ &= -E_{\mathrm{il}} \exp[-\mathrm{i}qL][[(\tau/2\kappa)^2 + (q/\kappa)^2 - 1 - q\tau/\kappa^2]/2\{(q\tau/\kappa^2)\cos qL \\ &\quad + \mathrm{i}[(\tau/2\kappa)^2 + (q/\kappa)^2 - 1]\sin qL\} \\ E_-^+ &= E_{\mathrm{il}} \exp[\mathrm{i}qL][[(\tau/2\kappa)^2 + (q/\kappa)^2 - 1 + q\tau/\kappa^2]/2\{(q\tau/\kappa^2)\cos qL \\ &\quad + \mathrm{i}[(\tau/2\kappa)^2 + (q/\kappa)^2 - 1]\sin qL\} \end{aligned} \tag{5.8}$$

The eigenwave amplitudes close to the stop-band edges are strongly oscillating functions of frequency (see Figs. 5.2 and 5.3 presenting the calculation results). At the points of maxima close to the stop-band edges, their values are much greater than the incident wave amplitude E_{il}. It turns out that the amplitude maxima frequencies coincide with the frequencies of zero reflection following from (5.6) for a non-absorbing CLC (see Figs. 5.2 and 5.3).

5.4 Edge Modes (Nonabsorbing LC)

We examine the formulas of the preceding section for a non-absorbing CLC in more detail. In a non-absorbing CLC, $\gamma = 0$ in the general expression for the dielectric constant $\varepsilon_0 = \varepsilon_0(1 + \mathrm{i}\gamma)$ (note that in real situations $|\gamma| \ll 1$). The calculations of the reflection, R, and transmission, T, coefficients as functions of the frequency in accordance with (5.6) (Figs. 5.2a and 5.3a) give the well-known results [21–25]: strong reflection inside the stop band, frequency oscillations of T and R outside the stop-band edges with $0 \leq R \leq 1$, and preservation of the relationship $T + R = 1$ for all frequencies. This means that $T = 1$ at the frequencies corresponding to $R = 0$ (see Fig. 5.3).

The corresponding calculations of the amplitudes E_+^+ and E_-^+ of the eigenwaves excited in the layer (Figs. 5.2 and 5.3b, c) reveal a nontrivial frequency dependence of E_+^+ and E_-^+. In particular, close to the stop-band edges (outside the stop-band edges), frequency oscillations of the amplitudes are accompanied by an essential enhancement of their magnitude relative to the incident wave amplitude (in the calculations, the incident wave amplitude is assumed to be equal to 1). The thicker layer, the greater the enhancement (see Figs. 5.2 and 5.3). As Figs. 5.2 and 5.3 show, the positions of the maxima of the amplitude oscillations just coincide with (or are very close to, for an absorbing or amplifying CLC) the positions of the reflection coefficient minima corresponding to $R = 0$ for a non-absorbing CLC.

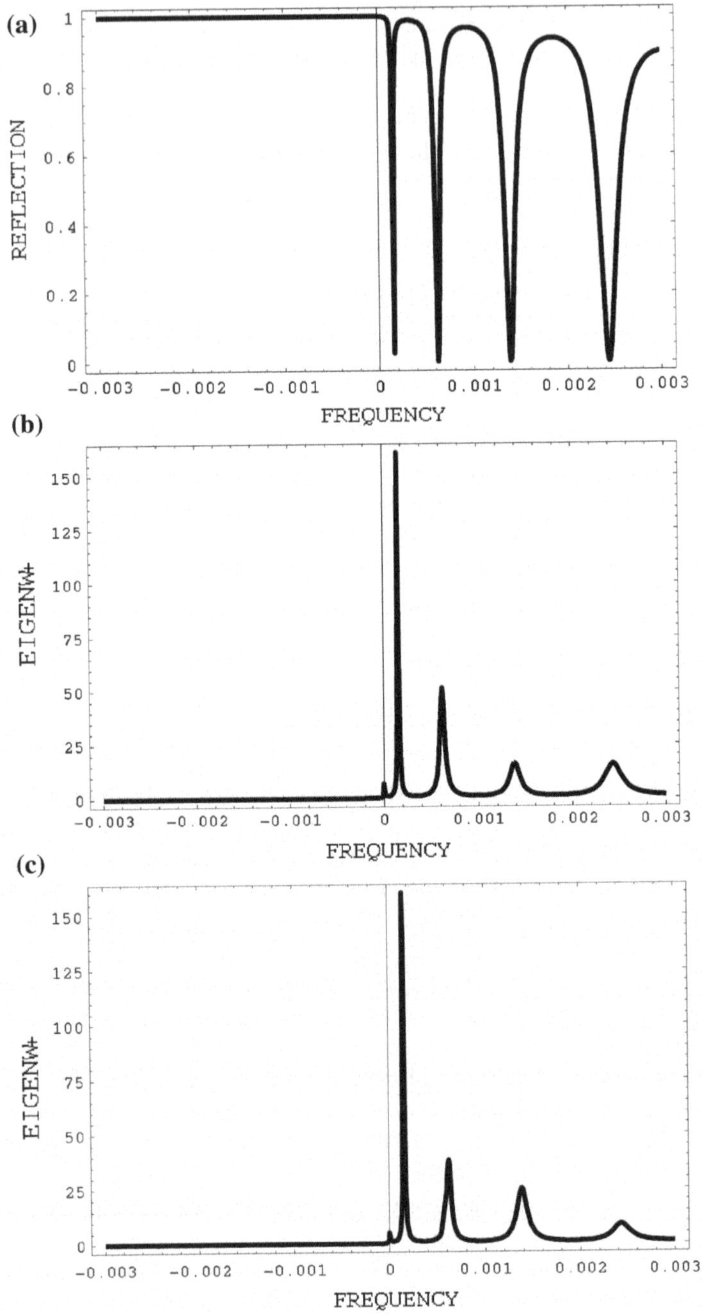

Fig. 5.2 Reflection coefficient R (**a**), and the squared E_+^+ (**b**) and E_-^+ (**c**) eigenmode amplitudes calculated versus frequency for a non-absorbing CLC layer ($\delta = 0.05$, $N = L/p = 250$). Here and in all figures below, $\delta(\upsilon - 1)$ is plotted along the frequency axis; i.e., the frequency deviation from the stop-band edge is plotted [normalized by the Bragg frequency, see (5.19) and (5.20)]

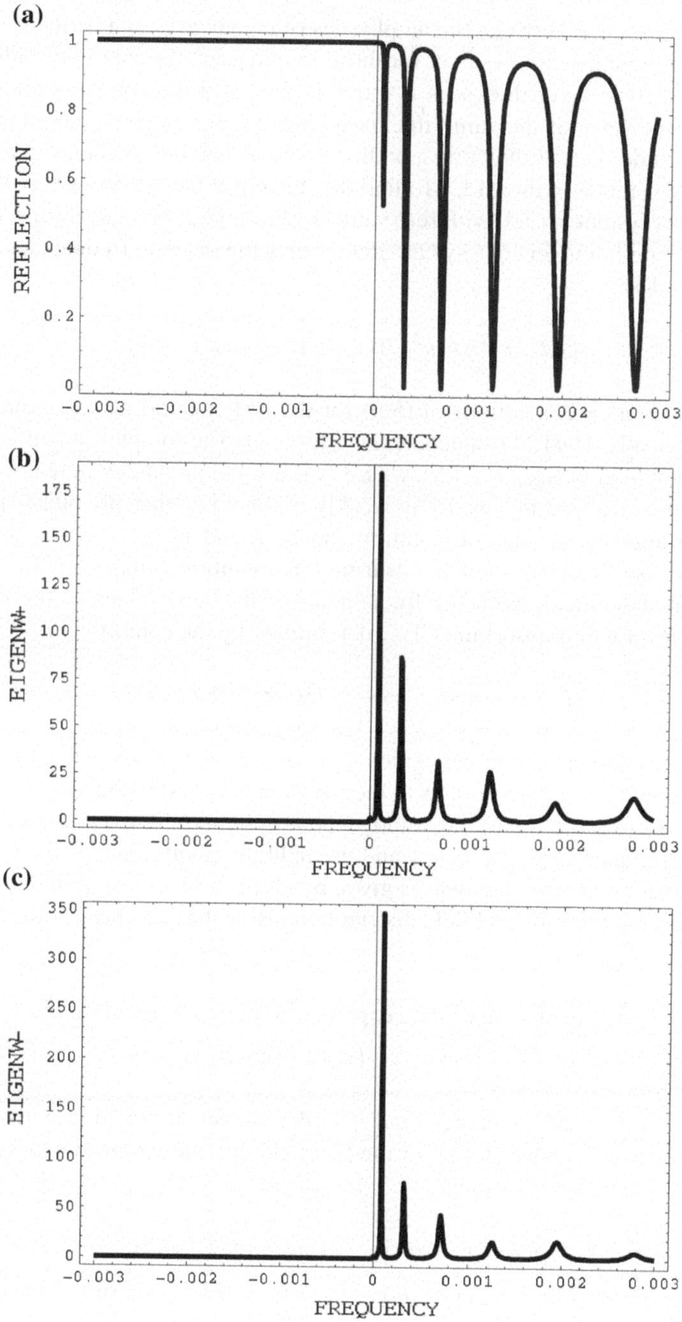

Fig. 5.3 Reflection coefficient R (**a**) and the squared E_+^+ (**b**) and E_-^+ (**c**) eigenmode amplitudes calculated versus frequency for a non-absorbing CLC layer ($\delta = 0.05, N = 350$)

The above relation between the amplitudes of eigenwaves and incident waves at the specific frequencies shows that, for these frequencies, the energy of radiation in the CLC at a fixed layer thickness is much higher than the corresponding energy of the incident wave at the same thickness. Hence, in complete accordance with [1], we conclude that, at the corresponding frequencies, the incident wave excites some localized mode in the CLC. To find this localized mode, we have to the solve homogeneous system of (5.3) with zero values of E_{ir} and E_{il}. The solvability condition for the resulting homogeneous system determines the discrete frequencies of these localized modes:

$$\mathrm{tg}qL = \mathrm{i}\left(q\tau/\kappa^2\right)/\left[(\tau/2\kappa)^2 + (q/\kappa)^2 - 1\right] \tag{5.9}$$

In the general case, solutions of (5.9) for the EM frequencies ω_{EM} can only be found numerically. The EM frequencies ω_{EM} turn out to be complex quantities, which can be written $\omega_{\mathrm{EM}} = \omega^0_{\mathrm{EM}}(1 + \mathrm{i}\Delta)$, where Δ is a small parameter in real situations. Therefore, the localized modes decay weakly in time; i.e., they are quasi-stationary modes. Fortunately, an analytic solution can be found in a certain limiting case, namely, for a sufficiently small Δ ensuring the condition $L\mathrm{Im}q \ll 1$. In this case, the ω^0_{EM} values coincide with the frequencies of the zero values of the reflection coefficient R for a non-absorbing CLC, determined by the conditions

$$qL = n\pi \text{ and } \Delta = -1/2\delta(n\pi)^2/(\delta\, L\tau/4)^3, \tag{5.10}$$

where n is the edge mode number, which increases as the frequency departs from the stop-band edge ($n = 1$ corresponds to the frequency closest to the stop-band edge).

In the resulting solution of the homogeneous system (5.3), the ratio of the eigensolution amplitudes is $E^+_-/E^+_+ = -1$ and the field distribution inside the CLC layer is a superposition of two eigenwaves given by (1.10) with this amplitude ratio. The explicit expression for the EM field distribution inside the CLC layer following from (1.10) is

$$\begin{aligned}\mathbf{E}\left(\omega_{\mathrm{EM},z,t}\right) &= \mathrm{i}\exp(-\mathrm{i}\omega_{\mathrm{EM}}t)\{\mathbf{n}_+\exp(\mathrm{i}\tau z/2)\sin qz + (\mathbf{n}_-/\delta)\exp(-\mathrm{i}\tau z/2)\\ &\left\{\left[(\tau/2\kappa)^2 + (q/\kappa)^2 - 1\right]\sin qz - \mathrm{i}\left(\tau q/\kappa^2\right)\cos qz\right\}\},\end{aligned} \tag{5.11}$$

where ω_{EM} is the EM frequency, and q is determined by (5.7). For the analytic solution mentioned, above (5.11) for the EM field distribution inside the CLC layer becomes

$$\begin{aligned}\mathbf{E}(\omega_n, z, t) &= \mathrm{i}\exp(-\mathrm{i}\omega_n t)\{\mathbf{n}_+\exp(\mathrm{i}\tau z/2)\sin(n\pi z/L) + (\mathbf{n}_-/\delta)\exp(-\mathrm{i}\tau z/2)\\ &\left\{\left[(\tau/2\kappa)^2 + (n\pi/L\kappa)^2 - 1\right]\sin(n\pi z/L) - \mathrm{i}\left(\tau n\pi/L\kappa^2\right)\cos(n\pi z/L)\right\}\},\end{aligned} \tag{5.12}$$

where ω_n corresponds to a zero value of R (i.e., ω_n is determined by $qL = n\pi$). The eigensolution of the boundary problem given by (5.11) and (5.12) is a standing wave localized at the layer thickness L, with the amplitude modulated along the z-axis.

The number of modulation periods at the layer thickness L coincides with the EM number n.

The total EM field (the sum of the fields of the two eigenmodes in (5.11) and (5.12)) at any coordinate z in the layer makes a fixed angle to the local director direction; i.e., the field in the layer rotates together with the director as a function of z and performs the same number of rotations as the director does at the layer thickness. The field distributions following from (5.11 and 5.12) for the EM numbers $n = 1, 2, 3$ are presented in Fig. 5.4. This shows that the EM field is localized inside the CLC layer and its energy density experiences oscillations inside the layer with the number of oscillations equal to the EM number n. Note that the figure presents the total energy distribution in the layer. However, as is clear from (1.10, 5.11, and 5.12), the total field at each point of the CLC layer is represented by two plane waves propagating in opposite directions, and hence, the intensities of the waves propagating in the opposite directions can be calculated separately at any point in the layer. In general, the coordinate distribution of the intensities of the waves propagating in the opposite directions is similar to the distribution presented in Fig. 5.4. But these distributions are of special interest close to the layer surfaces. Figure 5.5 shows the intensity coordinate distributions of the waves propagating inside and outside the layer close to the layer surfaces. We can see that, at the layer surface, the intensity of the wave propagating inside the layer is strictly zero, while the intensity of the wave propagating outside the layer is nonzero (although small). This means that the EM energy is leaking from the layer through its surfaces. Equation (5.12) implies the expression

$$E_{\text{out}} = (\tau n\pi/\delta L)/\kappa^2 \approx np/L\delta, \tag{5.13}$$

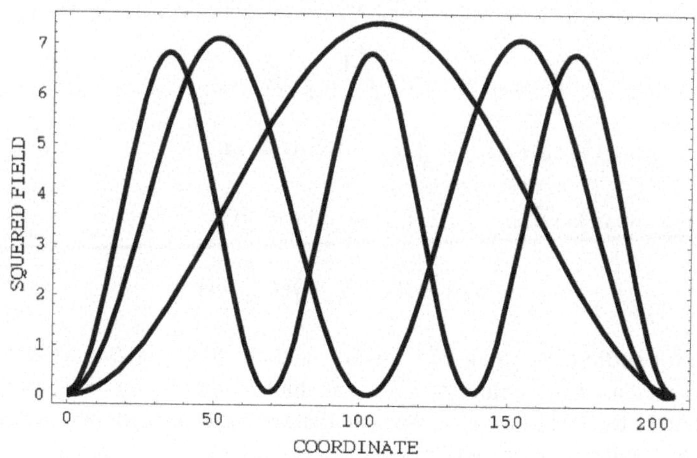

Fig. 5.4 Calculated EM energy (arbitrary units) distributions inside the CLC layer versus coordinate (in the dimensionless units $z\tau$) for the three first edge modes ($\delta = 0.05$, $N = 16.5$, $n = 1, 2, 3$)

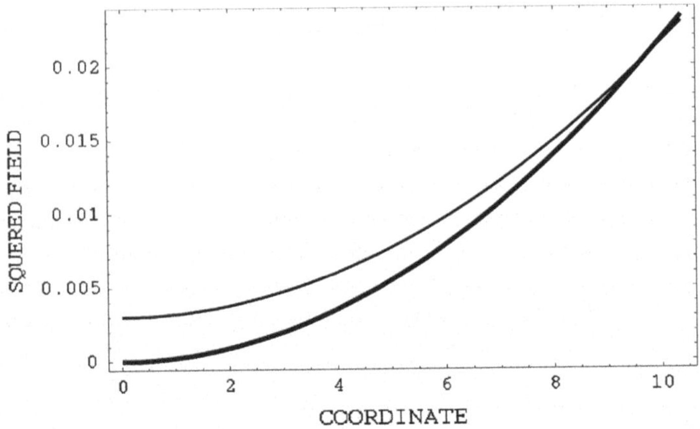

Fig. 5.5 Calculated EM energy (arbitrary units) distributions close to the CLC layer surface versus coordinate (in the dimensionless units $z\tau$) for a plane wave directed inside (bold line) and outside the layer for the first edge mode ($\delta = 0.05, N = 16.5, n = 1$)

for the leaking wave amplitude at the CLC layer surface, where p is the CLC pitch. Equation (5.13) shows that the EM energy leakage is inversely proportional to the square of the layer thickness L and proportional to the square of the EM number n. Hence, the longest-lived is the first EM in a CLC layer. If $L/p \gg 1$, the leaking wave amplitude E_{out} is small (i.e., $E_{\text{out}} < 1$), as (5.13) shows.

For a non-absorbing CLC layer, as assumed in this section, the only source of decay is the energy leakage through its surfaces; the decrease in the EM energy per unit time is equal to the energy flow of the leaking waves $(2c/\varepsilon_0^{1/2})|\mathbf{E}_{\mathbf{OUT}}|^2$, and therefore, using (5.11)–(5.13), we easily obtain the EM lifetime τ_{m} as

$$\begin{aligned}\tau_{\text{m}} &= \int |\mathbf{E}(\omega_{\text{EM}}, z, t)|^2 \text{d}z \Big/ \left\{ \text{d} \int |\text{E}(\omega_{\text{EM}}, z, t)|^2 \text{d}z]/\text{d}t \right\} \\ &\approx (5/16)\left(L\varepsilon_0^{1/2}/c\right)\left[1 + (4/5)(L\delta/pn)^2\right]. \end{aligned} \tag{5.14}$$

Under the condition $(L\delta/pn) \gg 1$, (5.14) reduces to

$$\tau_{\text{m}} \approx (1/4)\left(\varepsilon_0^{1/2} L/c\right)(L\delta/pn)^2. \tag{5.15}$$

Hence, for sufficiently thick CLC layers, as their thickness L increases, the EM lifetime τ_{m} increases as the third power of the thickness and is inversely proportional to the square of the EM number n. We note that the same dependence of the lifetime τ_{m} on n and L follows from (5.10):

$$\tau_{\text{m}} \approx 1/\text{Im}(\omega_{\text{EM}}) = (L/c)(\delta L/pn)^2. \tag{5.16}$$

5.5 Excitation of Edge Modes

The analysis of the localized EM (solution of the homogeneous system following from (5.3)) in the previous sections, together with the solution of the inhomogeneous (5.3) found in the previous sections, can be used to discuss ways of exciting EMs and the efficiency of this excitation. In a non-amplifying CLC layer, excitation requires an external wave (waves) whose frequency coincides with the EM frequency incident at the CLC layer. The general solution of the boundary problem found from system (5.3) in this case may be represented as a superposition of the particular solution corresponding to the inhomogeneous system (5.3) and the solution corresponding to the homogeneous system (5.3), i.e., corresponding to the EM, with the coefficient to be determined from the boundary conditions. It is easy to construct such a representation of the boundary problem solution determined by (5.8).

For this, it is sufficient to write $E_{\pm}^{+}$ in the form $E_{\pm}^{+} = E_{\pm}^{\mathrm{em}} + E_{\pm}^{p}$, where $E_{\pm}^{\mathrm{em}}$ and $E_{\pm}^{\mathrm{p}}$ are the respective amplitudes of eigenwaves in the EM and in the particular solution. Given that the amplitudes $E_{\pm}^{\mathrm{em}}$ satisfy (5.3) with zero right-hand sides, we obtain

$$E_{\pm}^{\mathrm{em}} = \pm E_{+}^{+}[1 + \mathrm{itg}(qL)]. \tag{5.17}$$

for frequencies close to ω_{EM}.

The $E_{\pm}^{\mathrm{em}}$ values calculated using (5.17) are presented in Fig. 5.6. Comparing Figs. 5.6, 5.2, and 5.3 shows that for a sufficiently thick LC layer, the amplitudes $E_{\pm}^{+}$ provide a very good approximation for $E_{\pm}^{\mathrm{em}}$.

The above results relate to a stationary process of EM excitation, i.e., to the situation of a plane wave of fixed amplitude incident on a CLC layer. The resulting formulas can be used to find the probability of EM excitation by a single photon.

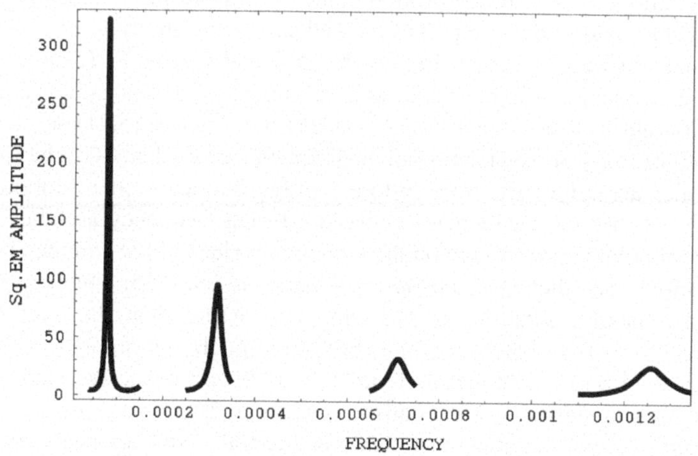

Fig. 5.6 Calculated squared amplitude of the EM close to the four differing EM frequencies ($\delta = 0.05$, $N = 350$)

This probability is given by

$$W_{\mathrm{em}} = (1 - R)\int \left|\mathbf{E}^{\mathrm{EM}}(\omega_{\mathrm{EM}}, z, t)\right|^2 \mathrm{d}z \bigg/ \left\{\left|\mathbf{E}^{\mathrm{EM}}(\omega, z, t)\right|^2 \mathrm{d}z + \int \left|\mathbf{E}^{\mathrm{P}}(\omega, z, t)\right|^2 \mathrm{d}z\right\} \tag{5.18}$$

where the integration is taken over the CLC layer thickness and $\mathbf{E}^{\mathrm{P}}(\omega_{\mathrm{EM}}, z, t)$ is the particular solution of (5.3).

It turns out that if a plane wave (with unit amplitude) is incident on the CLC layer, the eigenmode amplitudes in the excited EM are given by the expressions for $E_{\pm}^{\mathrm{em}}$ above. But in this case, it is impossible to excite only the EM. It is accompanied by the particular solution of (5.3) with nonzero amplitude determined by the relation $E_{\pm}^{+} = E_{\pm}^{\mathrm{em}} + E_{\pm}^{p}$.

Hence, the efficiency of the EM excitation by one plane wave (the ratio of the squared EM amplitude to the squared incident wave amplitude) for the specific values of the relevant parameters may be estimated from the squared E_{+}^{+} value in Figs. 5.2b and 5.3b. The values of $E_{\pm}^{\mathrm{em}}$ close to the EM frequencies are more accurately determined by (5.17) (see also Fig. 5.6).

5.6 Absorbing LC

We now examine EMs in an absorbing CLC. A key motivation for this study is DFB lasing in CLCs. It must be kept in mind that under lasing, a CLC is essentially an absorbing medium for the pumping wave. We examine the formulas in the previous sections in more detail with regard to their application to the pumping wave. We assume for simplicity that the absorption in the LC is isotropic. We denote the ratio of the imaginary part to the real part of the dielectric constant by γ, i.e., $\varepsilon = \varepsilon_0(1 + \mathrm{i}\gamma)$. In actual situations, $|\gamma| \ll 1$. In Figs. 5.7, 5.8, and 5.9, the R, T, and $1 - R - T$ frequency dependences are presented for several positive and negative values of γ, including its values close to the threshold values for lasing (see (5.21) below). Due to the assumed isotropy of the absorption, the frequency dependences of the calculated characteristics are symmetric relative to the Bragg frequency (the midpoint of the stop band), so only the frequencies above the Bragg frequency are shown in the figures. We should comment here on the numerical values of the parameters used in the calculations. The dielectric anisotropy is taken as $\delta = 0.05$, which corresponds to a typical value of this parameter. The same may be said about the layer thickness L. Because $\tau = 4\pi/p$, where p is the cholesteric pitch, the number of pitches N at the layer thickness L is equal to $1/4\pi(1 = \mathrm{L}\tau)$, and hence the value $l = 300$ adopted in the calculations corresponds to N close to 30, i.e., a typical number for such experiments. All the quantities mentioned reveal frequency beats close to the frequency edge of the selective reflection band. The positions of the corresponding maxima and minima are determined by the layer thickness L and by δ, and are slightly

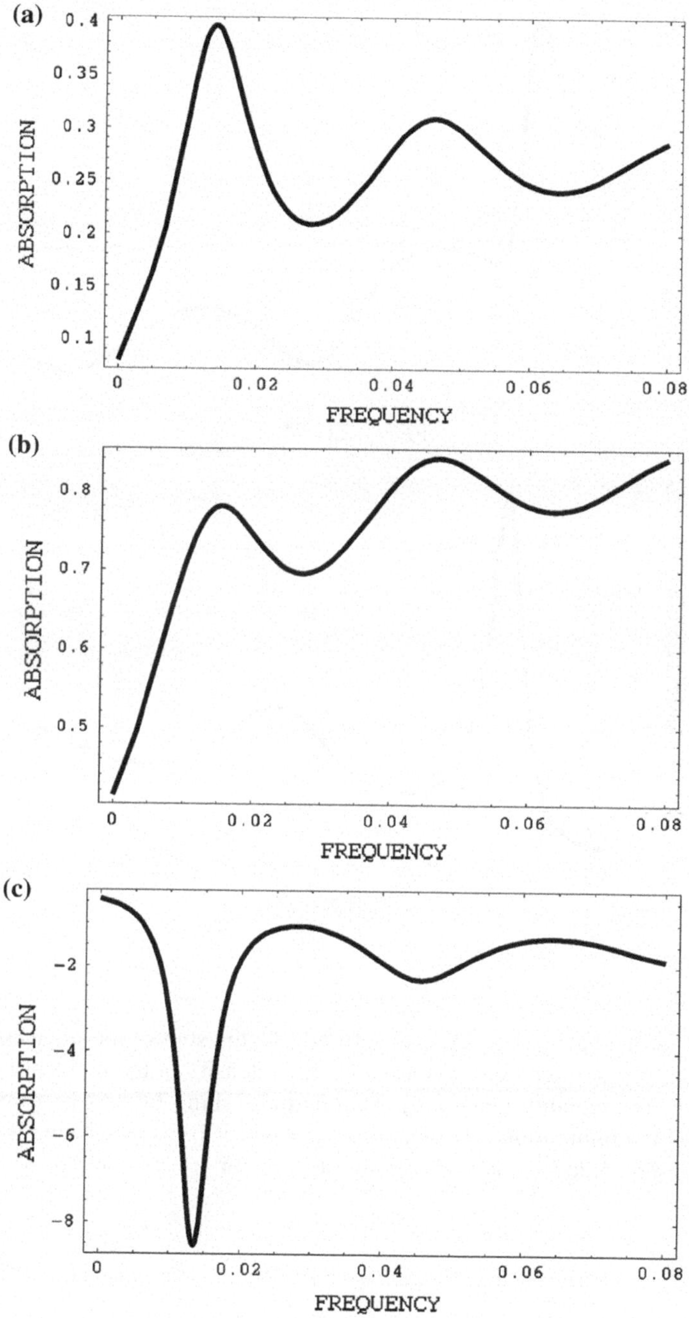

Fig. 5.7 Calculated frequency dependence for absorption $(1 - R - T)(l = 300, l = L\tau, \delta = 0.05)$: **a** $\gamma = 0.001$, **b** $\gamma = 0.005$, and **c** for a low-amplifying layer (below the threshold gain for the first lasing edge mode, $\gamma = -0.003$); $R(d)$ and $T(e)$ calculated versus the frequency $(l = 300, l = L\tau, \delta = 0.05)$ for a low-amplifying layer (below the threshold gain for the first edge mode, $\gamma = -0.003$)

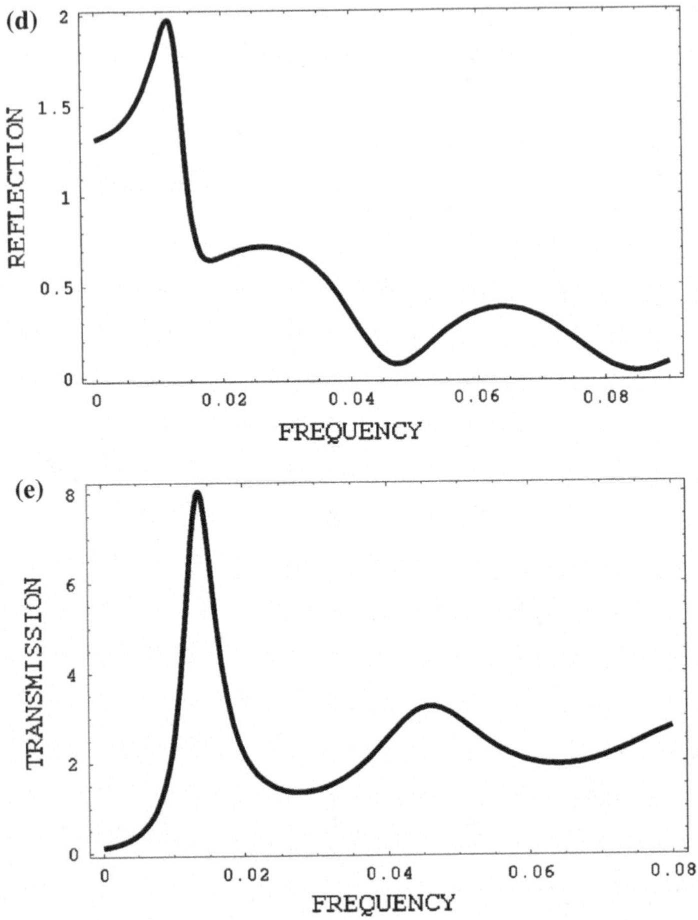

Fig. 5.7 (continued)

dependent on the value of γ. In an absorbing LC, the sum of the intensities of the reflected and transmitted beams is less than the intensity of the incident beam, i.e., $R + T < 1$. The equality holds only for a non-absorbing LC. As an example, the positions of the minima of the reflection coefficient R beats (following from (5.6)) are given in Fig. 5.2a for a non-absorbing LC, i.e., for $\gamma = 0$, corresponding to

$$qL = \pi n, \pm v = 1 + (n\pi/a)^2/2, \quad n = 1, 2, 3 \ldots,$$
$$v = 2(\omega - \omega_B)/\delta\omega_B, \quad \omega_B = c\tau/2\varepsilon_0^{3/2}, \quad a = \delta L\tau/4. \tag{5.19}$$

In a typical situation, $a \gg 1$.

The edges ω_ε of the selective reflection band are connected to the Bragg frequency ω_B by the relation

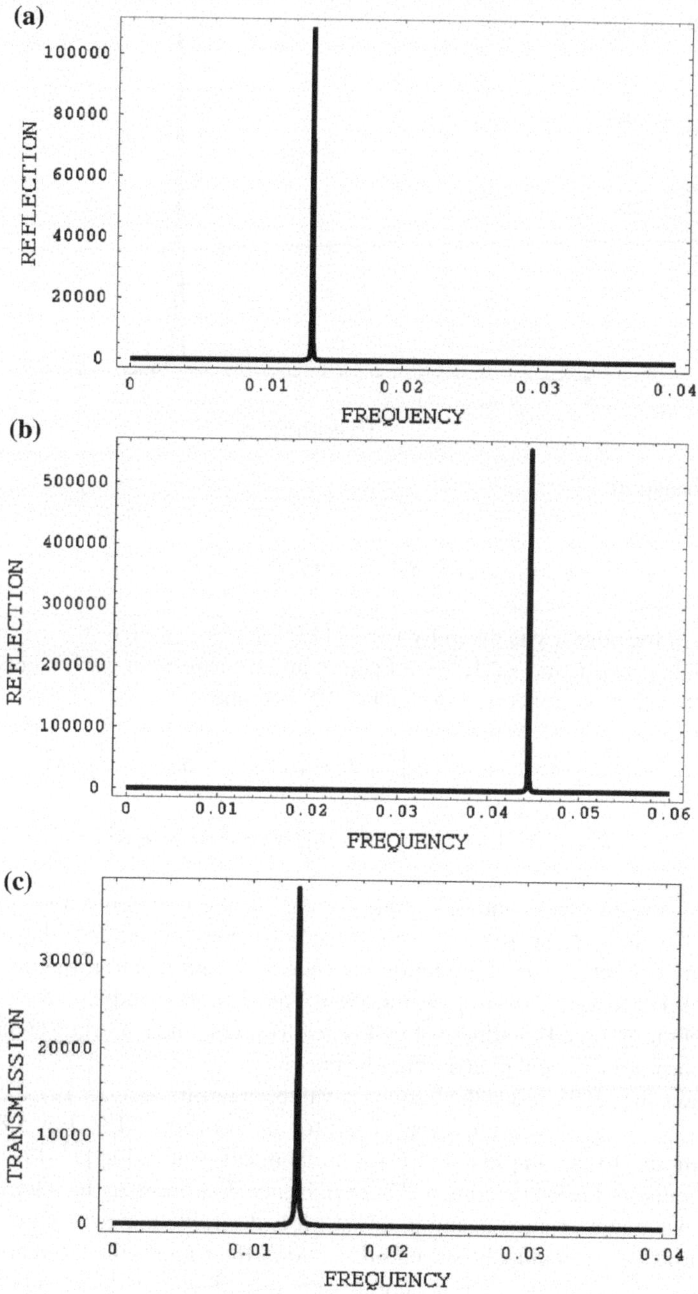

Fig. 5.8 Calculated frequency dependence of R ($l = 300, l = L\tau, \delta = 0.05$) **a** close to the threshold gain for the first lasing edge mode ($\gamma = -0.00565$), **b** close to the threshold gain for the second lasing edge mode ($\gamma = -0.0129$); calculated frequency dependence of T ($l = 300, l = L\tau, \delta = 0.05$) **c** close to the threshold gain for the first lasing edge mode ($\gamma = -0.00565$), and **d** close to the threshold gain for the second edge mode ($\gamma = -0.0129$)

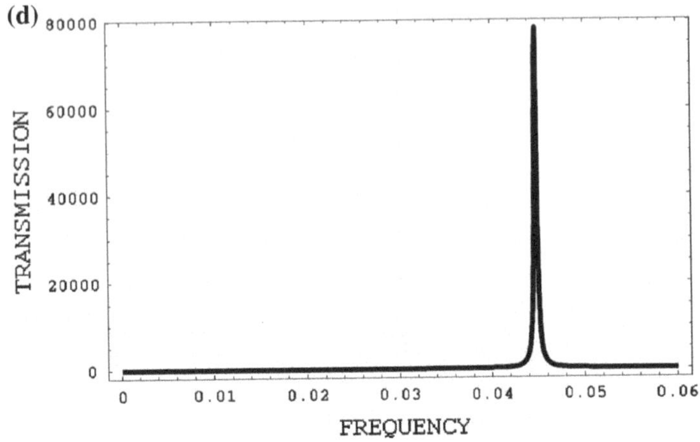

Fig. 5.8 (continued)

$$\omega_{\varepsilon} = \omega_{\mathrm{B}}/(1 \pm \delta)^{1/2} = c\tau/2[\varepsilon_0(1 \pm \delta)]^{1/2}. \tag{5.20}$$

Hence, at the edges, ν is given by $\nu_{\mathrm{e}} = 2\big(1/(1 \pm \delta)^{1/2} - 1\big)/\delta \approx -(\pm 1)$.

For small γ and $L\mathrm{Im}q \ll 1$, the reflection and transmission coefficients (5.6) at the frequencies of the reflection minima (5.19) become

$$R = \left(a^3\gamma/\delta\right)^2/\left[(n\pi)^2 + a^3\gamma/\delta\right]^2, \quad T = (n\pi)^4/\left[(n\pi)^2 + a^3\gamma/\delta\right]^2, \tag{5.21}$$

$$R + T = 1 - 2(n\pi)^2\alpha^3\gamma/\delta]/\left[(n\pi)^2 + a^3\gamma/\delta\right]^2.$$

It follows from (5.19) and 5.21) that for each n the maximum absorption, i.e., maximal $1 - R - T$, occurs for $(n\pi)^2 = a^3\gamma/\delta$. This means that the maximum absorption occurs for a special relation between δ, γ, and L, and if this relationship is satisfied, i.e., if $(n\pi)^2 = a^3\gamma/\delta$, then $R = 1/4$, $T = 1/4$, and $1 - R - T = 1/2$. Because of the assumed smallness of γ, this result corresponds to strong enhancement of the absorption for weakly absorbing layers.

As shown in [9, 10], just at the frequency values determined by (5.19), the effect of anomalously strong absorption reveals itself for an absorbing CLC (Fig. 5.8a, b) and the edge modes for an amplifying LC reveal themselves at lasing [1] (Fig. 5.9) (see similar results for layered media in [17, 18]). Hence, to minimize the intensity of the pumping wave that ensures lasing in a CLC, it is desirable to perform the pumping in conditions of anomalously strong absorption effect and with the realization of lasing at the EM frequency. These options were investigated in detail in [14, 15] and are briefly discussed in the following sections. To conclude this section, it should be noted that (see Figs. 5.7, 5.8 and 5.9) the absorption maxima in the frequency dependences are not so sharp as the intensity maxima for lasing.

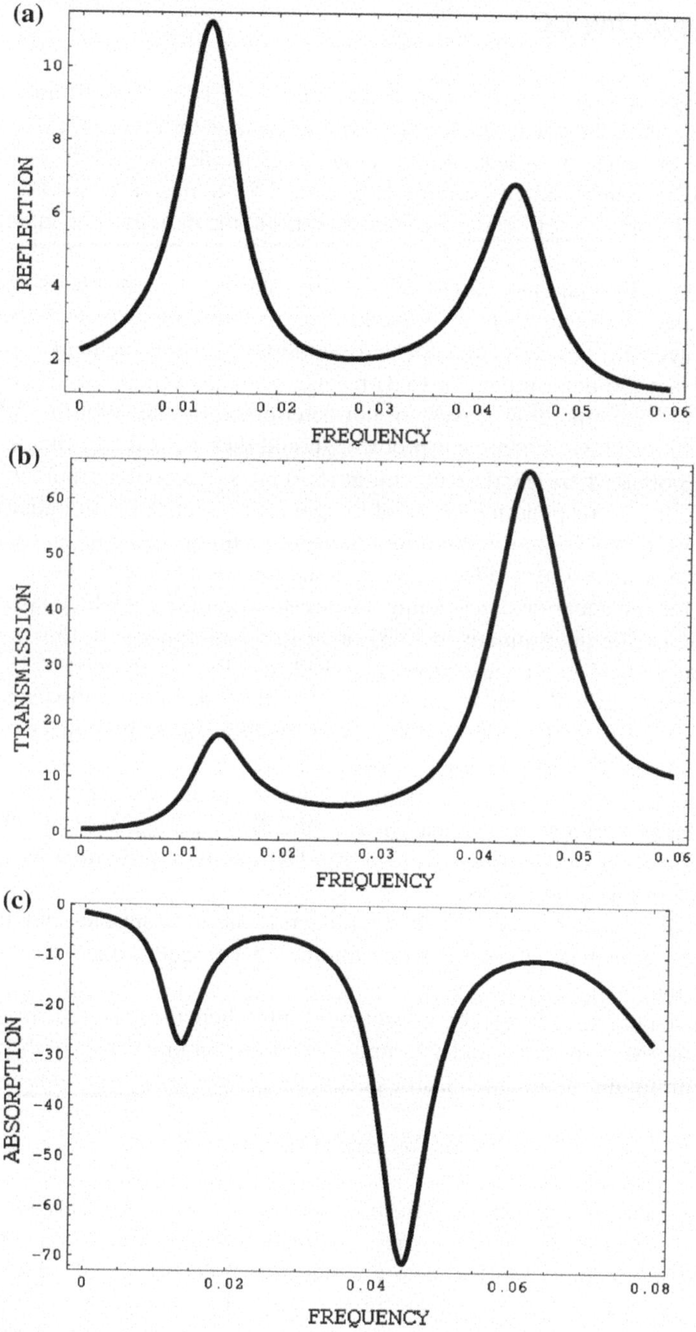

Fig. 5.9 R (**a**), T (**b**), $1-R-T$ (**c**) calculated as a function of frequency ($l = 300, l = L\tau, \delta = 0.05$) for $\gamma = -0.009$, i.e., for the gain between the thresholds for the first and the second edge modes

5.7 Amplifying LC

We now assume that $\gamma < 0$, which means that the CLC is amplifying. If $|\gamma|$ is sufficiently small, the waves emerging from the layer according to (5.3)–(5.6) exist only in the presence of at least one external wave incident on the layer, and their amplitudes are determined by solving (5.3) and (5.5). In this case (see Fig. 5.9c), $R + T > 1$ or $1 - R - T < 0$, which just corresponds to the definition of an amplifying medium.

However, if the imaginary part of the dielectric tensor, i.e., γ, reaches some critical negative value, the quantity $R + T$ diverges and the amplitudes of waves emerging from the layer are nonzero even for zero amplitudes of the incident waves. This happens when the determinant of (5.3) reaches zero. At this point, of course, the amplitudes of the emerging waves are not determined by the solution (5.5) of the linear equations (1.8) (a nonlinear problem should then be solved). But, as we saw above, the points where the determinant of (5.3) goes to zero determine the EM [1, 17, 18] and the corresponding values of the gain (or the negative imaginary part of the dielectric tensor), i.e., the minimum threshold gain at which lasing occurs (see the corresponding discussion for scalar periodic media in [17, 18]).

Therefore, the equation determining the threshold gain (γ) at which lasing occurs [zero value of the determinant of (5.3) or of the denominator in (5.6)] turns out to coincide with (5.9). It must now be solved, not for the frequency, but for the imaginary part of the dielectric constant (γ). In the general case, this equation has to be solved numerically. However, for a very small negative imaginary part of the dielectric tensor, the EM frequency values are pinned to the frequencies of the zeros of the reflection coefficient in its frequency beats outside the stop-band edge for the same layer with a zero imaginary part of the dielectric tensor [1, 9, 10]. This is why the threshold values of the gain for the EM can be represented by analytical expressions in this limiting case.

For small $|\gamma|$ and $L|\mathrm{Im}q| << 1$, the reflection and transmission coefficients in (5.6) at the frequencies of the reflection minima (5.19) reduce again to expressions (5.21), although with negative γ.

Hence, R and T may be divergent, and the points where they diverge correspond to the lasing at EM frequencies and determine the corresponding values of the threshold γ, i.e., a minimum $|\gamma|$ at which lasing occurs for the nth reflection minimum:

$$\gamma = -\delta(n\pi)^2/(\delta L\tau/4)^3. \tag{5.22}$$

As can be seen from (5.22), the threshold values of $|\gamma|$ are inversely proportional to the third power of the layer thickness and the minimum value of $|\gamma|$ corresponds to $n = 1$, i.e., to the EM closest to the selective reflection band edge (cf. analogous results for scalar layered media [17, 18]). The values of γ given by the (5.22) are convenient to use for estimating the threshold values of γ in the general case and as a zero approximation in the numerical solution of (5.9) for the threshold values. The frequency distance between the consequent edge lasing modes is equal to $\nu_{n+1} - \nu_n = ((\pi/a)^2/2)(2n + 1)$, i.e., inversely proportional to the second power of the layer thickness (cf. the corresponding distance between the lasing frequencies in a homogeneous layer, which is inversely proportional to the layer thickness).

It also follows from Fig. 5.8 that the different threshold values of γ correspond to the different edge lasing modes (divergent R and T) in Fig. 5.8a–d. This means that separate lasing modes can be excited by changing the gain (γ). If the value of γ is between the consequent threshold values of γ for neighboring EM, the lasing may not be achieved and the layer may reveal only amplifying properties (see Fig. 5.9a–c). This means that changing the pumping wave intensity allows achieving lasing at the individual EM and that the lasing intensity is not a monotonic function of the pumping intensity. Because the lasing frequency is determined by the EM frequencies, there is an option for some variation of the lasing frequency inside the width of the dye line by changing the CLC pitch by means of temperature variations [6, 7] or by application of an external electric or magnetic field to the layer [4]. We note that smooth variations of the external agent may result in jump-like variations of the lasing frequency [22] related to the jumps of the CLC pitch, which are sensitive to the surface anchoring [23].

5.8 Optimization of Pumping

The formulas in the preceding sections can be used to optimize the lasing threshold separately by reaching a coincidence of the lasing frequency with the frequency of the first EM and optimizing the pumping efficiency by reaching a coincidence of the pumping frequency with the frequency of the first maximum in the anomalously strong absorption effect [12]. We note that the same CLCs must simultaneously be an absorbing material at the pumping frequency and an amplifying one at the lasing frequency. In the present section, we discuss the possibilities for simultaneously reaching the highest pumping efficiency and the lowest value of the lasing threshold gain. The requirements for achieving the highest pumping efficiency and the lowest value of the lasing threshold gain are contradictory for a collinear geometry because they assume that the lasing frequency ω_ι and the pumping frequency ω_p practically coincide with the frequency edges of the selective reflection band. However, the lasing frequency ω_ι is less than the pumping frequency ω_p. We note that this contradiction for the collinear geometry may be overcome by pure chance or by very fine tuning of the lasing parameters if the difference $\omega_p - \omega_\iota$ is small, while ω_p coincides with the high frequency edge of the reflection band and ω_ι coincides with the low frequency edge of the reflection band. This results in the following frequencies of the lasing and pumping waves:

$$\begin{gathered}\omega_p = (c\tau/2)/\left[\varepsilon_{0p}(1-\delta_p)\right]^{1/2}, \omega_l = (c\tau/2)/[\varepsilon_{0l}(1+\delta_l)]^{1/2},\\ \omega_l/\omega_p = \left[\varepsilon_{0l}(1+\delta_l)/\varepsilon_{0p}(1-\delta_p)\right]^{1/2},\end{gathered} \tag{5.23}$$

where the dielectric constant ε_0 and the anisotropy δ carry subscripts p and l relating to the frequency dispersion of the dielectric properties and indicating that the respective parameters are to be taken at the pumping and lasing frequencies. We note that a decrease has been observed experimentally in the lasing threshold due to the anomalously strong absorption of the pumping wave in the collinear geometry [16].

Another possibility for reaching the lowest threshold in the collinear geometry may be realized by applying an external electric (or magnetic) field to the LC material. It is known [13] that in this case, due to the distortion of the LC helix, many diffraction orders exist for light propagating along the helix axis, and the pumping and lasing frequencies may be fitted to the frequencies of different diffraction orders. Nevertheless, the fitting still requires very fine tuning of the lasing parameters.

However, there is a regular way to reach optimization of the pumping intensity (under the assumption that lasing occurs along the helical axis). Non-collinear pumping can be used without any tuning of the lasing parameters; i.e., the pumping wave propagates at an angle to the helical axis, which allows it to satisfy the conditions for anomalously strong absorption. A rough estimate derived from the fact that the lasing and pumping waves experience Bragg scattering gives the following value for the angle θ between the pumping wave propagation direction and the helical axis:

$$\theta = \arccos[\omega_l/\omega_p]. \tag{5.24}$$

To obtain a more accurate expression for the pumping wave propagation direction, we must solve the Maxwell equations for light propagating at an angle to the helical axis and find the angle θ corresponding to the conditions for anomalously strong absorption [12, 13]. Unfortunately, no exact analytic solution of the Maxwell equations is known in this case, so a numerical approach must be used. The full power of the numerical approach is revealed if the frequency dispersions of the dielectric constant and of the LC dielectric anisotropy are taken into account. However, these quantities are not usually very well known, so in the experiment, even if the calculated angle θ is known, we must search experimentally for the actual angle at which the anomalously strong absorption effect occurs by changing the pumping wave propagation direction.

Under these circumstances, an approximate but more accurate expression for θ than (5.24) may be quite useful. The corresponding expression was found [14, 15] in the framework of the dynamical theory of diffraction applied to the case of light propagating at an angle to the helical axis [13, 25–29]. The dielectric anisotropy δ plays the role of a small parameter in this theory. Because the dielectric anisotropy δ is quite small in many practical cases ($\delta < 0.1$), the accuracy of the results found in the framework of the dynamical theory may be sufficient to describe the experimental results. An experimental observation of pumping optimization obtained by applying a non-collinear pumping geometry was reported in [27].

5.9 Calculation Results

To obtain the gain γ that corresponds to the onset of lasing in a CLC layer, we investigated the behavior of the R (reflection) and T (transmission) coefficients. The divergence of R and T corresponds to the lasing threshold value of γ. These divergences occur where the determinant of (5.3) vanishes. This condition gives a direct way to find the EM frequency corresponding to the solution of (5.9). Figure 5.10 presents the threshold gains found analytically and numerically for the first four EMs and for several values of the parameter $\delta L/p$. The comparison shows that the higher the value of $\delta L/p$, the closer the values of the threshold gains found analytically and numerically. For an amplifying CLC, the numerical solutions for the gain values normalized by δ [solutions of (5.9)] are presented in Fig. 5.11 for several values of the parameter $\delta L/p$ (the specific values of γ are also presented in Fig. 5.11 for $\delta = 0.05$ and 0.03). Zero values of the determinant of (5.3) occur at discrete values of γ and $\mathrm{Re}[\omega_\iota]$. Their values are found for the first few EMs; R and T were also calculated to control the solution procedure using the initial value of γ given by the analytic expression (5.22). The final values of ω_l and γ_i were found by graphical solution of the problem. The gain values normalized by δ for the fixed value of the parameter $\delta L/p$, as presented in Fig. 5.11, can be used to recalculate γ for various values of δ and L corresponding to the same value of $\delta L/p$.

The values of ω_i are very close to the values they have at the zeros of the reflection coefficient R for a non-absorbing CLC, as determined by (5.10). The values of γ_i are only close to the ones given by the analytic expression (5.22) for sufficiently high values of $\delta L/p$. On the whole, there is a general trend in the difference between the analytic and numerical solutions. The values of $\mathrm{Re}[\omega_{\iota\mathrm{i}}]$ found numerically are smaller than the analytic ones and the numerically found $|\gamma_\mathrm{i}|$ are greater than the analytic ones [see (5.22)].

5.10 Conclusion

The analytic and numerical theoretical examinations of the edge modes in LCs performed here can be used to represent the physics of these localized modes more clearly. We should also note that the results obtained here for spiral media are relevant to any periodic media, so the qualitative description of EMs in these media is the same as for CLCs and the analytic formula presented above can be taken as a useful guide when studying other periodic media.

As regards CLCs, in particular, it turns out that there are some real possibilities related to these localized modes for reaching a higher efficiency of DFB lasing in CLCs. Indeed, possibilities have been predicted for further reducing the lasing threshold relative to the one already achieved [14, 15], and these predictions have been partially observed experimentally [16]. There are also possibilities for varying

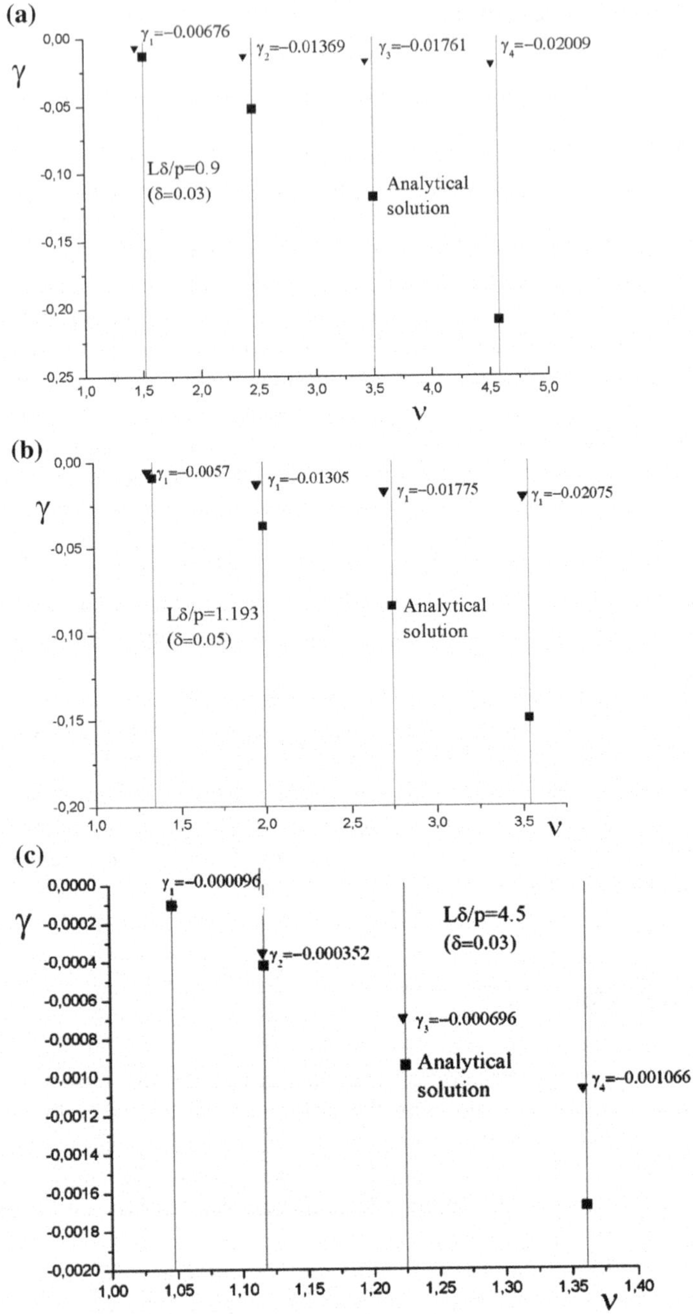

Fig. 5.10 Numerical and analytical solutions of the EM dispersion equation (5.9) for the parameter $\delta L/p = 0.9$ (**a**); 1.193 (**b**); and 4.5 (**c**)

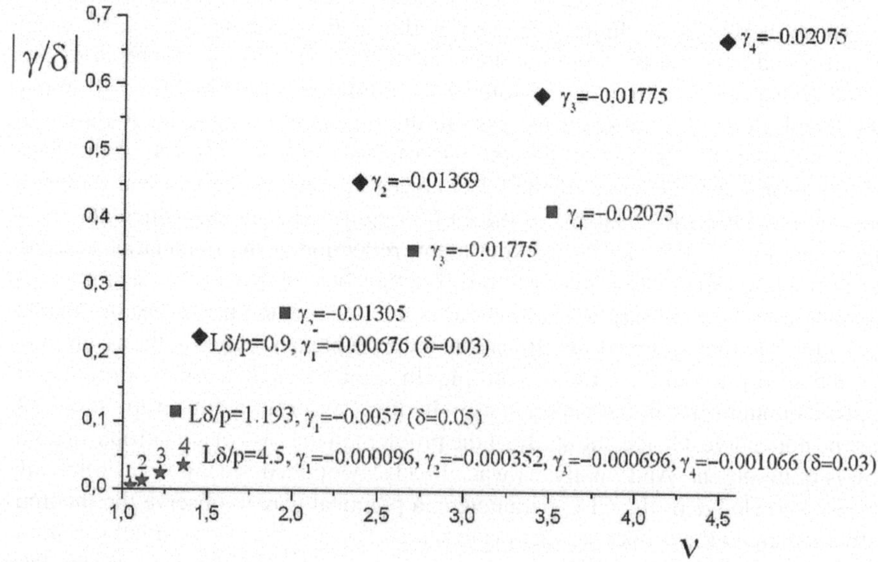

Fig. 5.11 Numerical solutions of the EM dispersion equation (5.9) for the gain normalized by δ for several values of the parameter $\delta L/p$ (the calculated values of γ are also given in Fig. 5.11 for $\delta = 0.05, 0.03$)

the lasing frequency in this kind of lasing. It should be mentioned that one advantage of CLCs over solid media consists in the fact that their parameters are easily variable. Hence, the CLC may be considered as a convenient model object for studying lasing in any periodic media, even in 3D periodic structures (see [30] on lasing in the LC blue phase). Another fortunate circumstance from the theoretical standpoint is the availability of an exact analytical solution of the Maxwell equations for light propagating along the helical axis. For other periodic media, no exact analytic solution is known and the coupled-wave approximation is usually applied to the problem [17, 18, 31].

Note that the equations in the previous sections, which formally assume a frequency-independent dielectric susceptibility and anisotropy, may be easily generalized for the experimentally realizable case of frequency-dependent dielectric susceptibility and anisotropy. To do this, the corresponding parameters in the equations are treated as functions of the frequency, but this raises the question of explicit expressions for these functions. It seems that the most pragmatic way to determine the corresponding frequency dependences is to obtain them from experimental measurements.

An essential point for the experimental observation of the examined anomalously strong absorption effect of the pumping wave is that the periodic structure has to be perfect enough to observe beats in the reflection coefficient at the edges of the reflection band. Insufficient perfection of the periodic structure leads to a decrease in this anomalously strong absorption. A similar decrease in the anomalously strong absorption is due to the finite frequency width of the pumping wave. The corresponding reduction in absorption comes about by averaging the expressions presented above over the frequency width of the pumping wave line [12, 13]. Imperfect samples can also cause an increase in the threshold lasing gain. It should be noted also that the assumption accepted above regarding the absence of dielectric reflection on the boundaries of the CLC layer (equality of the external dielectric and average dielectric constants of the CLC) requires special experimental care. If this assumption is not met in an experiment, reflection at the boundaries converts the diffracting polarization into a non-diffracting one, and this also decreases the anomalously strong absorption, and changes the polarization properties of the phenomenon. Another accepted simplification of the problem is related to the assumption that the absorption in the CLC is isotropic. In some cases, this assumption may be realistic enough. But in the general case, the local absorption anisotropy in a CLC may be noticeable. Hence, the study of the problem in the case of anisotropic absorption is quite urgent. And finally, as was already mentioned, owing to insufficiently precise knowledge of the CLC parameters, a practical way to observe the theoretically predicted effects in experiment is to search for the effect using small variations in the experimental parameters (propagation direction, temperature, etc.) around the values calculated using the formulas in this chapter.

It should also be kept in mind that the EMs studied here reveal themselves not only in the lasing, but also in other optical phenomena. For example, nonlinear optical harmonic generation [32] and Cherenkov radiation [33] in a periodic medium are enhanced at the EM frequencies (see also Figs. 29 and 32 in [13] and Figs. 5.10 and 6.2 in [25] relating to nonlinear optical harmonic generation and Cherenkov radiation in CLCs). In particular, note that the effect of enhanced absorption at the EM frequencies reveals itself in the luminescence and has recently been observed experimentally in several papers [34–36].

References

1. V.I. Kopp, Z.-Q. Zhang, A.Z. Genack, Prog. Quant. Electron. **27**(6), 369 (2003)
2. I.P. Il'chishin, E.A. Tikhonov, V.G. Tishchenko, M.T. Shpak, JETP Lett. **32**, 24 (1980)
3. B. Taheri, A.F. Muñoz, P. Palffy-Muhoray, R. Twieg, Mol. Cryst. Liq. Cryst. **358**, 73 (2001)
4. M. Ozaki, M. Kasano, T. Kitasho et al., Adv. Mater. **15**(12), 974 (2003)
5. M.I. Barnik, L.M. Blinov, V.V. Lazarev et al., J. Appl. Phys. **103**, 123113 (2008)
6. S.M. Morris, A.D. Ford, M.N. Pivnenko, H.J. Coles, J. Appl. Phys. **97**(2), 023103 (2005)
7. A. Chanishvili, G. Chilaya, G. Petriashvili et al., Appl. Phys. Lett. **86**, 055107 (2005)
8. H. Finkelmann, S.T. Kim, A. Muñoz et al., Adv. Mater. **13**(14), 1069 (2001)
9. V.F. Shabanov, S.Ya. Vetrov, A.V. Shabanov, *Optics of Real Photonic Crystals* (RAS, Sibirian Branch, 2005) (in Russian)
10. V.A. Belyakov, Localized optical modes in optics of chiral liquid crystals, in *New Developments in Liquid Crystals and Applications*, ed. by P.K. Choudhury (Nova Publishers, New York, 2013) Chapter 7, pp. 199–227
11. V.A. Belyakov, *Optics of Photonic Crystals* (MIPT Publishing House, 2013) (in Russian)
12. V.A. Belyakov, A.A. Gevorgian, O.S. Eritsian, N.V. Shipov, Zh. Tekhn. Fiz. **57**, 1418 (1987) [Sov. Phys. Technical Physics, **32** (n7), 843–845 (1987), English translation]

13. V.A. Belyakov, V.E. Dmitrienko, Optics of chiral liquid crystals, p. 80 in *Soviet Scientific reviews/Section A, Physics Reviews*, ed. by I.M. Khalatnikov (Harwood Academic Publisher), vol 13, pp. 1–203 (1989)
14. V.A. Belyakov, Mol. Cryst. Liq. Cryst. **453**, 43 (2006)
15. V.A. Belyakov, Ferroelectrics **364**, 33 (2008)
16. Y. Matsuhisa, Y. Huang, Y. Zhou et al., Appl. Phys. Lett. **90**, 091114 (2007)
17. H. Kogelnik, C.V. Shank, J. Appl. Phys. **43**, 2327 (1972)
18. A. Yariv, M. Nakamura, J. Quantum Electron. **QE-13**, 233 (1977)
19. J.P. Dowling, M. Scalora, M.J. Bloemer, C.M. Bowden, J. Appl. Phys. **75**, 1896 (1994)
20. V.A. Belyakov, S.V. Semenov, JETP **109**(4), 687–699 (2009)
21. H. De Vries, Acta Cryst. **4**, 219 (1951)
22. E.I. Kats, Sov. Phys. JETP **32**, 1004 (1971)
23. V.A. Belyakov, A.S. Sonin, *Optics of Cholesteric Liquid Crystals* (Moscow, "Nauka", 1982) (in Russian)
24. P.G. Gennes, J. Prost, *The Physics of Liquid Crystals* (Clarendon Press, Oxford, 1993)
25. V.A. Belyakov, *Diffraction Optics of Complex Structured Periodic Media* (Springer, New York, 1992) Chap. 4
26. K. Funamoto, M. Ozaki, K. Yoshino, Jpn. J. Appl. Phys. **42**, L1523 (2003)
27. M.F. Moreira, S. Relaix, W. Cao, B. Tahery, P. Palffy-Muhoray, in *Liquid Crystals Microlasers*, 2010 Transword Research Network, ed. by L. Blinov, R. Bartolino, p. 223
28. V.A. Belyakov, I.W. Stewart, M.A. Osipov, Phys. Rev. E **71**, 051708 (2005)
29. V.A. Belyakov, V.E. Dmitrienko, Sov.Phys.- Solid State, **15**(1811), 2364 (1974)
30. W. Cao, A. Munoz, P. Palffy-Muhoray, B. Taheri, Nat. Mater **1**(2), 111 (2002)
31. B.I. Mantstzov, *Coherent and Nonlinear Optics of Photonic Crystals* (Fizmatlit, Moscow, 2009). (in Russian)
32. V.A. Belyakov,N.V. Shipov, Zh. Eksp. Teor. Fiz, **82**,1159 (1982) [Sov. Phys. JETP, **55**, 674 (1982)]
33. N.V. Shipov, V.A. Belyakov, Zh. Eksp. Teor. Fi., **75**,1589 (1978). [Sov. Phys. JETP, **48**, 802 (1978)]
34. B.A. Umanskii, L.M. Blinov, S.P. Palto, Quantum Electron. **43**, 1078 (2013)
35. J. Etxebarria, J. Ortega, C.L. Folcia, Liq. Cryst. **45**, 122 (2017)
36. P.V. Dolganov, JETP Lett. **105**, 657 (2017)

Chapter 6
Optical Defect Modes in Chiral Liquid Crystals

6.1 Introduction

Recently, there has been very intense activity in the field of localized optical modes, in particular, defect modes in chiral liquid crystals (CLCs) mainly due to the possibilities for reaching a low lasing threshold for mirrorless distributed feedback (DFB) lasing [1–4], using defect modes as narrowband filters [5, 6], and enhancing nonlinear optical high harmonic generation [7] in chiral liquid crystals. Some reviews of the problem have been published recently [8–10]. The defect modes existing as a localized electromagnetic eigenstate with their frequencies in the forbidden band gap at the structure defect were first investigated in three-dimensional periodic dielectric structures [5]. The corresponding defect modes in chiral liquid crystals, and more generally in spiral media, are very similar to the defect modes in one-dimensional scalar periodic structures. They display abnormal reflection and transmission inside the forbidden band gap [1, 2] and allow DFB lasing at a low lasing threshold [3]. The qualitative difference with the case of scalar periodic media lies in the polarization properties. The defect mode in chiral liquid crystals is associated with a circular polarization of the electromagnetic field eigenstate whose chirality sense coincides with that of the chiral liquid crystal helix. There are two main types of defects in the chiral liquid crystals so far studied. One of them is a plane layer of some substance differing from the CLC, dividing a perfect cholesteric structure in two parts and lying perpendicular to the helical axes of the cholesteric structure [1]. The other is a jump in the cholesteric helix phase at some plane perpendicular to the helical axes (without the insertion of any substance at the location of this plane) [2]. Recently, many new types of defect layer have been studied [11–17], for example, a CLC layer whose pitch differs from the pitch of the two layers sandwiching the defect layer [11]. Clearly, there are many possible versions of the dielectric properties of the defect layer, but in the considerations below we restrict to the two main types of defects in chiral liquid crystals mentioned above.

V. Belyakov, *Diffraction Optics of Complex-Structured Periodic Media*, Springer Series in Optical Sciences 203,
https://doi.org/10.1007/978-3-319-43482-7_6

Almost all studies of the defect modes in chiral and scalar periodic media have been performed by numerical analysis, with the exception of [18, 19], where the exactly known analytical expression for the eigenmodes propagating along the helix axes [20, 21] was used to carry out a general study of the defect mode associated with a jump in the helix phase. The approach used in [18, 19] looks particularly fruitful because it provides an easy understanding of defect mode physics. It thus deserves further implementation in the study of defect modes and, in particular, in specific cases allowing essential simplification of the general relationships governing defect modes. In general, the helical media are the only periodic structure admitting a simple exact analytic solution of the Maxwell equations, so naturally this advantage of helical media over other periodic media should be fully exploited in solving specific boundary problems relating to DMs. In the present chapter, we present an analytic solution of the DM associated with insertion of an isotropic layer in the perfect cholesteric structure and consider some limiting cases simplifying the problem [22].

6.2 General Equations

To consider the defect mode associated with insertion of an isotropic layer in the perfect cholesteric structure, we have to solve the Maxwell equations and a boundary problem for an electromagnetic wave propagating along the cholesteric helix in the structure depicted in Fig. 6.1.

The Maxwell equations for a wave propagating along the helix axes have the form given by (1.8), with the dielectric tensor of the chiral liquid crystal given by (1.6) [17–21].

As is well known (see also Chap. 3), in the perfect cholesteric structure, i.e., in each cholesteric layer depicted in Fig. 6.1, there are four eigensolutions of the Maxwell equations (first studied in [20, 21]), each of them being a superposition of two plane waves. Two of these eigensolutions correspond to the wave non-diffracting in the

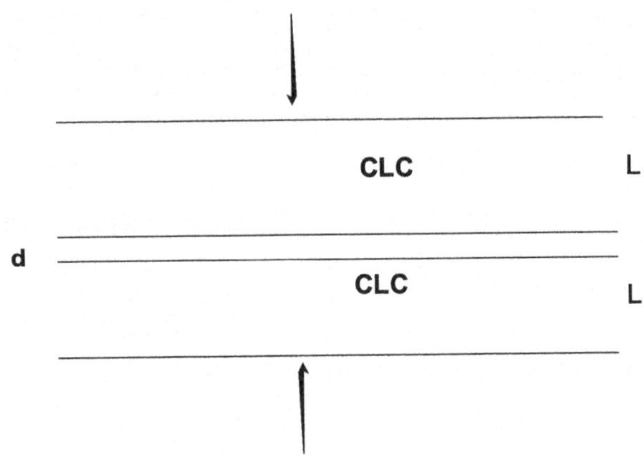

Fig. 6.1 Schematic of the CLC defect mode structure with an isotropic defect layer

cholesteric, and two correspond to the wave diffracting in the cholesteric and having one frequency band (around the Bragg frequency) that is forbidden for propagation of the wave (the so-called stop band). The diffracting eigensolutions are related to the circular polarization with the same sense of chirality as the cholesteric helix, while the non-diffracting eigenstate solutions are related to the circular polarization with the opposite sense of chirality. Therefore, solving the boundary problem involves expressing the electromagnetic wave in the CLC as a linear superposition of the four eigensolutions in each cholesteric layer of the structure depicted in Fig. 6.1. For the entire structure, the boundary problem reduces to the four problems related to the four interfaces of the structure. In addition to the parameters entering in the eigensolutions (see below), the system also involves a parameter given by the ratio of the dielectric constant of the defect layer to the average dielectric constant of the cholesteric. Therefore, if we assume simple dielectric properties for the defect layer, e.g., isotropy, the exact solution of the boundary problem can be presented analytically as a solution of the system of eight linear equations. The zero value of the determinant of this system determines the eigensolution of the boundary problem, i.e., the DM. However, the corresponding analytical solution and the equation determining the eigenmode are rather cumbersome in the general case.

An example of the exact solution of a simpler boundary problem for a cholesteric layer surrounded by an isotropic medium (one half of Fig. 6.1) is presented in the preceding chapters [24–26].

As a result of the general analysis of the solutions of the present boundary problem and the boundary problem for a single cholesteric layer, we conclude that no truly localized defect mode exists for the boundary problem under consideration. This is because, at the isotropic layer boundaries, there is a conversion of the diffracting eigensolutions into the non-diffracting eigensolutions escaping from the external cholesteric layer surfaces. In the general case, we can therefore speak of a quasi-localized DM (see also [18, 19] for the DMs related to the helix phase jump). However, in the most favorable case, where the dielectric constant of the layer is equal to the average dielectric constant of the CLC, the polarization conversion is low (proportional to the dielectric anisotropy of the cholesteric). This is why, in this case, for thick CLC layers, the DM is practically completely localized (see below for the case of infinitely thick CLC layers).

In what follows, to clarify the general results, we present the explicit form of the eigenstate solutions and the specific cases simplifying solution of the boundary problem, namely the case where the dielectric constant of the isotropic defect layer is equal to the average dielectric constant of the cholesteric, and in particular, the case of diffractionally thick cholesteric layers in Fig. 6.1.

6.3 Eigenwaves in CLC

It is known [21–25] (see also Chap. 3) that the eigenwaves corresponding to propagation of light along the spiral axis in a CLC, i.e., the solutions of the Maxwell

(1.8), are a superposition of two plane waves of the form (1.10) with wave vectors $\mathbf{K}^{\pm}$ satisfying the condition (5.4). The ratio of amplitudes (E^-/E^+) in (1.10) is given by the (5.2).

Two of the eigenwaves, corresponding to a circular polarization whose chirality coincides with that of the liquid crystal spiral, experience strong diffraction scattering at the frequencies in the stop band. The other two eigenwaves, corresponding to the opposite circular polarization, are almost unaffected by diffraction scattering, even at the frequencies of the stop band for the diffracting polarization.

Because the specific features of the DMs in CLCs are related to the eigenwaves with the diffracting polarization (see [2, 3, 18, 19]), we restrict ourself in what follows to the consideration of the case of light with the diffracting polarization only.

6.4 Boundary Problem

In this section, we consider the boundary problem discussed above in general form under the assumption that the specific parameters allow a simplification of the problem. We assume (see Fig. 6.1) that the CLC is given by a planar layer with spiral axes perpendicular to the layer surfaces. We also assume that the average CLC dielectric constant ε_0 coincides with the dielectric constant of the isotropic external medium and also that of the isotropic layer inserted between the two cholesteric layers. This assumption allows us to neglect the conversion of one circular polarization to another at the layer surfaces [24, 25] because it is proportional to the small parameter δ, the cholesteric dielectric anisotropy; it also allows us to take into account only two eigenwaves [and correspondingly only two wave vectors from (5.6)] with diffracting circular polarization.

We first consider a linear boundary problem in the formulation where two plane waves with the diffracting polarization and with the same frequency are assumed to be incident along the spiral axis from opposite sides on both CLC layers (see Fig. 6.1), and the dielectric tensor may have a nonzero imaginary part of any sign. Two diffracting eigensolutions with the structure determined by (1.10) are excited in each cholesteric layer. The amplitudes of the two diffracting eigenwaves are denoted by $E_+^{+\mathrm{u}}$ and $E_-^{+\mathrm{u}}$ for the upper layer and $E_+^{+\mathrm{d}}$ and $E_-^{+\mathrm{d}}$ for the bottom layer; they must satisfy the system of four linear equations [24, 25] (following from the continuity of the tangential components of the electric and magnetic fields at the four boundaries of the cholesteric layers depicted in Fig. 6.1):

$$
\begin{aligned}
&E_+^{+\mathrm{u}} + E_-^{+\mathrm{u}} = E_{\mathrm{iu}}\\
&\exp[\mathrm{i}\,kd]\,\exp\left[\mathrm{i}\,K_+^+ L_-\right]E_+^{+\mathrm{u}} + \exp[\mathrm{i}\,kd]\exp\left[\mathrm{i}\,K_-^+ L_-\right]E_-^{+\mathrm{u}}\\
&\quad = \exp\left[\mathrm{i}K_+^+ L_+\right]E_+^{+\mathrm{d}} + \exp\left[\mathrm{i}\,K_+^+ L_+\right]E_-^{+\mathrm{d}}\\
&\xi^+ \exp\left[\mathrm{i}\,K_+^- L_-\right]E_+^{+\mathrm{u}} + \xi^- \exp\left[\mathrm{i}K_-^- L_-\right]E_-^{+\mathrm{u}}\\
&\quad = \exp[\mathrm{i}kd]\xi^+ \exp\left[\mathrm{i}\,K_+^- L_+\right]E_+^{+\mathrm{d}}
\end{aligned}
$$

$$+\exp[ikd]\xi^-\exp\left[iK_-^-L_+\right]E_-^{+d}$$
$$\exp\left[i\,K_+^-L\right]\xi^+E_+^{+d}+\exp\left[i2K_-^-L\right]\xi^-E_-^{+d}=E_{id} \qquad (6.1)$$

where E_{iu} and E_{id} are the respective amplitudes of the waves of diffracting polarization incident at the cholesteric layers from the top (E_{iu}) and the bottom (E_{id}) of the structure in Fig. 6.1, 2L is the total CLC structure thickness, d is the isotropic layer thickness, $L_\pm = L \pm d/2$, and

$$K_\pm^+=\tau/2\pm q,\;\; K_\pm^-=K_\pm^+-\tau=-\tau/2\pm q,$$
$$\xi^\pm=E_\pm^-/E_\pm^+=\delta/\left[\left(K_\pm^+-\tau\right)^2/\kappa^2-1\right], \qquad (6.2)$$

where q is given by (5.7).

We note that for the further study of DMs, it is essential that q given by (5.7) is purely imaginary at the DM frequencies (located inside the stop band). The frequency at the stop-band center (the Bragg frequency ω_B) is given by

$$\omega_B=2\pi\,c/p\varepsilon_0^{1/2}=\tau\,c\varepsilon_0^{-1/2}/2$$

and the band edge frequencies $\omega_e^\pm$ are

$$\omega_e^\pm=\omega_B/[1-(\pm\delta)]^{1/2}.$$

If we assume E_{iu} (E_{id}) to be the only nonzero amplitudes, then (6.1) describes the reflection and transmission of light incident on the structure (Fig. 6.1) from above (downward).

In this case, the amplitude reflection, R, and transmission, T, coefficients of the defect structure (Fig. 6.1), as solutions of (6.1), are given by the formulas

$$R(d,L)=\xi^+E_+^{+u}+\xi^-E_-^{+u}, \qquad (6.3)$$

$$T(d,L)=\exp\left[i\left(2K_+^+L+kd\right)\right]E_+^{+d}+\exp\left[i\left(2K_-^+L+kd\right)\right]E_-^{+d}, \qquad (6.4)$$

obtained by solving (6.1) under the assumption that only the wave incident from above exists ($E_{iu}\neq 0$, $E_{id}=0$).

We now consider in more detail the solutions of the system (6.1) for some specific situations.

6.5 Perfect Cholesteric Layer

The case of a perfect cholesteric layer corresponds to two limits in (6.1) and is considered here for completeness (the corresponding results can also be found in

[24, 25]). One option corresponds to $d = 0$ and the other also to $d = 0$, and the removed one CLC layer in Fig. 6.1. The first and second options correspond to perfect CLC layers of thickness 2*L* and *L*, respectively. Solving system (6.1) yields the following expressions for the amplitude reflection, *R*, and transmission, *T*, coefficients of a CLC layer of thickness *L* (obtained with a slight modification of the notation in (3.32)):

$$
\begin{aligned}
R(L) &= \mathrm{i}\delta \sin qL/\{(q\,\tau/\kappa^2)\cos qL + \mathrm{i}[(\tau/2\kappa)^2 + (q/\kappa)^2 - 1]\sin qL\}\\
T(L) &= \exp[\mathrm{i}\tau L/2](q\tau/\kappa^2)/\{(q\tau/\kappa^2)\cos qL\\
&\quad + \mathrm{i}[(\tau/2\kappa)^2 + (q/\kappa)^2 - 1]\sin qL\},
\end{aligned}
\tag{6.5}
$$

where the phases of *T* and *R* correspond to the assumption that the coordinate $z = 0$ at the entrance surface and correspondingly the director orientation at the entrance surface are determined by the expression (1.6) for the dielectric tensor $\varepsilon(z)$ of the cholesteric liquid crystal (see also [21–25]) at $z = 0$.

6.6 Reflection and Transmission for DM Structure

As noted, the system (6.1) determines the amplitude light transmission *T*(*d*, *L*) and reflection *R*(*d*, *L*) coefficients for the DM structure (DMS) (see 6.3, 6.4) if one of the amplitudes E_{iu} or E_{id} is assumed to be zero. For a finite value of *L*, we have to solve the system (6.1) and use (6.3, 6.4) to find the transmission and reflection coefficients.

But there is another way to obtain formulas for the optical properties of the structure depicted in Fig. 6.1. If we use the expressions (6.5) for the amplitude transmission, *T*(*L*), and reflection, *R*(*L*), coefficients for a single cholesteric layer (see also [24, 25]), then the corresponding intensity transmission, $|T(d, L)|^2$, and reflection, $|R(d, L)|^2$, coefficients for the entire structure can be written as

$$
|T(d, L)|^2 = \left|[T_{\mathrm{e}}T_{\mathrm{d}}\exp(\mathrm{i}kd)]/[1 - \exp(2\mathrm{i}kd)R_{\mathrm{d}}R_{\mathrm{u}}]\right|^2, \tag{6.6}
$$

$$
|R(d, L)|^2 = \left|\{R_{\mathrm{e}} + R_{\mathrm{u}}T_{\mathrm{e}}T_{\mathrm{u}}\exp(2\mathrm{i}kd)/[1 - \exp(2\mathrm{i}kd)R_{\mathrm{d}}R_{\mathrm{u}}]\}\right|^2, \tag{6.7}
$$

where $R_{\mathrm{e}}(T_{\mathrm{e}})$, R_{u} (T_{u}), and $R_{\mathrm{d}}(T_{\mathrm{d}})$ are, respectively, the amplitude reflection (transmission) coefficients of the CLC layer (6.5) (see Fig. 6.1) for light incident on the outer (top) layer surface, for light incident on the inner top CLC layer surface from the inserted defect layer, and for light incident on the inner bottom CLC layer surface from the inserted defect layer. It is assumed in deriving (6.6, 6.7) that the external beam is only incident on the structure (Fig. 6.1) from above.

It also easy to find expressions for the eigenmode amplitudes excited by the incident wave on the two CLC layers of the structure depicted in Fig. 6.1 by using the expressions for the amplitude transmission, *T*(*d*, *L*), and reflection, *R*(*d*, *L*), coefficients in (6.6, 6.7). The eigenmode amplitudes at the CLC entrance layer are expressed in terms of $R(d, L)$ by

$$E_+^{+u} = (E_{iu})[\xi - R(d, L)](\xi^- - \xi^+), \quad E_-^{+u} = -(E_{iu})[\xi^+ - R(d, L)](\xi^- - \xi^+). \tag{6.8}$$

Similarly, the eigenmode amplitudes in the CLC exit layer are expressed in terms of $T(d, L)$ by

$$\begin{aligned} E_+^{+d} &= (E_{iu})[\xi^- T(d, L)] \exp[-i(2K_+^+ L + kd)]/(\xi^- - \xi^+), \\ E_-^{+d} &= -(E_{iu})[\xi^+ T(d, L)] \exp[-i(2K_-^+ L + kd)]/(\xi^- - \xi^+). \end{aligned} \tag{6.9}$$

The corresponding calculations of the amplitudes E_+^{+u}, E_-^{+u}, E_+^{+d}, and E_-^{+d} of the eigenwaves excited in the layer reveal (see Sect. 6.7) a non-trivial frequency dependence of these amplitudes. In particular, close to the DM frequencies (inside the stop band, see below), the eigenmode amplitudes change sharply. However, in contrast to the corresponding amplitude changes for localized edge modes, where the changes are huge [26] (see Chap. 5), the amplitude changes are of the order of unity for DMs. Nevertheless, there is an essential enhancement of the field magnitude relative to the incident wave amplitude (in calculations, the incident wave amplitude is assumed to be equal to (1) close to the defect layer). The reason is an exponential increase in the field for increasing distance between the CLC layer external surface and the defect layer surface. As a result, the thicker the layer, the greater the enhancement of the field at the defect layer.

6.7 Non-absorbing LC

Here we examine in more detail the formulas of the preceding section for non-absorbing cholesteric layers. Figures 6.2 and 6.3 show the calculated reflection $|R(d, L)|^2$ and transmission $|T(d, L)|^2$ spectra at normal incidence inside the stop band for the structure sketched in Fig. 6.1. The figures show maxima in $|T(d, L)|^2$ and minima in $|R(d, L)|^2$ at some frequencies inside the stop band, at positions which depend on the defect layer thickness d. As is well known [1–3, 18, 19], the corresponding minima of $|R(d, L)|^2$ and maxima of $|T(d, L)|^2$ correspond to the defect mode frequencies. For the layer thickness $d = p/4$, which is just one half of the dielectric tensor period in the cholesteric, these maxima and minima are situated just at the center of the stop band. In the interval $0 < d/p < 0.5$, the defect mode frequency value moves from the high-frequency stop-band edge to the low-frequency stop-band edge. For a further increase in the defect layer thickness, the defect mode frequency oscillates between the high-frequency and low-frequency stop-band edges. However, this is only true if Δkd is less than approximately 2π,

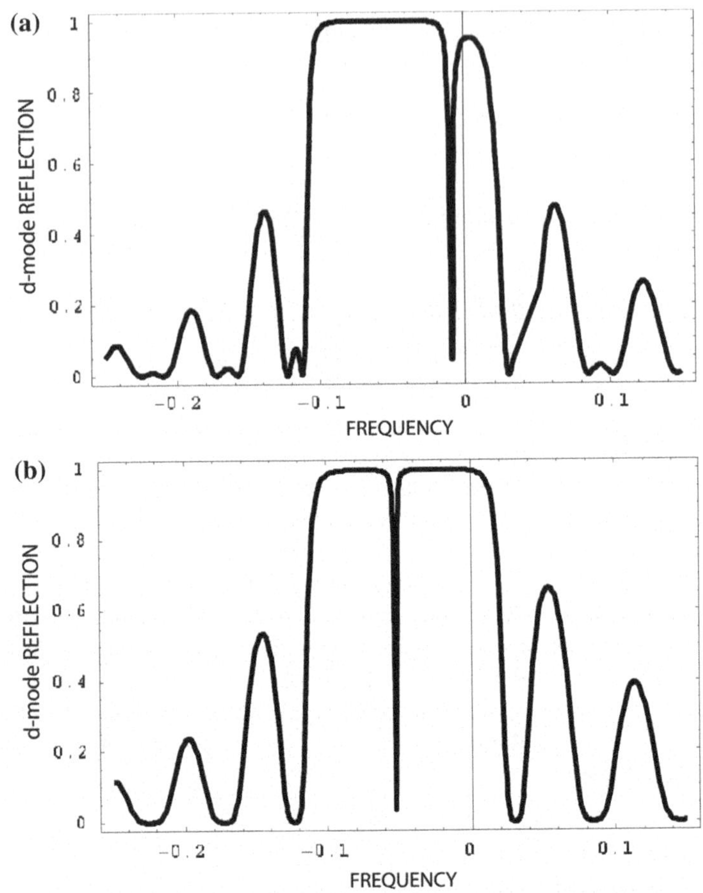

Fig. 6.2 Reflection coefficient $R(d)$ versus the dimensionless frequency for a non-absorbing CLC ($\gamma = 0$) at $d/p = 0.1$ (**a**) and $d/p = 0.25$ (**b**); $\delta = 0.05$, $1 = 200$, $1 = L\tau = 2\pi N$, where N is the director half-turn number at the CLC layer thickness L. Here and in all remaining figures of Chap. 6, the frequency is defined in the same way as in Chap. 5 (see Fig. 5.2)

where Δk is the change in the wave vector at the frequency width of the stop band. When Δkd exceeds 2π, the second defect mode frequency appears in the stop band. Additional defect mode frequencies appear for a further increase in d, and their number may be estimated as $\Delta kd/2\pi$. The described appearance of many defect mode frequencies inside the stop band is illustrated in Fig. 6.3, where only $T(d)$ or $R(d)$ need be shown since, for a non-absorbing structure, $|R(d, L)|^2 + |T(d, L)|^2 = 1$.

Figures 6.2 and 6.3 show that the reflection coefficient $R(d) = 0$ at some frequency. We find from (6.7) the equation determining the frequencies of the reflection coefficient zeros:

$$R_e\left[1 - \exp(2ikd)R_d R_u\right] + R_u T_e T_u \exp(2ikd) = 0. \tag{6.10}$$

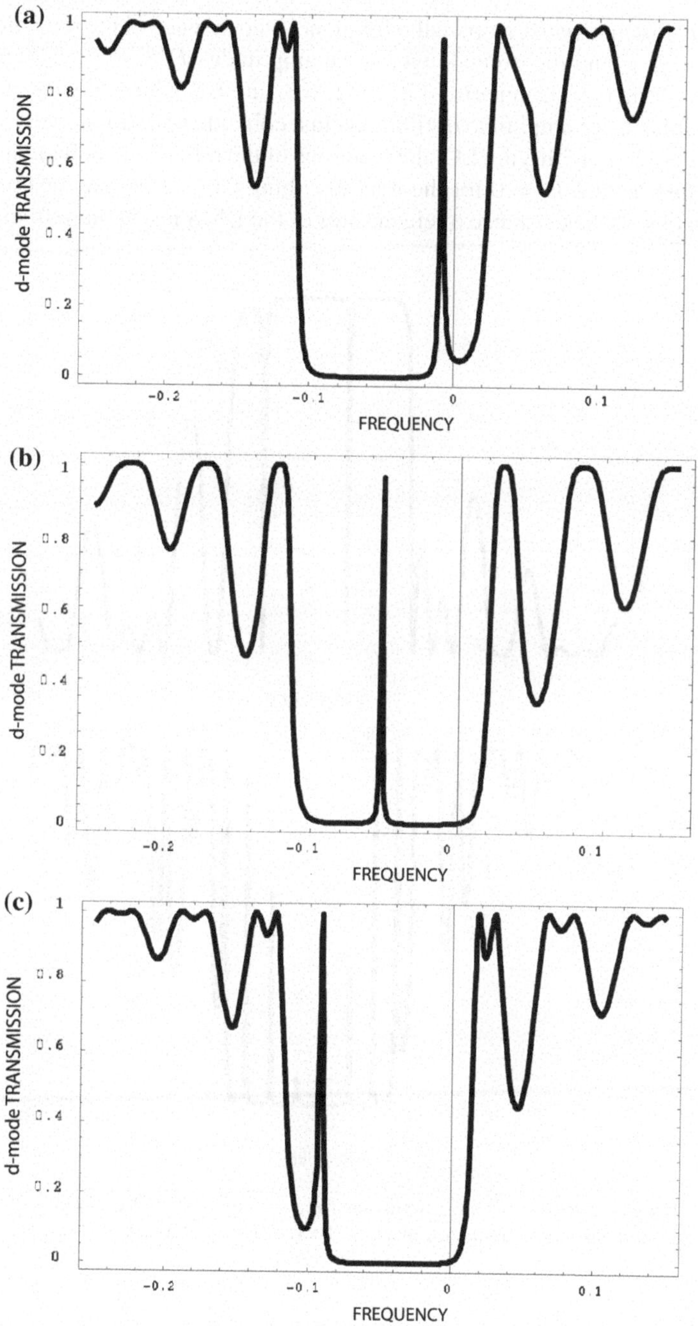

Fig. 6.3 Transmission coefficient $T(d)$ versus the dimensionless frequency for a non-absorbing CLC ($\gamma = 0$) at $d/p = 0.1$ (**a**), $d/p = 0.25$ (**b**), and $d/p = 0.4$ (**c**); $\delta = 0.05$, $\mathrm{l} = 200$, $\mathrm{l} = L\tau = 2\pi N$, where N is the director half-turn number at the CLC layer thickness L

It is quite instructive to compare the frequency dependence of the reflection coefficient $|R(d, L)|^2$ and the excited eigenmode amplitudes E_{+}^{+u}, E_{-}^{+u}, E_{+}^{+d}, and E_{-}^{+d} in the defect structure. Comparing Figs. 6.2, 6.3, and 6.3 with 6.5, we see that the positions of the sharp amplitude oscillations just coincide with (or are very close to, for an absorbing or amplifying LC) the positions of the reflection coefficient minima corresponding to $R(d, L) = 0$ for the non-absorbing CLC. This may be considered as an indication of the existence of eigenstates of the DMS just at these frequencies.

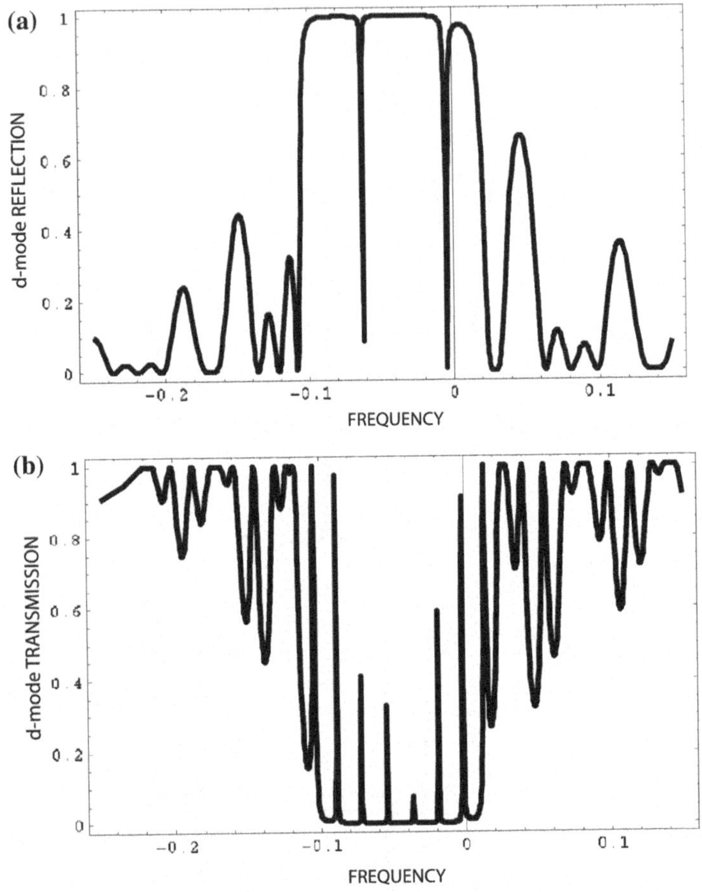

Fig. 6.4 $R(d)$ versus the dimensionless frequency for a non-absorbing CLC at $d/p = 200.1$ (**a**) and $T(d)$ versus the dimensionless frequency at $d/p = 1000.6$ (**b**); $\delta = 0.05$, $N = 33$

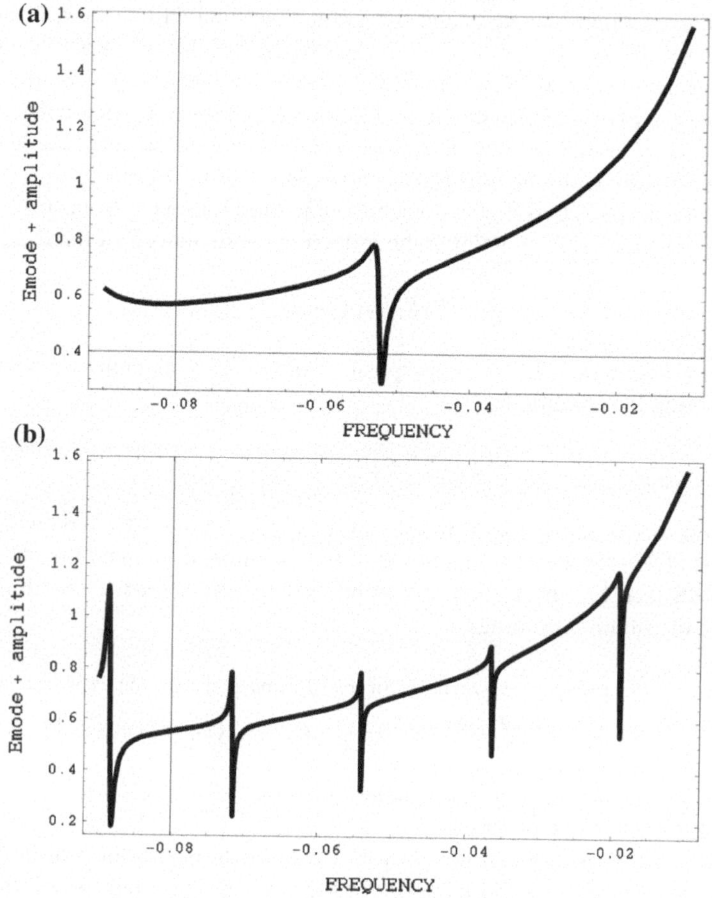

Fig. 6.5 Calculated eigenmode amplitude (E_{+}^{+u}) in the DMS excited by an incident wave of unit amplitude: **a** for $d/p = 0.25$, **b** for $d/p = 1000.6$, $\delta = 0.05$, $N = 33$

6.8 Defect Mode (A Non-absorbing LC)

The solution of (6.1) in the general case is a linear superposition of a propagating wave and a pure defect mode, i.e., a standing wave of complicated structure that cannot simply be reduced to two plane counter-propagating waves. The pure defect mode (DM) is determined by (6.1) with $E_{\text{iu}} = E_{\text{id}} = 0$, i.e., by the solution of (6.1) if no waves are incident from outside on the structure shown in Fig. 6.1.

The defect mode frequency ω_{D} is determined by the zero value of the determinant of the system (6.1):

$$\begin{aligned}\text{Det}(d, L) = 4\{&\exp(2ikd)\sin^2 qL - \exp(-\text{i}\,\tau\, L)[(\tau\, q/\kappa^2)\cos qL \\ &+ \text{i}((\tau/2\kappa)^2 + (q/\kappa)^2 - 1)\sin qL]^2/\delta^2]\}.\end{aligned} \tag{6.11}$$

Note that, at a finite length L and for a non-absorbing CLC, Det(d, L) does not reach zero for a real value of ω, while it does reach zero for a complex value of ω. The greater the thicknesses L of the CLC layers in the DMS (Fig. 6.1), the smaller the imaginary part of ω, and in the limit of infinite L, it reduces to zero in the accepted approach (see Sect. 6.9 below). Therefore, the DM is a quasi-stable mode, and its lifetime is determined by the imaginary part of ω_D.

With the help of (6.6, 6.7), the dispersion equation following from (6.11) can be reduced to an expression containing the reflection coefficients R of the CLC layers:

$$1 - R_d R_u \exp(2ikd) = 0 \tag{6.12}$$

The DM field in the CLC is a superposition of two CLC eigenmodes with amplitude satisfying the condition:

$$E_+^{+u} + E_-^{+u} = 0. \tag{6.13}$$

Relation (6.13) allows us to find the DM field inside the CLC layers using the expression (1.10) for the CLC eigenmodes. For example, in an individual CLC layer of the DMS (see Fig. 6.1), the corresponding expression for the coordinate field amplitude distribution becomes

$$\begin{aligned}\mathbf{E}(\omega_\mathbf{D}, z, t) = {} & \mathrm{i}\exp(-\mathrm{i}\omega_\mathrm{D}t)\{\mathbf{n}_+ \exp(\mathrm{i}\tau z/2)\sin qz + (\mathbf{n}_-/\delta)\exp(-\mathrm{i}\tau z/2) \\ & \{[(\tau/2\kappa)^2 + (q/\kappa)^2 - 1]\sin qz - \mathrm{i}(\tau q/\kappa^2)\cos qz\}\},\end{aligned} \tag{6.14}$$

where ω_D is the DM frequency, q is determined by (5.7), and $z = 0$ corresponds to the external surface of CLC layer.

In particular, Figs. 6.5 and 6.6 show the coordinate dependence of the squared modulus of the total field and its parts for the wave propagating toward the defect layer and in the opposite direction, respectively, close to the external surface of CLC layer. In Fig. 6.6, we see that the greater the dielectric anisotropy δ, the sharper the growth of the DM field toward the defect layer. Figure 6.7 shows that, at the external surfaces of the DMS, only the amplitude of the wave directed toward the defect layer reduces strictly to zero. The amplitude of the wave directed outward is small, but does not reduce to zero. This is why there is a leakage of the DM energy outward through the external surfaces of the DMS. The ratio of the corresponding energy flow to the whole DM energy accumulated in the DMS determines the inverse lifetime (and correspondingly the DM frequency width), which may be given by an analytical expression.

For non-absorbing CLC layers (as considered in this section), the only source of decay is the energy leakage through their external surfaces. The decrease in the DM energy per unit time is equal to the energy flow of the leaking waves $\left(2c/\varepsilon_0^{1/2}\right)|E_{\mathrm{out}}|^2$, where E_{out} is the amplitude of the wave exiting the DMS through the external surfaces of the CLC layers. Therefore, using (6.14), we easily obtain the DM lifetime τ_{m} as

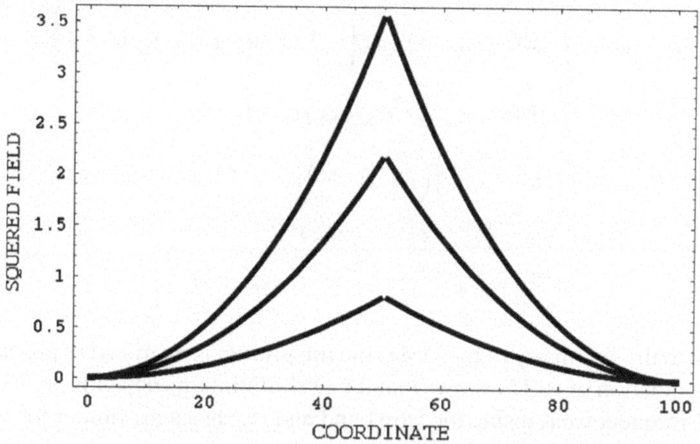

Fig. 6.6 Coordinate dependence ($x = z/p$) of the squared DM field amplitude at the DM frequency when it lies at the stop-band center for various dielectric anisotropy values (from top to bottom $\delta = 0.05, 0.04, 0.025$) and the defect layer thickness $d = p/4$ for the cholesteric layer thickness $L = 50(p/2)$

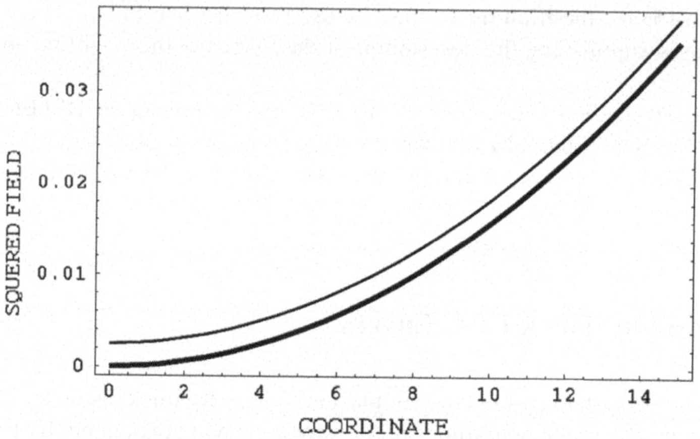

Fig. 6.7 Coordinate dependence ($x = z/p$) of the squared amplitude of the DM waves inside the CLC layer directed toward the defect layer (bold line) and out of the DM structure (narrow line) at the DM frequency when it lies at the stop-band center for dielectric anisotropy $\delta = 0.05$ close to the external surface of the CLC layer

$$\tau_m = \int |\mathbf{E}(\omega_{EM}, z, t)|^2 dz \left\{ d \int |\mathbf{E}(\omega_{EM}, z, t)|^2 dz]/dt \right\}$$
$$= \int \left| \mathbf{E}(\omega_{EM}, z, t)^2 dz / \left(2c/\varepsilon_0^{1/2}\right) \right|$$
$$E_{out}^2| = \left[\left(\delta\kappa^2/\tau\, q \right)^2 \right] \left(L\varepsilon_0^{1/2}/2c \right) \{ |1 - (2/qL) \sin 2qL|$$
$$\left(1 + \delta^{-2}\left[(\tau/2\kappa)^2 + (q/\kappa)^2\right) - 1\right]^2$$
$$+\delta^{-2} \left| \left(\tau q/\kappa^2 \right)^2 (1 + (2/qL) \sin 2qL) \right| \Big\}, \tag{6.15}$$

where, due to the symmetry of the DMS, the integration is restricted to just half of the DMS. Examination of (6.15) shows that the DM lifetime τ_m depends on the position of the DM frequency ω_D inside the stop band and reaches a maximum for ω_D just at the middle of the stop band, i.e., at $k = \tau/2$.

In the general case, description of the DM requires numerical solution of the dispersion (6.11). But because Reω_D is determined by the frequencies corresponding to the zeros of the reflection coefficient for non-absorbing CLC layers, it is easier to investigate DMs in general. In particular, because the DM lifetime can be written as 1/Imω_D, (6.15) for the lifetime τ_m may be used to calculate Imω_D. There are also limiting cases simplifying the description of the DM, and these will be considered below.

As the analysis shows, the thicker the CLC layers, the greater the DM lifetime and correspondingly less the DM frequency width. In the limit of infinitely thick CLC layers, the DM lifetime is infinite in our model (this is shown directly in the next section).

6.9 Infinitely Thick CLC Layers

Let us study the system (6.1) in the simplest case for very thick cholesteric layers in Fig. 6.1. In this case, the amplitudes of the waves growing exponentially toward the external surfaces of the cholesteric layers have to be zero. Formally, we may set L equal to infinity.

Hence, in the cholesteric layers nonzero amplitudes correspond only to the eigenwaves propagating toward the isotropic defect layer. This means that in the system (6.1), in the absence of waves incident on the DMS from outside, the amplitudes E_-^{+u} and E_+^{+d} are equal to zero and the system reduces to the following two linear equations:

$$\exp[\mathrm{i}\,kd] \exp\left[\mathrm{i}\, K_{++}^{+} \mathrm{L}_-\right] E_+^{+u} = \exp\left[\mathrm{i}K_-^{+} \mathrm{L}_+\right] E_-^{+d}$$
$$\xi^+ \exp\left[\mathrm{i}K^- + L_-\right] E_+^{+u} = \exp[\mathrm{i}kd]\xi^- \exp\left[\mathrm{i}K_-^- L_+\right] E_-^{+d} \tag{6.16}$$

The DM frequency ω_D in this case is determined by a zero value of the determinant of the system (6.16), reduced to the following relation:

$$-(2/\delta)\exp[[i(k-q)d]\{(\tau q/\kappa^2\}\cos(\tau/2-k)d \\ +i[(\tau/2\kappa)^2+(q/\kappa)^2-1]]\sin(\tau/2-k)d\}=0. \quad (6.17)$$

For light frequencies inside the selective reflection band, the value of the determinant (6.17) becomes zero if the isotropic layer thickness d is related to the light frequency according to

$$(d/p)(1-k/(\tau/2))=(1/2\pi)\mathrm{arctg}[2(\tau/2\kappa)\{[4(\tau/2\kappa)^2+\delta^2]^{1/2} \\ -1-(\tau/2\kappa)^2\}^{1/2}/[2(\tau/2\kappa)^2-[4(\tau/2\kappa)^2+\delta^2]^{1/2}]] \quad (6.18)$$

This implies that, for any frequency inside the selective reflection band, the defect mode exists, but the existence condition requires a specific value of the isotropic layer thickness for each chosen frequency value.

The relation between the defect mode frequency and the isotropic layer thickness d for infinitely thick CLC layers given in (6.18) is shown in Fig. 6.8 (where the frequency scale is shifted so that $\nu=1$ corresponds to the stop-band center). Again the defect mode frequency (in Fig. 6.8) at the stop-band center corresponds to a layer thickness $d/p=1/4$. The same result follows directly from (6.18) if we assume that ω_D is located in the middle of the stop band, i.e., $\tau/2\kappa=1$. In that case, (6.18) implies

$$d/p=(1/2\pi)\mathrm{arctg}\left[2/\left\{[4+\delta^2]^{\frac{1}{2}}-2\right\}^{\frac{1}{2}}\right]=(1/2\pi)\mathrm{arctg}[4/\delta] \quad (6.19)$$

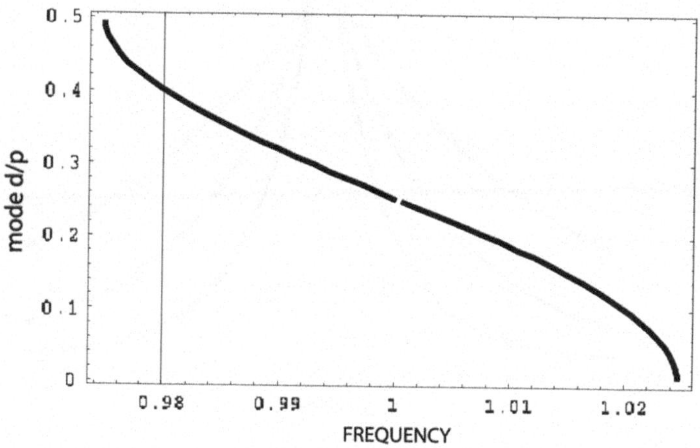

Fig. 6.8 Calculated relation between *d/p* and the DM frequency for infinitely thick non-absorbing CLC layers ($\delta=0.05$)

which corresponds approximately to $d/p = 1/4 + n/2$, where n is zero or an integer.

The shift of the defect mode frequency $\Delta\omega$ in the interval $|\Delta\omega/\omega_B| < \delta$ due to a small variation in $d(\Delta d)$ close to $d/p = 1/4$ is given approximately by the relation

$$\Delta\omega/\omega_B = -4\Delta d/p, \tag{6.20}$$

where the Bragg frequency (frequency at the stop-band center) is determined by (5.19).

As the calculations show, the defect mode frequency is only slightly dependent on the CLC layer thickness L. This is why the corresponding dependence of ω_D on d/p for an infinitely large L may be regarded as a good approximation for ω_D at any L.

It is known that the defect mode field is localized in the defect layer and nearby [1–3, 15]. An illustration of such a localization is presented by the coordinate field distribution of the defect mode for infinitely thick CLC layers in Fig. 6.9. The maximum of the field amplitude is located at the defect layer, and the field amplitude attenuates exponentially in the CLC outside the defect layer. The strongest attenuation occurs for the layer thickness $d/p = 1/4$, i.e., for the DM frequency just at the stop-band center, with the attenuation decreasing as the DM frequency approaches the stop-band edges. The attenuation also grows with increasing dielectric anisotropy δ of the CLC layers.

As the (6.17, 6.18 and 6.19) show, for infinitely thick CLC layers, the DM frequency is a real quantity in the model under consideration and hence the DM lifetime is infinite. This is not the case for a limited CLC layer thickness. For non-absorbing CLC layers of finite thickness, the determinant (6.11) reaches zero for a complex

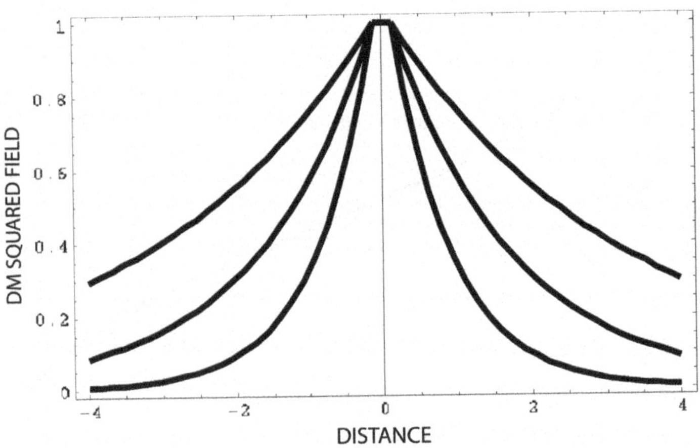

Fig. 6.9 Calculated distribution of the squared field modulus versus the distance from the defect layer center ($x = z/p$) for non-absorbing infinitely thick CLC layers ($\delta = 0.05, 0.1, 0.2$ from top to bottom); $d/p = 1/4$

frequency. The physical meaning of the complex frequency is quite clear. As mentioned above, for non-absorbing CLC layers of finite thickness, there is a leakage of the defect mode electromagnetic field through the external surfaces of the CLC layers which results in the decay of the defect mode. The imaginary part of the frequency and the finite lifetime of the defect mode are determined by this leakage.

6.10 Thick CLC Layers

In the case of a DMS with thick CLC layers ($|q|L \gg 1$), some analytic results can also be obtained for DMs. These results may also be obtained from the dispersion (6.11) or from expressions (6.6 and 6.7) for the DMS transmission and reflection coefficients.

Instead of directly solving the equation for a complex frequency following from the expression (6.11) (Det(d, L) = 0), we can use (6.6 and 6.7) for the DM structure transmission and reflection coefficients to estimate the imaginary part of the defect mode frequency. We have to allow a nonzero imaginary addition to the frequency (defined by the relation $\omega/\mathrm{Re}[\omega] = (1 + \mathrm{i}\Delta)$, where Δ is a small quantity) and search for extrema of (6.6 and 6.7) relative to this imaginary addition iΔ. The results of the corresponding calculations are presented in Figs. 6.10 and 6.11. These calculations of the transmission, $|T(d, L)|^2$, and reflection, $|R(d, L)|^2$, intensity coefficients show that the imaginary addition to the defect mode frequency ω_D decreases as the CLC layer thickness increases and the addition also decreases as $\mathrm{Re}\omega_\mathrm{D}$ approaches the stop-band center for fixed CLC layer thickness. In particular, the calculations show that, for $d/p = 1/4$, corresponding to ω_D being at the stop-band center, Δ is the smallest at a fixed CLC layer thickness. The DM lifetime τ determined from $\tau = 1/|\mathrm{Im}\omega_D|$ increases as the CLC layer thickness increases and reaches a maximum for the DM frequency at the stop-band center for fixed CLC layer thickness. These results are in complete agreement with the numerical solutions of the dispersion (6.21) performed in Sect. 6.13.

We now examine analytically how the lifetime τ increases with the CLC layer thickness, i.e., we find the value of Δ (in the limit $|q|L \gg 1$) corresponding to divergence of the DM structure transmission and reflection coefficients as functions of Δ, by expanding the denominators in (6.6) and (6.7) in the small parameter Δ.

The corresponding expression for Δ is

$$\Delta = \Delta q/\left[\mathrm{i}q F\left(\delta^2\right)\right] \tag{6.21}$$

where Δq is the change in q due to the imaginary addition to the DM frequency ω_D, ensuring the divergence of the DMS transmission and reflection coefficients:

$$\begin{aligned} F\left(\delta^2\right) = &\left\{1 + \left\{1/\left(2[(\tau/\kappa)^2 + \delta^2]^{1/2}\right)\right\}\right. \\ &- (\tau/2\kappa)^2/\left\{\left(1 - \left[(\tau/\kappa)^2 + \delta^2\right]^{1/2} + (\tau/2\kappa)^2\right)\right\} \end{aligned}$$

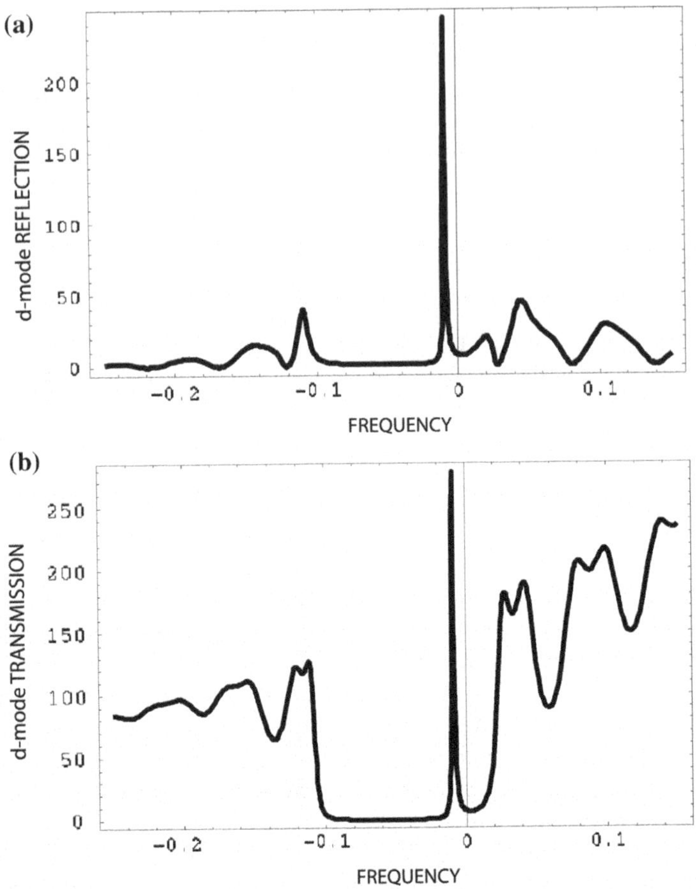

Fig. 6.10 $R(d)$ **a** and $T(d)$ **b** versus the frequency for a non-absorbing CLC when the complex addition to the frequency is given by $\Delta = -0.012$ in $\omega/\mathrm{Re}[\omega] = (1 + \mathrm{i}\Delta))$; $d/p = 0.1$, $\delta = 0.05$, $N = 33$

and

$$\Delta q = \left[2\kappa^2/(q\tau L)\right] \exp[-|q|L].$$

Because the DM lifetime is $\tau_D = 1/\mathrm{Im}\omega_D$, the expression (6.21) reveals an exponential increase in τ_m with increasing CLC thickness L, and also a strong dependence of the increased rate on the position of the DM frequency ω_D inside the stop band. When ω_D lies in the middle of the stop band, the expression (6.21) for Δ becomes

$$\Delta = -(2/3\pi)(p/\delta\, L)\exp[-2\pi\delta(L/p)] \tag{6.22}$$

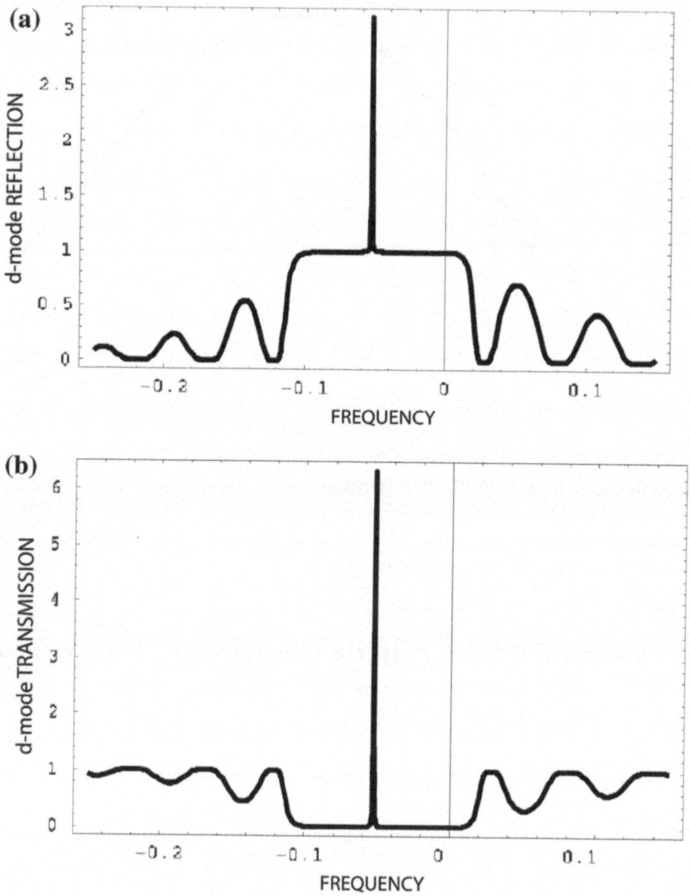

Fig. 6.11 $R(d)$ (**a**) and $T(d)$ (**b**) versus frequency for a non-absorbing CLC when the complex addition to the frequency is given by $\Delta = -0.00466$ in $\omega/\mathrm{Re}[\omega] = (1 + \mathrm{i}\Delta)$; $d/p = 1/4$, $\delta = 0.05$, $N = 33$

The dependence of the DM lifetime on the position of ω_D inside the stop band in the limit of thick CLC layers $|q|L \gg 1$) is shown in Fig. 6.12, which presents the results of calculations using (6.21) for the frequency range inside the stop band where the condition $|q|L \gg 1$ holds. The value of Δ found from (6.21 and 6.22) can be regarded as an initial approximation in numerical calculations in the case of arbitrary CLC layer thickness L.

The same result for the DM lifetime τ_m for thick CLC layers in a DM structure ($|q|L \gg 1$) follows from (6.15), determining the finite lifetime due to electromagnetic wave leakage from the DMS:

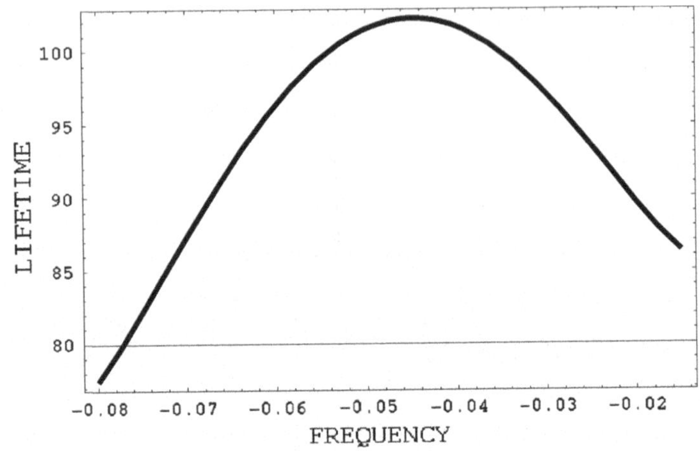

Fig. 6.12 Dependence of the DM lifetime (normalized by the time of flight of light through the DMS $2L\varepsilon_0^{1/2}/c$) on the location of the DM frequency ω_D inside the stop band, calculated for thick CLC layers according to (6.21) ($\delta = 0.05$, $N = 40$, the frequency of the midpoint of the stop band in the figure corresponds to the abscissa value −0.05)

$$\tau_m = \left[\left(\delta\kappa^2/\tau q\right)^2\right]\left(L\varepsilon_0^{1/2}/2c\right)\left[1/\left(\delta^2 q L\right)\right]\big)\left\{\left[(\tau/2\kappa)^2 + (q/\kappa)^2 - 1\right]^2 + \left|\left(\tau q/\kappa^2\right)^2\right|\right\}\exp[2|q|L]. \quad (6.23)$$

For the DM frequency in the middle of the stop band, (6.23) gives

$$\tau_m = 3\pi\left(\varepsilon_0^{1/2}/\tau c\right)(\delta L/p)\exp[2\pi\,\delta L/p].$$

Again, as mentioned above, the maximum of the DM lifetime τ_m corresponds to the location of the DM frequency in the middle of the stop band (i.e., at $k = \tau/2$), where $|q|$ reaches a maximum.

6.11 Absorbing LC

We now examine the formulas (6.6, 6.7) for absorbing CLC layers. This case is directly related to lasing in CLCs: At the lasing frequency, the CLC has to be amplifying, whereas at the frequency of the pumping wave, it is absorbing. To take the absorption into account, we define the ratio of the imaginary part of the dielectric constant to the real part of ε as γ, i.e., $\varepsilon = \varepsilon_0(1 + i\gamma)$. Note that in real situations $\gamma \ll 1$. A natural consequence of the nonzero absorption ($\gamma > 0$) is a reduction in the transmission, $T(d)$, and reflection, $R(d)$, coefficients. However, there are some interesting peculiarities in the optical properties of the structure under considera-

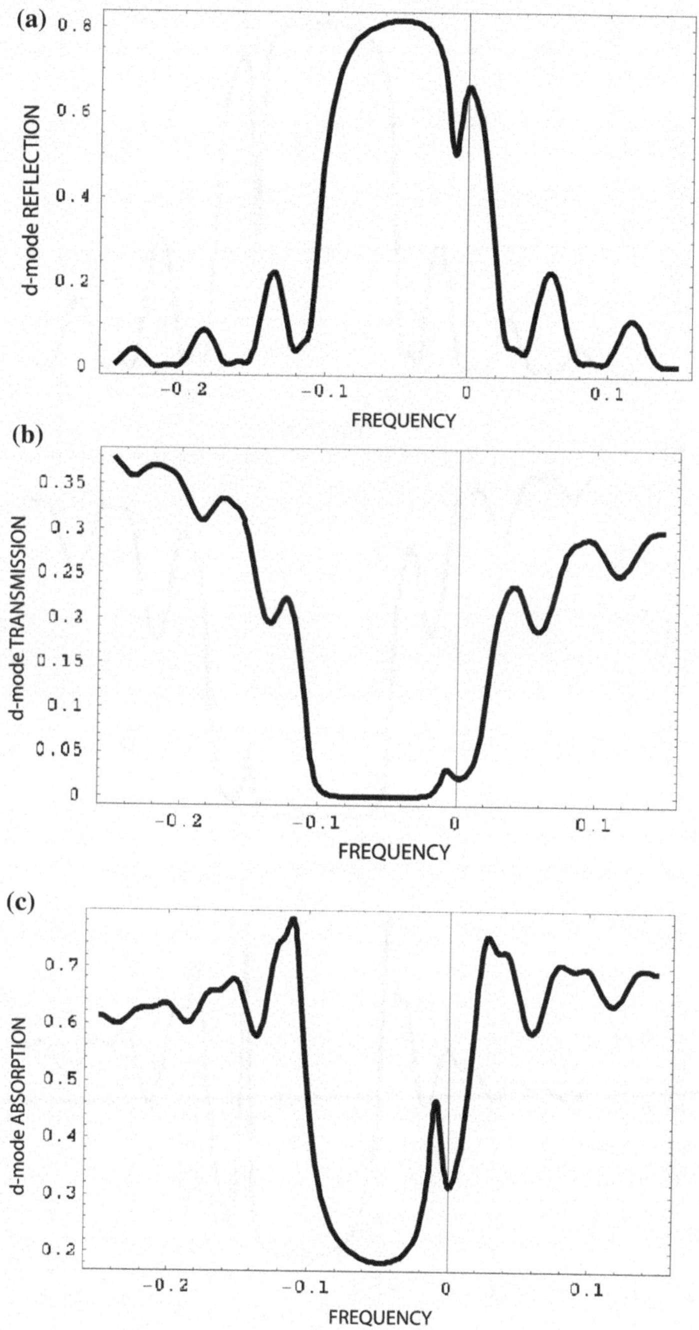

Fig. 6.13 $R(d)$ (**a**), $T(d)$ (**b**), and the total absorption $(1 - R(d)–T(d))$ (**c**) versus the dimensionless frequency for an absorbing CLC $\gamma = 0.005$; $d/p = 0.1$, $\delta = 0.05$, $N = 33$

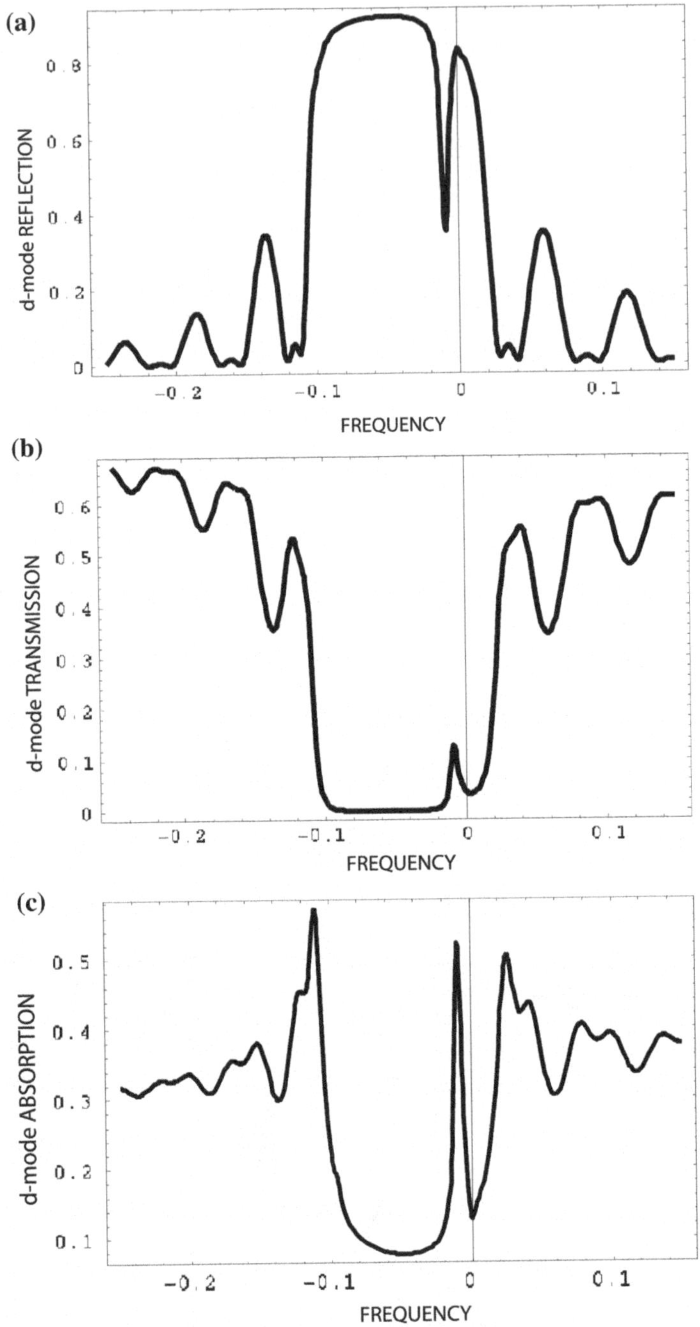

Fig. 6.14 $R(d)$ (**a**), $T(d)$ (**b**), and the total absorption (**c**) for an absorbing CLC versus the dimensionless frequency $\gamma = 0.002$; $d/p = 0.1$, $\delta = 0.05$, $N = 33$

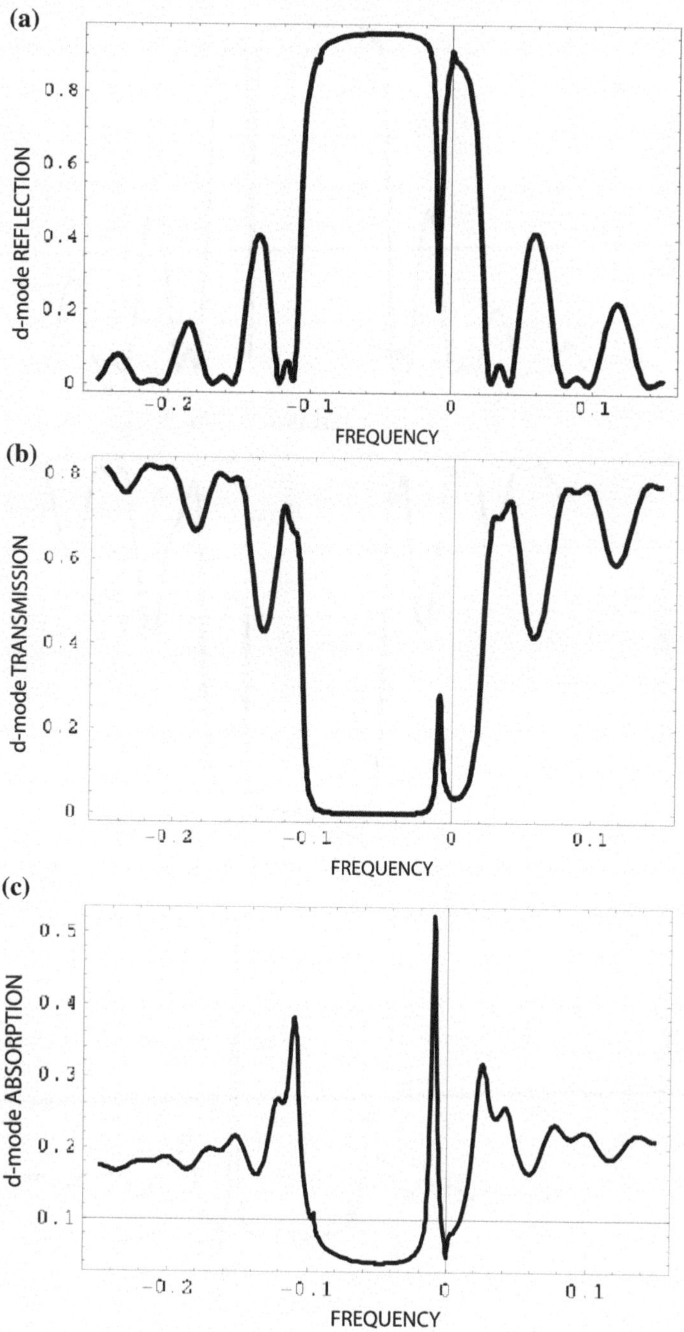

Fig. 6.15 $R(d)$ (**a**), $T(d)$ (**b**), and the total absorption (**c**) for an absorbing CLC versus the dimensionless frequency $\gamma = 0.001$; $d/\mathrm{l} = 0.1$, $\delta = 0.05$, $N = 33$

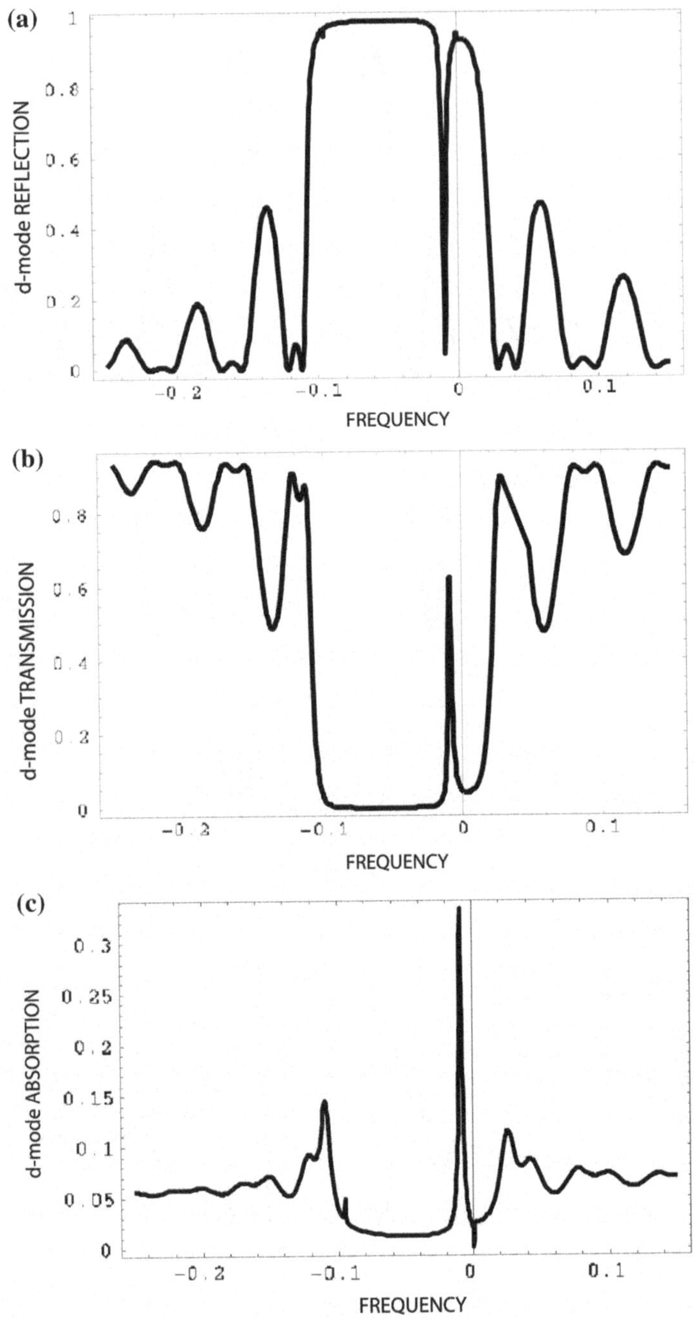

Fig. 6.16 $R(d)$ (**a**), $T(d)$ (**b**), and the total absorption (**c**) for an absorbing CLC versus the dimensionless frequency $\gamma = 0.0003$; $d/p = 0.1$, $\delta = 0.05$, $N = 33$

tion (Fig. 6.1). The calculation results presented in Figs. 6.13, 6.14, 6.15, and 6.16 reveal these peculiarities. For absorbing structures, $|T(d, L)|^2 = |R(d, L)|^2 < 1$, and the quantity $1 - |T(d, L)|^2 - |R(d, L)|^2$ shown in Figs. 6.13, 6.14, 6.15, and 6.16 gives the total absorption in the structure. Up to a relatively high level of absorption ($\gamma = 0.005$ in Fig. 6.13), the spectral shapes of reflection and transmission are typically conserved for the defect mode minima and maxima in the reflection and transmission coefficients, respectively, although they deviate from the case of a non-absorbing CLC (see Figs. 6.2 and 6.3). As γ decreases to small values, the spectral shapes of reflection and transmission almost approach the corresponding shapes for non-absorbing CLCs (see Fig. 6.16a, b corresponding to $\gamma = 0.0003$), but the total absorption $(1 - |T(d, L)|^2 - |R(d, L)|^2)$ at the DM frequency behaves in an unusual way.

As regards the total absorption, it displays a non-conventional frequency dependence. At small γ, for some frequencies, the absorption turns out to be much greater than the absorption outside the stop band (see Figs. 6.13, 6.14, 6.15, and 6.16). If γ is not too small (Fig. 6.13c, $\gamma = 0.005$), the total absorption increase reveals itself at the stop-band edges (at the frequencies of the edge modes). This is a manifestation of the so-called anomalously strong absorption effect, known for perfect CLC layers at the edge mode frequency [25, 27]. For smaller γ, the total absorption at the defect mode frequency ω_D begins to exceed the absorption outside the stop band, and the absorption has the same value as for the edge modes (Fig. 6.14c, $\gamma = 0.002$). As γ decreases further, the anomalously strong absorption effect becomes more pronounced at the DM frequency than at the EM frequencies (Fig. 6.15c, $\gamma = 0.001$, Fig. 6.16c, $\gamma = 0.0003$). It follows that, at the DM frequency ω_D, there is an anomalously strong absorption effect similar to the one for EMs [24, 27], and furthermore, the absorption enhancement for DMs at small γ is higher than for EMs. It is clear that the anomalously strong absorption effect at the DM frequency is solely due to the localized DM, i.e., to the defect layer in the structure. Its realization assumes some relation between γ and the other LC parameters. This relation is determined by the condition

$$\partial(1 - |T(d, L)|^2 - |R(d, L)|^2)/\partial\gamma = 0. \tag{6.24}$$

In the general case, (6.24) can only be solved numerically. But in the case of thick CLC layers ($|q|L \gg 1$), the dependence of γ on L and other parameters ensuring maximal absorption can be found analytically:

$$\gamma = -2\left[2\kappa^2/(q\tau L)\right]\exp[|q|L](1/\mathrm{i}q)/\Big\{1 + \Big\{1/\Big(2\left[(\tau/\kappa)^2 + \delta^2\right]^{1/2}\Big) - (\tau/2\kappa)^2\Big\}/ \\ \Big(1 - \left[(\tau/\kappa)^2 + \delta^2\right]^{1/2} + (\tau/2\kappa)^2\Big)\Big\} \tag{6.25}$$

The value of γ given by (6.25) may be regarded as an initial approximation in numerical calculations in the case of an arbitrary CLC layer thickness L. Figure 6.14 shows the frequency dependence of γ that corresponds to the maximum absorption

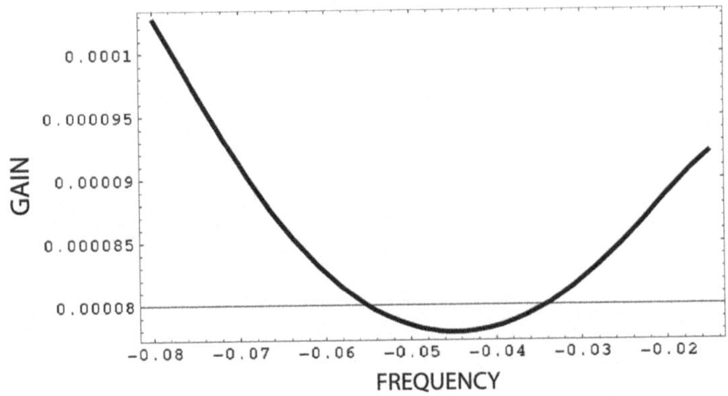

Fig. 6.17 Gain γ corresponding to maximal absorption versus the DM frequency ω_D located inside the stop band is calculated for thick CLC layers using (6.25) ($\delta = 0.05$, $N = 40$, and the frequency of the midpoint of the stop band in the figure corresponds to the abscissa value -0.05)

for a thick CLC layer in the limit of $|q|L \gg 1$. Figure 6.17 shows that maximal absorption enhancement occurs just in the center of the stop band.

When ω_D lies in the middle of the stop band, expression (6.25) for γ becomes

$$\gamma = (4/3\pi)(p/\delta L)\exp[-2\pi\,\delta(L/p)] \tag{6.26}$$

We note that the "anomalously strong absorption effect" at the DM frequency and its realization under some relation between γ and other LC parameters reveal themselves in the calculations of the total absorption at the DM frequency as a function of γ performed in [18] (the absorption reaches a maximum at a small finite value of γ; see Fig. 8 in [18]).

6.12 Amplifying LC

We examine formulas (6.6, 6.7) for amplifying cholesteric layers. As previously, we assume that the dielectric constant is given by the same formula $\varepsilon = \varepsilon_0(1 + i\gamma)$, but with $\gamma < 0$. The calculation results for the transmission $|T(d, L)|^2$ and reflection $|R(d, L)|^2$ coefficients at $\gamma < 0$ are presented at Figs. 6.18, 6.19, 6.20, and 6.21. For small absolute value of γ, the shape of the transmission, $|T(d, L)|^2$, and reflection, $|R(d, L)|^2$, coefficients is qualitatively the same as for zero amplification ($\gamma = 0$) (Figs. 6.18, 6.20, and 6.21a, b). But the absorption is a small negative quantity (which means amplification) at all frequencies with some amplification enhancement at the DM frequency and at the EM frequencies (Fig. 6.18c). As the absolute value of γ increases, the shape of the reflection coefficient $|R(d, L)|^2$ changes at some value of γ [a typical minimum in $|R(d, L)|^2$ is superseded by a small maximum close to 1, and the transmission $|T(d, L)|^2$ significantly exceeds 1 (Fig. 6.19)]. As the absolute

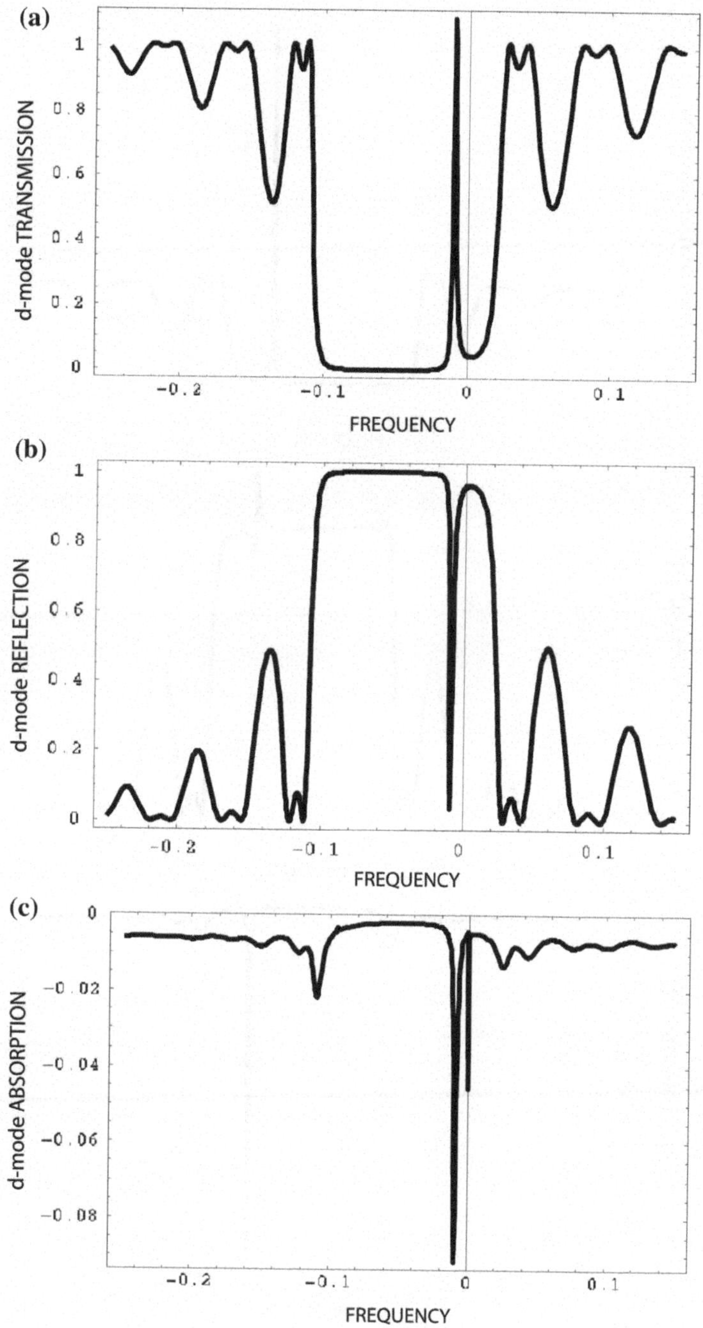

Fig. 6.18 $T(d)$ (**a**), $R(d)$ (**b**), and the total absorption (**c**) for an amplifying CLC versus the dimensionless frequency $\gamma = -0.00005$; $d/p = 0.1$, $\delta = 0.05$, $N = 33$

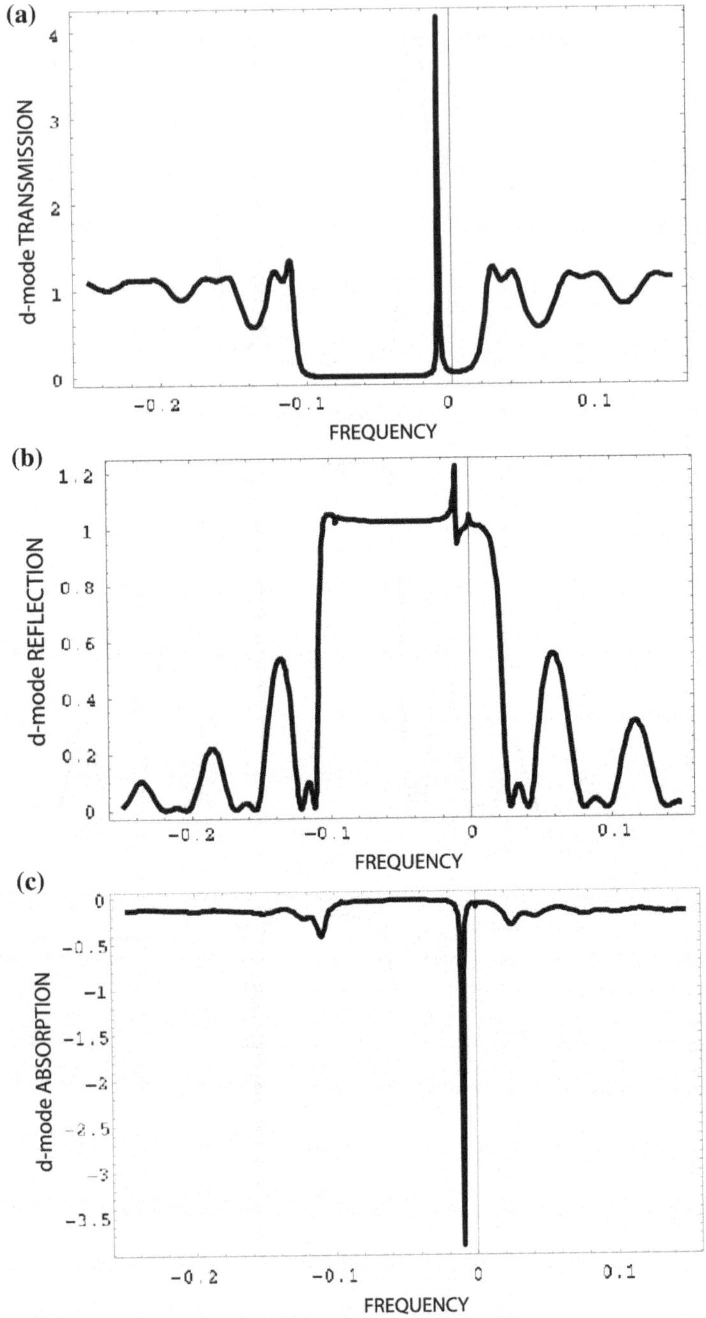

Fig. 6.19 $T(d)$(**a**), $R(d)$ (**b**), and the total absorption (**c**) for an amplifying CLC versus the dimensionless frequency $\gamma = -0.0006$; $d/p = 0.1$, $\delta = 0.05$, $N = 33$

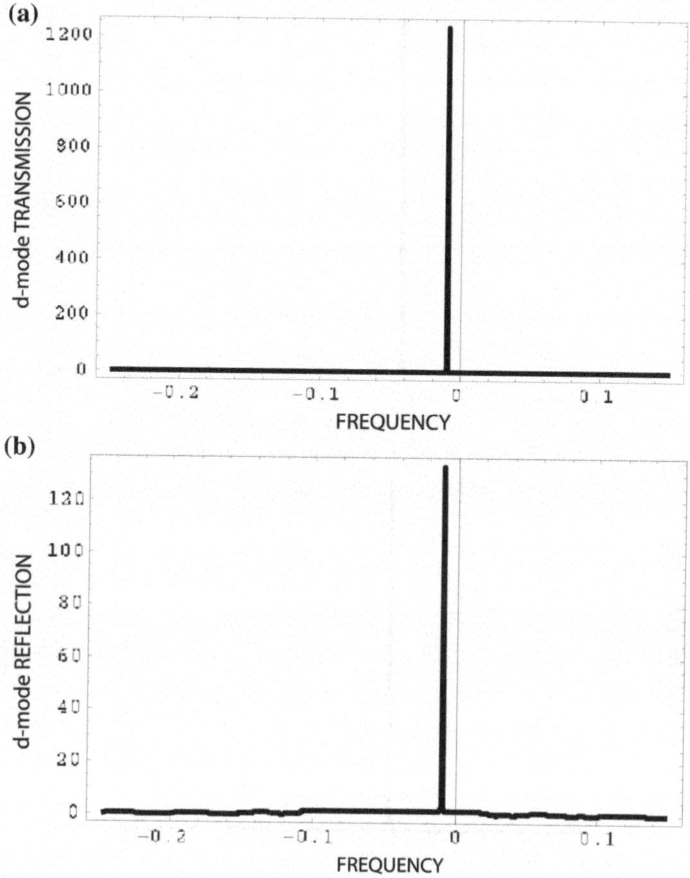

Fig. 6.20 $T(d)$ **a** and $R(d)$ (**b**) for an amplifying CLC versus the dimensionless frequency $\gamma = -0.00117$; $d/p = 0.1$, $\delta = 0.05$, $N = 33$

value of γ increases further, the reflection and transmission coefficients at the DM frequency for the chosen values of the problem parameters exceed 100 (Fig. 6.20) with no signs of noticeable maxima at other frequencies. The corresponding value of γ may be regarded as being close to the threshold value of the gain (γ) for DFB lasing at the DM frequency. With still further increase in the absolute value of γ, diverging maxima of $|R(d, L)|^2$ appear at the edge mode frequencies (with no trace of a maximum at the DM frequency) for a gain five times greater than the threshold gain for the DM (Fig. 6.21). At even greater absolute values of γ, we find that new edge mode frequencies appear, further away from the stop-band edge [26, 28]. The observed result shows that the DM lasing threshold gain is lower than the corresponding threshold for the EM. The study also reveals the existence of some interconnection between the LC parameters at the lasing threshold which, for thick CLC layers, was found analytically for DMs (see 6.27) and for edge modes (see [22,

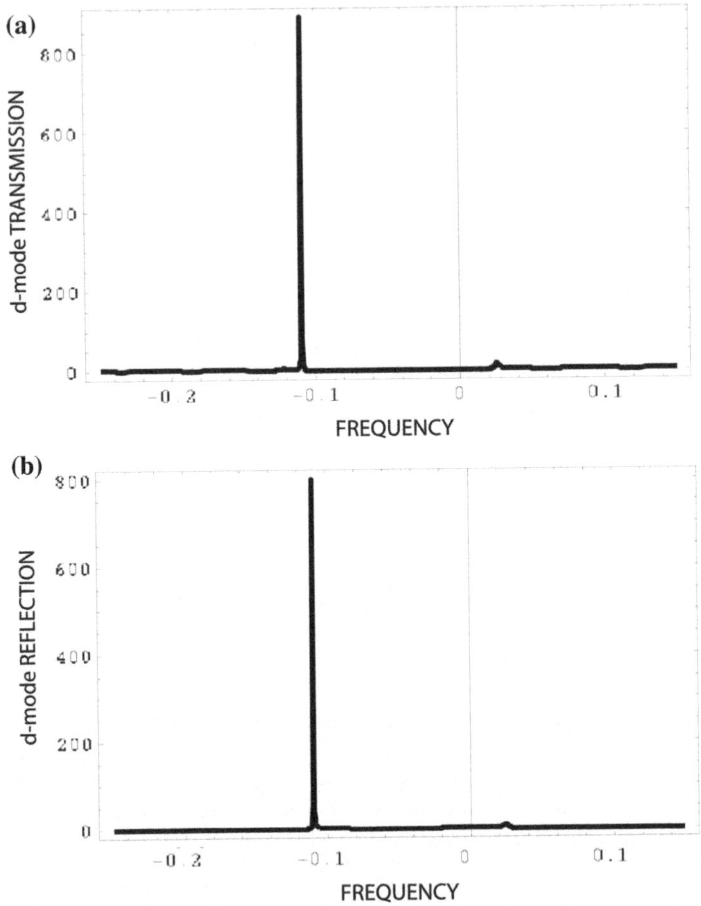

Fig. 6.21 $T(d)$ (**a**) and $R(d)$ (**b**) for an amplifying CLC versus the dimensionless frequency $\gamma = -0.0045$; $d/p = 0.1$, $\delta = 0.05$, $N = 33$

24] and Chap. 5). In fact, a continuous increase in the gain results in the appearance of lasing at new modes, with the disappearance of lasing at the previous modes corresponding to lower thresholds (this was observed experimentally in [3]).

To find the interconnection between the LC parameters at the lasing threshold mention above, we have to solve the DM dispersion equation following from (6.11) under the assumption that the CLC layers are amplifying ($\gamma < 0$). In the general case, this should be done numerically. But in the case of thick CLC layers ($|q|L \gg 1$), the dependence of the threshold γ on L and other parameters can be found analytically. For example, if the DM frequency ω_D is located at the stop-band center, the corresponding relation for the threshold gain (γ) is

$$Y = -(4/3\pi)(p/\delta\, L)\exp[-2\pi\, \delta(L/p)]. \tag{6.27}$$

The exponentially small value of $|\gamma|$ following from (6.27) for thick CLC layers confirms the above statement about the lower lasing threshold for the DM compared to the EM. In fact, the lasing threshold for EM for thick CLC layers does not decrease exponentially with L, but is inversely proportional to only the third power of L.

6.13 Calculation Results

The above plots obtained analytically in limiting cases may be compared with numerical calculations performed for the problem parameters corresponding to their typical values in experiment. Figure 6.22 presents the calculated values of the DM lifetime as a function of the defect layer thickness (d/p) at a fixed CLC layer thickness L. Figure 6.23 presents the calculated values of the lasing threshold $|\gamma|$ as a function of the defect layer thickness (d/p) at a fixed CLC layer thickness L. Figure 6.24 presents the calculated values of the lasing threshold $|\gamma|$ as a function of the CLC layer thickness L. Within the range of applicability of the analytical approach, the analytical and calculated values are in a good agreement. In particular, Fig. 6.24 reveals that the threshold ($|\gamma|$) decreases exponentially with increasing CLC layer thickness L for thick layers in the middle of the stop band ($d/p = 0.25$), but at the same thicknesses L close to the stop-band edge ($d/p = 0.1$), where the limit of thick layers is inapplicable, there are deviations from this exponential decrease.

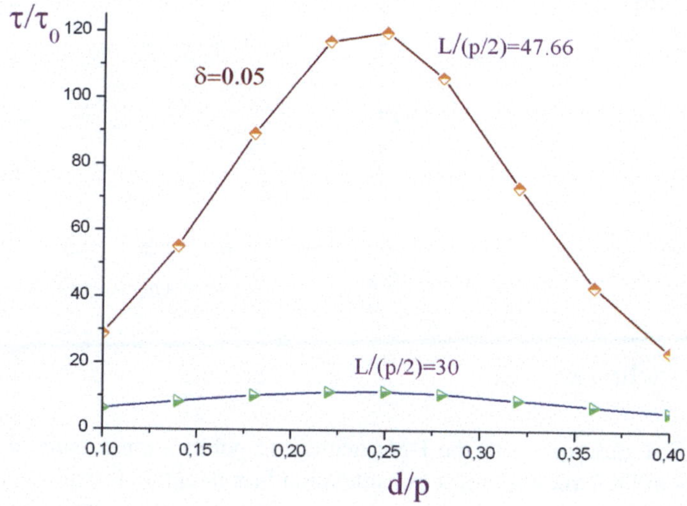

Fig. 6.22 Dependence of the DM lifetime (normalized by the time of flight of light $\tau_0 = 2L\varepsilon_0^{1/2}/c$ through the DMS) on the defect layer thickness found numerically for two values of the CLC layer thickness L

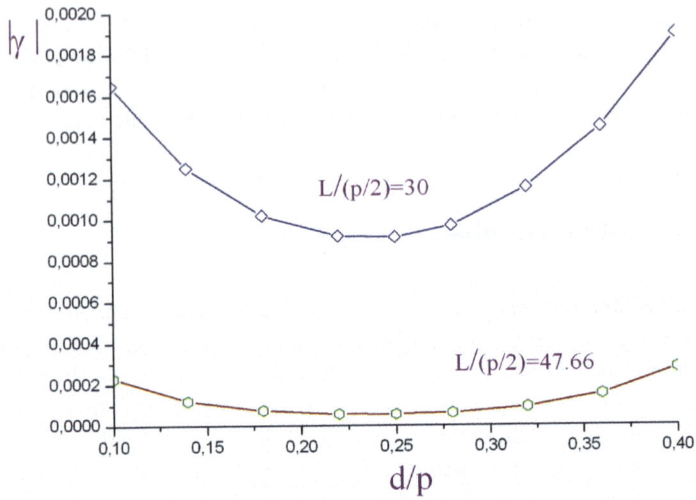

Fig. 6.23 Lasing threshold at the DM frequency versus the defect layer thickness found numerically for two values of the CLC layer thickness L ($\delta = 0.05$)

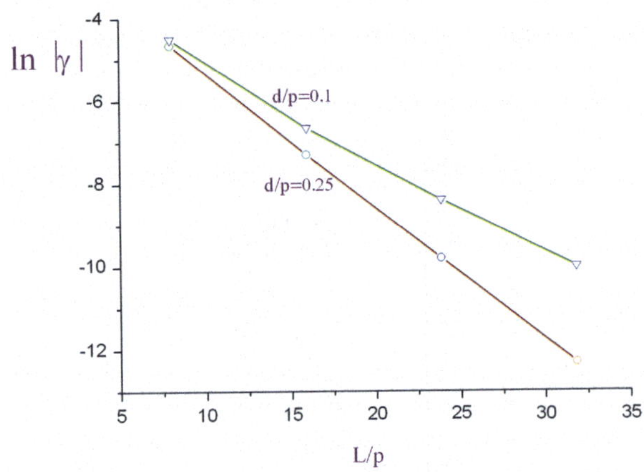

Fig. 6.24 Lasing threshold at the DM frequency versus the CLC layer thickness found numerically for two values of the defect layer thickness L ($\delta = 0.05$)

6.14 Conclusion

Our analytical description of the DMs neglecting polarization mixing at the CLC boundaries in the structure under consideration brings out a clear physical picture of these modes, which is applicable to DMs in general. For example, a lower lasing threshold and stronger absorption (under the conditions of the anomalously strong absorption effect) at the DM frequency compared to the EM frequencies are features of any periodic media. We note that experimental studies of the lasing threshold [3] agree with the corresponding theoretical result obtained above. Moreover, experiment

[3] also confirms the existence of some interconnection between the gain and other LC parameters at the threshold pumping energy for lasing at the defect (as well at the EM) mode frequency. Specifically, this was demonstrated by the observation that an increase in the pumping energy above the threshold value results in a decrease in the lasing intensity (see Fig. 5 in [3]).

For a particular choice of the parameters in the experiment, the resulting formulas may be directly applied to the experiment. Nevertheless, it should be kept in mind that direct comparison of theory and experiment requires some conditions to be met. In the general case, one has take into account a mutual transformation at the boundaries of the two circular polarizations of opposite sense. For example, the circular polarization sense of the wave emitted from the defect structure above the lasing threshold, as observed in the experiment [3], may be opposite to the polarization sense responsible for the existence of the defect mode. The obvious explanation of the "lasing" for the opposite (non-diffracting) circular polarization is as follows. Due to the polarization conversion of the generated wave into a wave of opposite circular polarization, the converted wave with non-diffracting polarization freely escapes from the structure. This polarization conversion phenomenon also contributes to the frequency width of the defect mode. This means that polarization mixing must also be taken into account in calculations of the DM lifetime (frequency width). In the general case, the defect mode field leakage from the structure is also determined by the finite CLC layer thickness, and hence by the leakage due to polarization conversion. Only for sufficiently thin CLC layers or in the case where the DM frequency is very close to the stop-band frequency edges, the main contribution to the frequency width of the DM is determined by the thickness effect, and the model developed above may be directly applied to the description of experimental data.

There is an opportunity to obtain results in the present model which can simulate the polarization conversion and may be quantitatively applicable to the description of experiments. Because the polarization conversion at the CLC surfaces is of the order of the dielectric anisotropy δ (in the absence of reflection at a dielectric boundary) [24, 25], the analysis of the problem in the framework of the present model and its results will correspond to a real situation for very thick CLC layers if the CLC layer thickness L is accepted to be less than the actual CLC layer thickness, and will correspond to the transmission coefficient of the CLC layer inside the stop band being approximately equal to the CLC anisotropy δ. More generally, for a real structure with layer thickness L, a structure with CLC layer thickness less than L has to be considered in our approach to simulation of the polarization conversion. It should be noted that the applied analytical approach helped to reveal the "anomalously strong absorption effect" at the DM frequency. The corresponding "observation" would be much more difficult to do in a purely numerical approach.

The defect type considered above is a homogeneous layer. The approach developed here is also applicable to a defect of "phase jump" type [2, 3, 15, 16], and the corresponding results are practically the same as above. In particular, the equations related to the case of a "phase jump" defect can be obtained from the equations presented above by substituting the quantity $2\Delta\varphi$ for $2kd$ in the factor $\exp(2ikd)$, where $\Delta\varphi$ is the spiral phase jump at the defect plane. Based on the present results, we can already point out one difference between the two types of defects: for a "phase jump" defect, in contrast to the considered homogeneous layer defect, only one defect mode frequency is possible inside the stop-band frequency range, because the value of phase jump is limited by the condition $\Delta\varphi| \leq 2\pi$. In particular, the DM frequency for a "phase jump" defect is located at the stop-band center at $|\Delta\varphi| = \pi/2$.

We also note that localized DMs (as well as EMs) reveal themselves in an enhancement of some inelastic and nonlinear optical processes in photonic liquid crystals. For example, we mention the experimentally observed effects of the enhancement of nonlinear optical second harmonic generation [29] and lowering of the lasing threshold [30] in photonic liquid crystals, along with the theoretically predicted enhancement of Cherenkov radiation (see Sect. 4 in [24] and Chap. 5 in [25]).

To conclude this chapter, we would like to point out that the results obtained here for defect modes (see also [26, 28, 31] for the edge modes) clarify the physics of these modes and entirely agree with the corresponding results from previous investigations performed using a numerical approach [32]. The results presented are qualitatively applicable to the corresponding localized electromagnetic modes in any periodic media and may be regarded as a useful guide in studies of localized modes in general.

References

1. Y.-C. Yang, C.-S. Kee, J.-E. Kim et al., Phys. Rev. E, **60**, 6852 (1999)
2. V.I. Kopp, A.Z. Genack, Phys. Rev. Lett. **89**, 033901 (2003)
3. J. Schmidtke, W. Stille, H. Finkelmann, Phys. Rev. Lett. **90**, 083902 (2003)
4. P.V. Shibaev, V.I. Kopp, A.Z. Genack, J. Phys. Chem. B **107**, 6961 (2003)
5. E. Yablonovi tch, T.J. Gmitter, R.D. Meade et all., Phys. Rev. Lett. **67**, 3380 (1991)
6. I.J. Hodgkinson, Q.H. Wu, K.E. Torn et all., Opt. Commun. **184**(57) (2003)
7. H. Hoshi, K. Ishikava, H. Takezoe, Phys. Rev. E **68**, 020701(R) (2003)
8. V.F. Shabanov, S.Ya. Vetrov, A.V. Shabanov, Optics of Real Photonic Crystals, RAS, Sibirian Branch, 2005 (in Russian)
9. V.A. Belyakov, Localized optical modes in optics of chiral liquid crystals, in *New Developments in Liquid Crystals and Applications*, Chap. 7, ed. by P.K. Choudhury (Nova Publishers, New York, 2013) pp. 199–227
10. V.A. Belyakov, *Optics of Photonic Crystals*. (MIPT Publishing House, 2013) (in Russian)
11. A.V. Shabanov, S.Ya. Vetrov, A.Yu. Karneev, JETP Lett. **80**(#3), 181 (2004))
12. I.J. Hodkinson, Q.H. Wu, K.E. Thorn, A. Lakhtakia, M.W. McCall, Optics Communications **184**, 57 (2000)
13. F. Wang, A. Lakhtakia, Opt. Express **13**(n19), 7319 (2005)
14. M.H. Song, N.Y. Ha, K. Amemiya, B. Park, Y. Takanishi, K. Ishikaw, J.W. Wu, S. Nishimura, T. Toyooka, H. Takezoe, Adv. Mater. **18**, 193 (2006)
15. H. Yoshida, C.H. Lee, A. Fuji, M. Ozaki, Appl. Phys. Lett. **89**, 231913 (2006)
16. H. Yoshida, R. Ozaki, K. Yoshino, M. Ozaki, Thin Solid Films **509**, 197 (2006)
17. A.H. Gevorgyan, M.Z. Haratyunyan, Phys. Rev. **E76**, 031701 (2007)
18. M. Becchi, S. Ponti, J.A. Reyes, C. Oldano, Phys. Rev. B **70**, 033103 (2004)
19. J. Schmidtke, W. Stille, Eur. Phys. J. **90**, 353 (2003)
20. H. Vries, Acta Crystallogr **4**, 219 (1951)
21. E.I. Kats, Sov. Phys., *JETP* **32**, 1004 (1971)

22. V.A. Belyakov, S.V. Semenov, Mol. Cryst. Liq. Cryst. **543**, 123 (2011)
23. P.G. de Gennes, J. Prost, *The Physics of Liquid Crystals* (Clarendon Press, Oxford, 1993)
24. V.A. Belyakov, V.E. Dmitrienko, Optics of Chiral Liquid Crystals, p. 54 in *Soviet Scientific reviews/Section A, Physics Reviews* ed. by I.M. Khalatnikov, Harwood Academic Publisher), v. 13, p. 1–203 (1989)
25. V.A. Belyakov, *Diffraction Optics of Complex Structured Periodic Media* (Springer, New York, 1992). Chap. 4
26. V.A. Belyakov, S.V. Semenov, JETP, **109**, 687 (2009); Mol. Cryst. Liq. Cryst., **507**, 209 (2009).
27. V.A. Belyakov, A.A. Gevorgian, O.S. Eritsian and N.V. Shipov, *Zhurn.Tekhn.Fiz.,* **57**, 1418 (1987) [*Sov. Phys. Technical Physics*, **32** (n7), 843 (1987), English translation]; *Sov. Phys. Crystalography*. **33** (n3), 337 (1988)
28. V.A. Belyakov, Mol. Cryst. Liq. Cryst., **453**, 43 (2006); Ferroelectrics, **344**, 163 (2006)
29. K. Shin, H. Hoshi, D. Chang, K. Ishikawa, H. Takezoe, Optics Lett. **27**, 128 (2002)
30. Y. Matsuhisa, Y. Huang, Y. Zhou et al., Appl. Phys. Lett. **90**, 091114 (2007)
31. V.A. Belyakov, Mol. Cryst. Liq. Cryst. **494**, 127 (2008)
32. V.I. Kopp, Z.-Q. Zhang, A.Z. Genack, Prog. Quant. Electron. **27**(6), 369 (2003)

Chapter 7
Optical Defect Modes at Active Defect Layer in Photonic Liquid Crystals

7.1 Introduction

The field of mirrorless distributed feedback (DFB) lasing in photonic structures consisting of many layers of chiral liquid crystals has recently attracted much attention, mainly due to the possibilities of reaching a low lasing threshold for DFB lasing [1–8]. For definiteness, we study photonic liquid crystals with the example of the best-known type of photonic liquid crystals, i.e., cholesteric liquid crystals (CLC). The related theory is mainly based on numerical calculations [9] whose results are not always interpreted in the framework of a clear physical picture. Several recent papers [10–17] have shown that an analytic theoretical approach to the problem (sometimes limited by the introduction of approximations) can be used to build a clear physical picture of the linear optics and lasing in the relevant structures. In particular, the physics and the role of localized optical modes (edge and defect modes) in the structures under consideration was clearly demonstrated. The most promising results in DFB lasing relate to defect modes (DM) [12–14]. The defect modes existing at a structure defect as a localized electromagnetic eigenstate with its frequency in the forbidden band gap were investigated initially in the three-dimensional periodic dielectric structures [18]. The corresponding defect modes in chiral liquid crystals, and more generally, in spiral media, are very similar to the defect modes in one-dimensional scalar periodic structures. They reveal abnormal reflection and transmission inside the forbidden band gap [1, 2] and allow DFB lasing at a low lasing threshold [3]. The qualitative difference from the case of scalar periodic media consists in the polarization properties. The defect mode in chiral liquid crystals is associated with the circular polarization of an electromagnetic field eigenstate whose chirality sense coincides with that of the chiral liquid crystal helix. Two main types of defects have been studied in chiral liquid crystals up to now. One is a plane layer

V. Belyakov, *Diffraction Optics of Complex-Structured Periodic Media*, Springer Series in Optical Sciences 203,
https://doi.org/10.1007/978-3-319-43482-7_7

of some substance, differing from the CLC, dividing a perfect cholesteric structure into two parts, and lying perpendicular to the helical axis of the cholesteric structure [1]. The other defect type is a jump in the cholesteric helix phase at some plane perpendicular to the helical axis (without insertion of any substance at the location of this plane) [2]. Recently, many new types of defect layers have been studied [19–26], such as a CLC layer whose pitch differs from the pitch of two layers sandwiched between these layers [8]. Clearly, there are many versions of the dielectric properties of the defect layer, but considerations below are limited to the first type of defect, a layer inserted in a chiral liquid crystal. Our focus is on the active defect layers (absorbing, amplifying, or changing the light polarization). The reason for this is connected with both experimental research on DFB lasing in CLCs where dyes are placed in a defect layer [26] and the general idea that the unusual properties of DMs manifest themselves most clearly just at the middle of the defect structure, i.e., at the defect layer, where the DM field intensity reaches its maximum. We therefore assume that there is no absorption in the CLC layers of the DMS, and that absorption, amplification, or changes in the light polarization occur only in the defect layer. The analytic approach to studying a DMS with an active defect layer is very similar to the DM studies discussed in Chap. 5 [12, 13], and we therefore present below the final results of the present investigation, referring the reader to [12, 13] for full details of the investigation.

In the present chapter, we discuss an analytical solution of the defect mode associated with insertion of an active defect layer in the perfect cholesteric structure, for light propagating along the helical axis, and we consider some limiting cases simplifying the problem [27].

7.2 Defect Mode at Amplifying (Absorbing) Defect Layer

To consider the defect mode associated with the insertion of an isotropic layer in the perfect cholesteric structure, we have to solve Maxwell equations and a boundary problem for an electromagnetic wave propagating along the cholesteric helix for the layered structure depicted in Fig. 6.1. This investigation was performed in [12, 13] under the assumption that the CLC layers can be absorbing or amplifying in Fig. 6.1. It is possible to use the results of [12, 13] in the present case of an amplifying (absorbing) isotropic defect layer and non-absorbing CLC layers, introducing only some physically clear changes in the formulas obtained in [12, 13]. In this section, we retain the assumptions in [12, 13] that the average dielectric constant ε_0 of the CLC coincides with the dielectric constant of the defect layer and the external medium, so that polarization conversion is absent at the interfaces and only light with the diffracting circular polarization has to be taken into account. The main notations of the papers [12, 13] are also maintained in this chapter. As is well known [9], much information about the DM is available from the spectral properties of the DMS transmission, $T(d, L)$, and reflection, $R(d, L)$, coefficients.

Formulas for the optical properties of the structure depicted in Fig. 6.1 can be obtained using the expressions for the amplitude transmission, $T(L)$, and reflection, $R(L)$, coefficients for a single cholesteric layer (see Chap. 1 [28, 29]). The transmission, $|T(d, L)|^2$, and reflection, $|R(d, L)|^2$, intensity coefficients (of light with diffracting circular polarization) for the whole structure may be presented in the form

$$|T(d, L)|^2 = |[T_e T_d \exp(ikd(1+ig))]/[1-\exp(2ikd(1+ig))R_d R_u]|^2, \quad (7.1)$$

$$|R(d, L)|^2 = \{R_e + R_u T_e T_u \exp(2ikd(1+ig))/[1-\exp(2ikd(1+ig))R_d R_u]\}|^2, \quad (7.2)$$

where $R_e(T_e)$, $R_u(T_u)$, and $R_d(T_d)$ are the amplitudes of the reflection (transmission) coefficients of the individual CLC layers (see Fig. 6.1) for light incident on the outer top layer surface, on the inner top CLC layer surface from the inserted defect layer, and on the inner bottom CLC layer surface from the inserted defect layer, respectively. It is assumed in deriving (7.1) and (7.2) that the external beam is only incident on the structure (Fig. 6.1) from above. The factor $(1 + ig)$ is related to the defect layer alone and corresponds to the dielectric constant of the defect layer having the form $\varepsilon_0(1 + 2ig)$, with a small g that is positive for an absorbing defect layer and negative for an amplifying one.

The expressions for the amplitude transmission, $T(L)$, and reflection, $R(L)$, coefficients for a single non-absorbing cholesterol layer of thickness L for light with the diffracting circular polarization are given by (6.1) (see also [28–32]).

The defect mode frequency ω_D is determined by the dispersion equation [compare with (6.18)]:

$$\begin{aligned}\{&\exp(2ikd(1+ig))\sin^2 qL - \exp(-i\tau L)[(\tau q/\kappa^2)\cos qL \\ &+ i((\tau/2\kappa)^2 + (q/\kappa)^2 - 1)\sin qL]^2/\delta^2]\} = 0 \end{aligned} \quad (7.3)$$

For CLC layers of finite thickness L, the DM frequency ω_D is a complex quantity which may be found by solving (7.3) numerically. For very small values of the parameter g, the reflection and transmission spectra of the DMS with an active defect layer are similar to the spectra studied in Chap. 6 [12, 13] (see Fig. 7.1). In particular, the frequency positions of dips in reflection and spikes in transmission inside the stop band just correspond to $\mathrm{Re}[\omega_D]$, and this observation is very useful for solving the dispersion equation numerically. For its part, the DM lifetime is shorter for absorbing defect layers than for a non-absorbing defect layer [12, 13].

7.2.1 *Absorbing Defect Layer*

As in the case investigated in Chap. 6 [12, 13], i.e., DMs with non-absorbing CLC layers, the effect of anomalously strong absorption also occurs in DMSs with absorbing

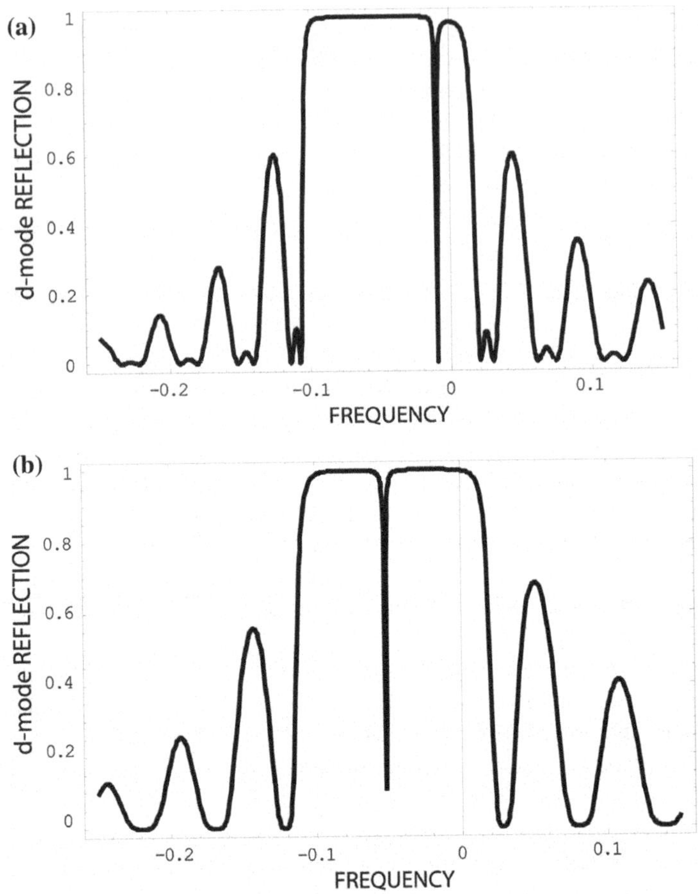

Fig. 7.1 Reflection $|R(d, L)|^2$ versus frequency for a non-absorbing defect and CLC layers ($g = 0$) at $d/p = 0.1$ (**a**) and $d/p = 0.25$ (**b**); $\delta = 0.05$, $l = 200$, $l = L\tau = 2\pi N$, where N is the director half-turn number at the CLC layer thickness L. Here and in all further figures in this chapter, the frequency is defined in the same way as in Chaps. 5 and 6 (see Fig. 5.2). Furthermore, $\delta = 0.05$ and the director half-turn number at the CLC layer thickness is $N = 33$

defect layers. The effect reveals itself at the DM frequency and reaches its maximum, viz. $1 - |T(d, L)|^2 - |R(d, L)|^2$, for a certain value of g that can be found using the expressions (7.1), (7.2) for $|T(d, L)|^2$ and $|R(d, L)|^2$. Figure 7.2 demonstrates the existence of the anomalously strong absorption effect at the DM frequency, where it can be seen that the maximum value of the anomalous absorption [28, 33] $(1 - |T(d, L)|^2 - |R(d, L)|^2)$ at two differing values of d/p are reached for $g = 0.04978$ and $g = 0.00008891$. (Taken with the opposite sign of g, these are the approximate values of the lasing threshold gain for the same DMSs found in the next section.)

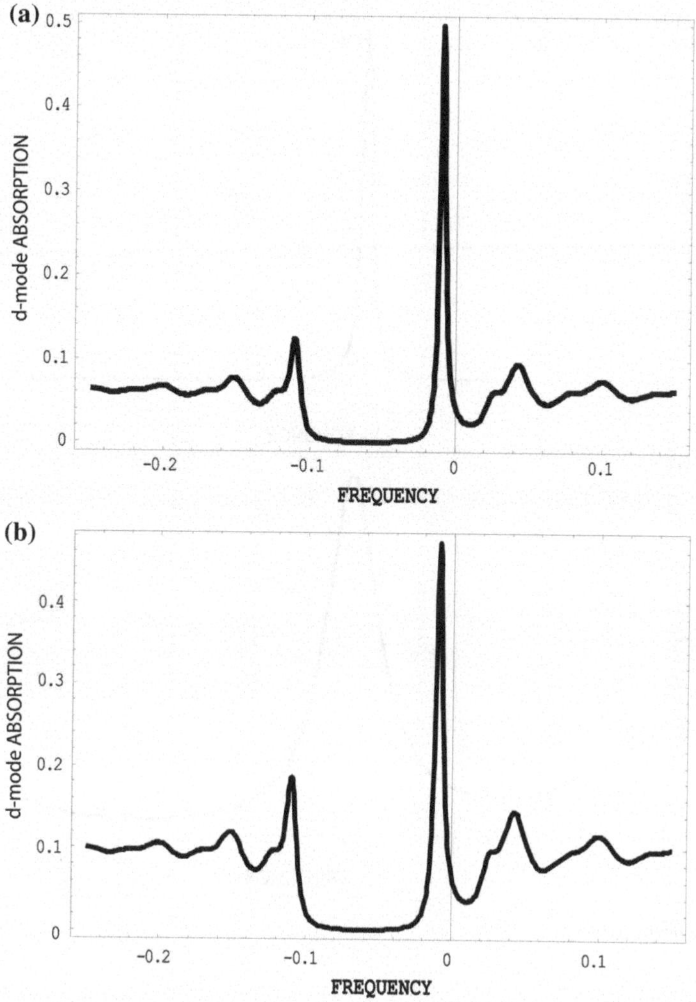

Fig. 7.2 Total absorption $(1 - |T(d, L)|^2 - |R(d, L)|^2)$ versus frequency for an absorbing defect layer and non-absorbing CLC layers at $g = 0.04978$ (**a**) and $g = 0.08$ (**b**) for $d/p = 0.1$; at $g = 0.00008891$ (**c**) and at $g = 0.0008891$ (**d**) for $d/p = 22.25$

In the case of thick CLC layers ($|q|L \gg 1$) in the DMS, the g value ensuring maximum absorption can be found analytically:

$$g_t = (L/d)|[2\kappa^2/(q^2 L\tau)]\exp[-2|q|L]\{1 + \{1/(2[(\tau/\kappa)^2 + \delta^2]^{1/2}) - (\tau/2\kappa)^2\}/(1 - [(\tau/\kappa)^2 + \delta^2]^{1/2} + (\tau/2\kappa)^2]\}^{-1}| \quad (7.4)$$

For the defect mode frequency ω_D in the middle of the stop band, the maximal absorption corresponds to

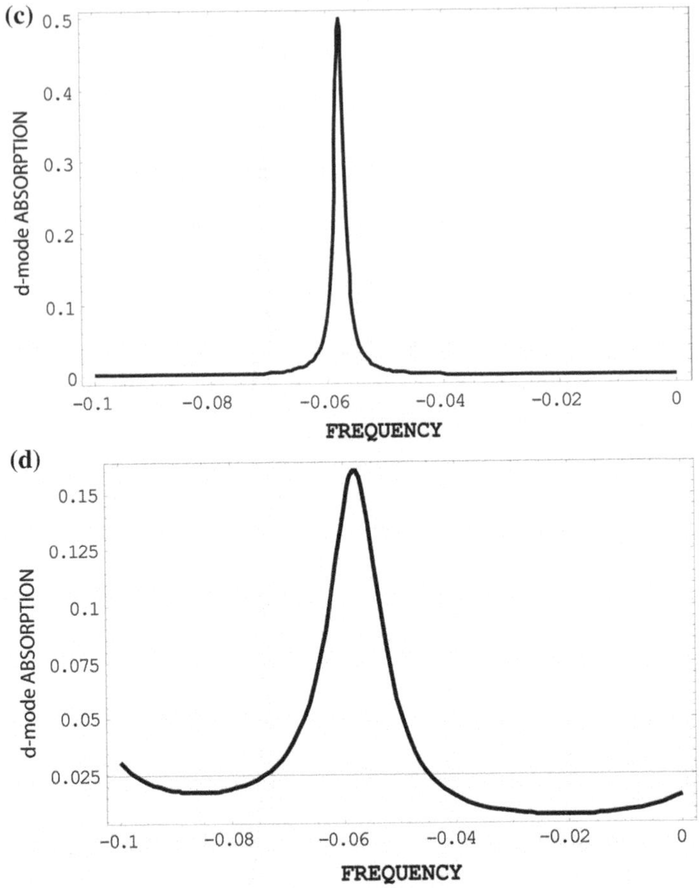

Fig. 7.2 (continued)

$$g_t = (2/3\pi)(p/\delta d)\exp[-2\pi\delta(L/p)]. \tag{7.5}$$

As the calculations and the formulas (7.4), (7.5) show, the gain g corresponding to the maximal absorption is approximately inversely proportional to the defect layer thickness d.

7.2.2 Amplifying Defect Layer

In the case of DMSs with an amplifying defect layer ($g < 0$), the reflection and transmission coefficients diverge at some value of $|g|$. The corresponding values of g are the gain lasing thresholds. They can be found by solving the dispersion Equation (7.3)

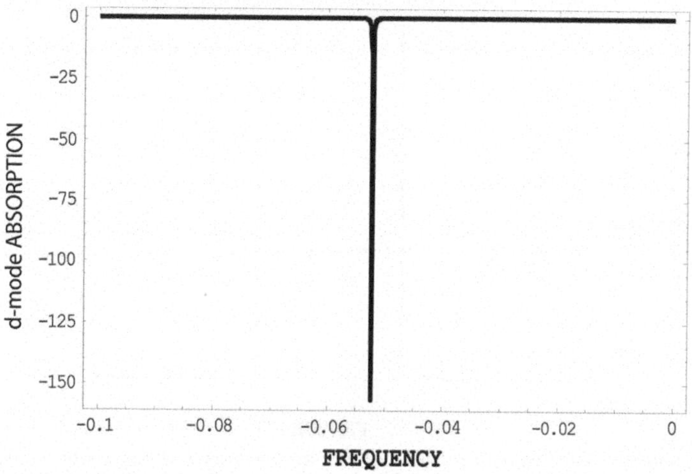

Fig. 7.3 Total absorption $(1 - |T(d, L)|^2 - |R(d, L)|^2)$ versus frequency for an amplifying defect layer and non-absorbing CLC layers at $g = -0.0065957$ for $d/p = 0.25$

for g or numerically using the expressions (7.1), (7.2) for $|T(d, L)|^2$ and $|R(d, L)|^2$, or they can be found approximately by plotting $|T(d, L)|^2$ and $|R(d, L)|^2$ as functions of g. The third option is illustrated in Figs. 7.3, 7.4, and 7.5 where "almost divergent" values of $|T(d, L)|^2$, $|R(d, L)|^2$, or the absorption $(1 - |T(d, L)|^2 - |R(d, L)|^2)$ are shown. The values of g used in Figs. 7.3, 7.4, and 7.5 are close to the threshold values ensuring divergence of $|T(d, L)|^2$ and $|R(d, L)|^2$. The calculation results show that the minimal threshold $|g|$ occurs when ω_D lies precisely in the middle of the stop band and $|g|$ is almost inversely proportional to the defect layer thickness. Figures 7.3 and 7.4 actually correspond to the situation where the defect mode frequency ω_D is located close to the midpoint of the stop band, and show that there is a decrease in the lasing threshold gain with increasing thickness of the defect layer. Figure 7.6 corresponds to the situation where the defect mode frequency ω_D is located close to the stop-band edge and shows that there is an increase in the lasing threshold gain as the defect mode frequency ω_D approaches the stop-band edge.

The analytic approach for thick CLC layers ($|q|L \gg 1$) results in similar predictions, namely the gain threshold value is given by (7.4) "with a negative sign on the right-hand side of this expression." For thick CLC layers with ω_D in the middle of the stop band, the threshold gain is given by the expression

$$g_t = -(2/3\pi)(p/\delta d) \exp[-2\pi\delta(L/p)]. \tag{7.6}$$

Hence, as can be seen from (7.6), the thinner the amplifying defect layer, the higher the threshold gain g.

As mentioned above, the same result is also valid for the absorption enhancement (formulas (7.4) and (7.5)). The thinner the absorbing defect layer, the higher the g value ensuring maximal absorption.

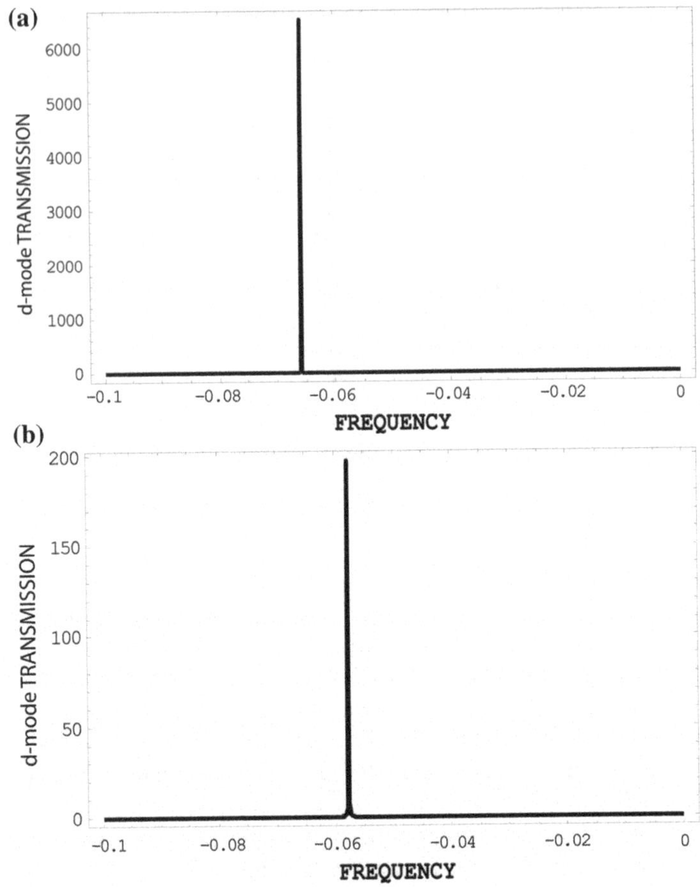

Fig. 7.4 Transmission $|T(d, L)|^2$ versus frequency for an amplifying defect layer and non-absorbing CLC layers at $g = -0.001000$ for $d/p = 2.25$ (**a**); at $g = -0.00008891$ for $d/p = 22.25$ (**b**)

An important result relating to DFB lasing in the DMS with amplifying (absorbing) defect layer can be formulated as follows. The lasing threshold gain in a defect layer decreases as the amplifying layer thickness increases, being almost inversely proportional to the thickness. A similar result holds for the anomalously strong absorption phenomenon, where the value of g in the defect layer ensuring maximal absorption is almost inversely proportional to the defect layer thickness. We note that this decrease in the lasing threshold gain with increasing thickness of the amplifying defect layer cannot be regarded directly as the corresponding reduction in the lasing energy threshold of the pumping wave pulse. The situation depends on the specifics of the pumping arrangement. This question requires separate and more thorough consideration. For example, if we assume that the pumping is arranged such that the product of the gain g and the defect layer thickness d is proportional to the pumping

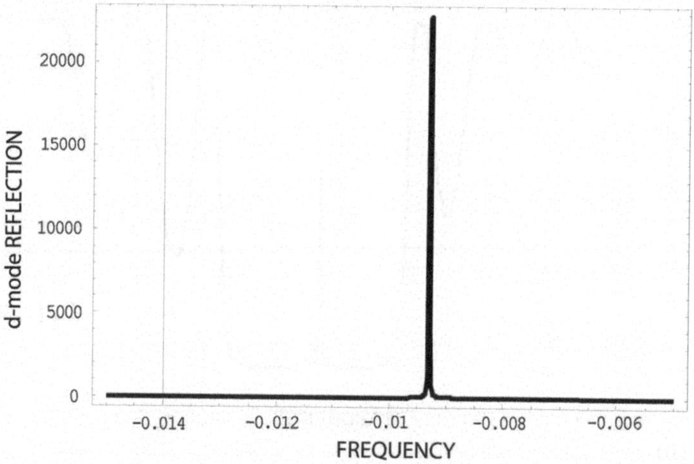

Fig. 7.5 Reflection $|R(d, L)|^2$ versus frequency for an amplifying defect layer and non-absorbing CLC layers at $g = -0.04978$ for $d/p = 0.1$

pulse energy, then the threshold pumping pulse energy is almost independent of the defect layer thickness because of the almost inverse proportionality of the threshold gain to the defect layer thickness found above.

7.3 Defect Mode at Birefringent Defect Layer

In this section, we focus on a birefringent defect layer and, in particular, the case of low birefringence. As mentioned above, the reason for this is connected with both experimental research on DFB lasing in CLCs where the defect layer is birefringent [26] and the general idea that the unusual properties of DMs manifest themselves most clearly just at the middle of the defect structure, i.e., at the defect layer, where the DM field intensity reaches its maximum. We also assume at the outset that there is no absorption in the CLC and the birefringent defect layer. The analytic approach used to study a DMS with a birefringent defect layer is very similar to the DM studies carried out for an isotropic defect layer in Chap. 6 [12, 13], and we therefore present the final results, referring the reader to Chap. 6 for the details.

7.3.1 Non-absorbing CLC Layers

In this section, we present an analytic solution for the DM associated with the insertion of a birefringent defect layer in the perfect cholesteric structure for light propagating along the helical axes and we consider some limiting cases simplifying the

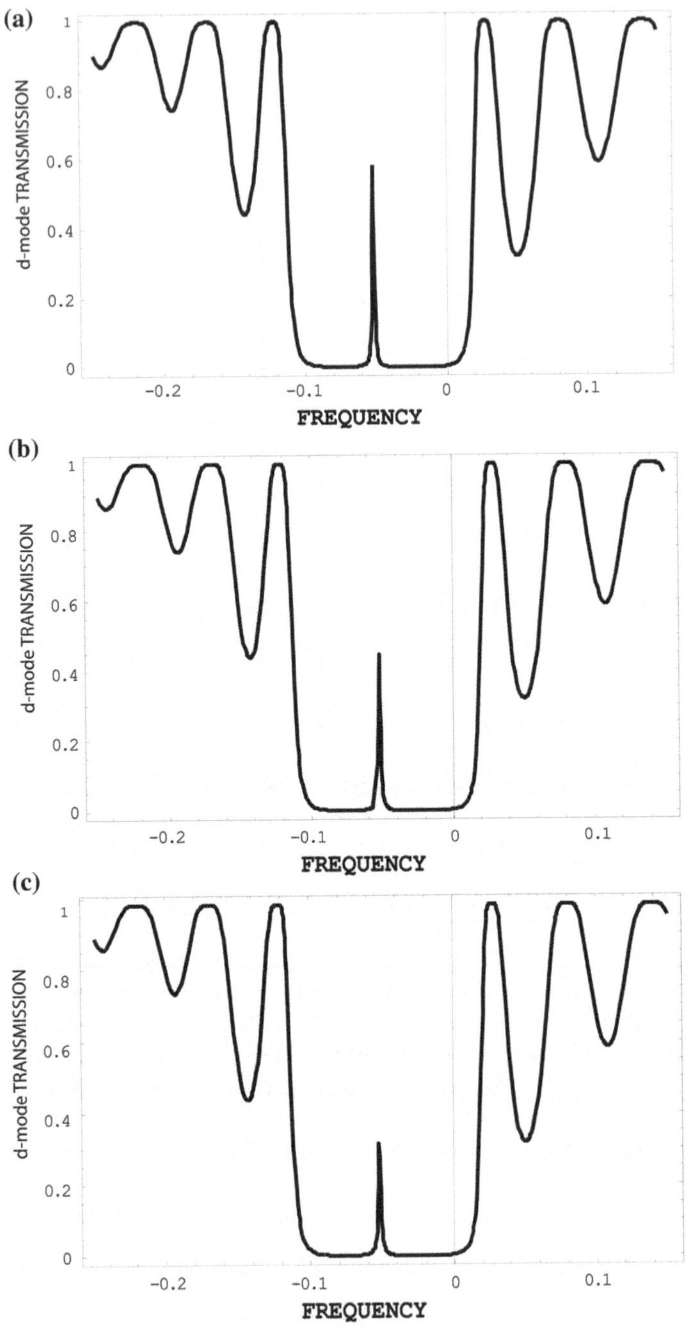

Fig. 7.6 Calculated diffracting polarization intensity transmission coefficient $|T(d, L)|^2$ for a low birefringent defect layer versus frequency for diffracting incident polarization and for a birefringent phase shift at the defect layer thickness equal to $\Delta\varphi = \pi/20$ (**a**), $\pi/16$ (**b**), $\pi/12$ (**c**), $\pi/8$ (**d**), $\pi/6$ (**e**), $\pi/4$ (**f**), $\pi/2$ (**g**), and $\Delta\varphi = 0$ (**h**) [figure (**h**) corresponds to the isotropic defect layer] for a non-absorbing CLC with $d/p = 0.25$

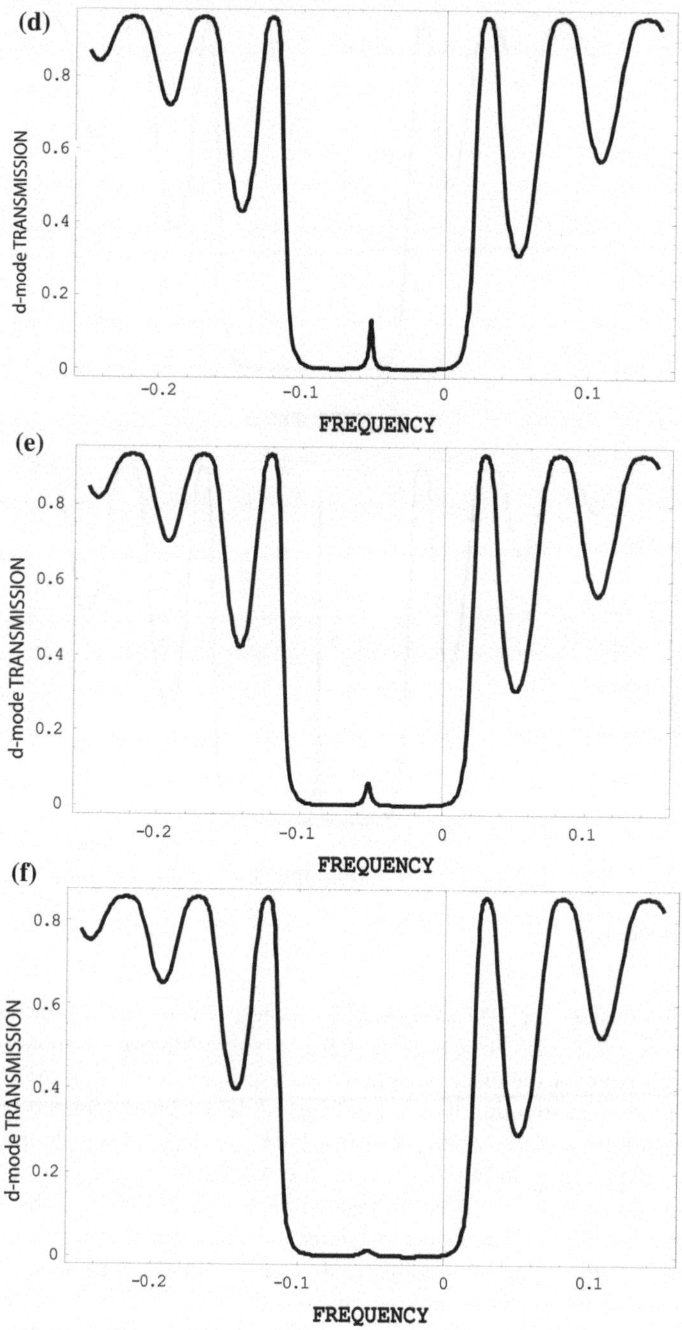

Fig. 7.6 (continued)

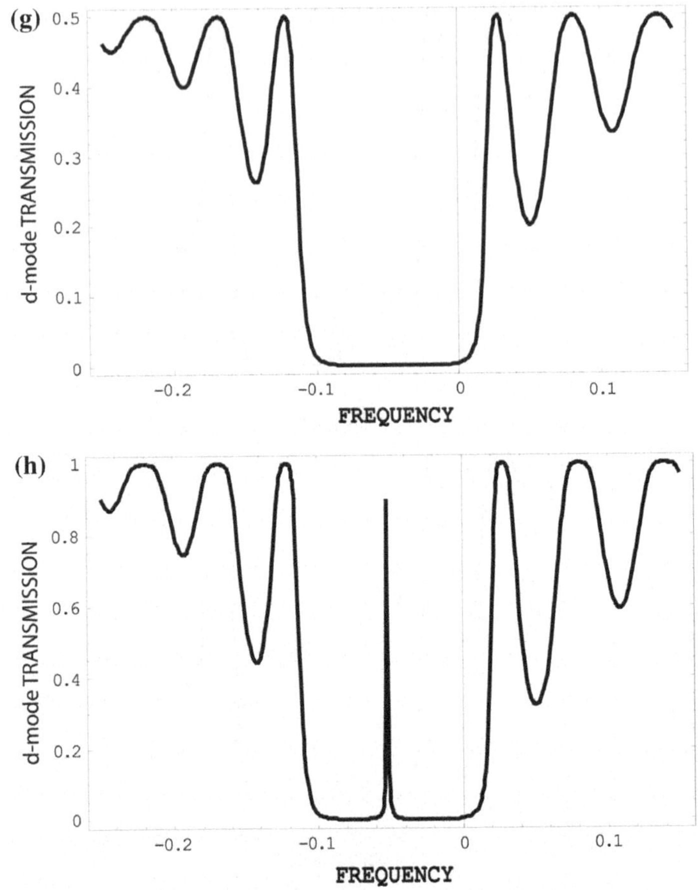

Fig. 7.6 (continued)

problem. To consider the DM associated with the insertion of a birefringent layer in the perfect cholesteric structure, we have to solve Maxwell's equations and a boundary problem for the electromagnetic wave propagating along the cholesteric helix for the layered structure with a birefringent defect layer depicted in Fig. 6.1. This investigation has already been performed in Chap. 6 [12, 13] under the assumption that the defect layer in Fig. 6.1 is isotropic. We can therefore use the results of [12, 13] for the present case of a birefringent defect layer and non-absorbing and amplifying (absorbing) CLC layers (keeping the main notation of Chap. 6 here), but introducing just a few physically clear changes in the formulas obtained in Chap. 6. We can no longer assume, as we did in [12, 13], that polarization conversion is absent and only light with the diffracting circular polarization need be taken into account (due to the assumption that the average CLC dielectric constant ε_0 coincides with the dielectric constant of the defect layer and the external medium).

In fact, due to the birefringence of the defect layer, the light polarization changes in the course of its propagation through the defect layer, from one of its surfaces to the other, and hence, generally speaking, the polarization of the light after crossing the defect layer will differ from the polarization at the first defect layer surface. This is why the polarization component differs from the diffracting polarization, generally speaking, and there is leakage of the correspondingly polarized light from the DMS. The obvious consequence of this leakage is a reduction in the DM lifetime for the case of a birefringent defect layer.

Formulas for the optical properties of the structure with a birefringent defect layer depicted in Fig. 6.1 can be obtained using the expressions for the amplitude transmission, $T(L)$, and reflection, $R(L)$, coefficients for a single cholesteric layer in the presence of dielectric interfaces (see Chap. 3 and [28, 33]). If we neglect multiple scattering of light with the non-diffracting polarization, the transmission, $|T(d, L)|^2$, and reflection, $|R(d, L)|^2$, intensity coefficients (of light with the diffracting circular polarization) for the whole structure may be presented in the following form:

$$|T(d, L)|^2 = |[T_e T_d M(k, d, \Delta n)(\boldsymbol{\sigma}_e \boldsymbol{\sigma}^*_{ed})][1 - M^2(k, d, \Delta n)(\boldsymbol{\sigma}_r \boldsymbol{\sigma}^*_{ed})^2 (R_d R_u]|^2, \tag{7.7}$$

$$|R(d, L)|^2 = |\{R_e + R_d T_e T_u M^2(k, d, \Delta n)|(\boldsymbol{\sigma}_e \boldsymbol{\sigma}_{ed}*)|^2/[1 - M^2(k, d, \Delta n)|(\boldsymbol{\sigma}_r \boldsymbol{\sigma}_{ed}*)^2 R_d R_u]\}|^2, \tag{7.8}$$

where the meaning of $R_e(T_e)$, $R_u(T_u)$, and $R_d(T_d)$ is the same as in (7.1) and (7.2) and $\boldsymbol{\sigma}_e$, $\boldsymbol{\sigma}_r$, and $\boldsymbol{\sigma}_{ed}$ are the polarization vectors of light exiting the CLC layer inner surface, light reflected at the inner bottom CLC layer surface after incidence from the inserted defect layer, and light whose polarization vector $\boldsymbol{\sigma}_{ed}$ transforms to the polarization vector $\boldsymbol{\sigma}_e$ upon crossing the birefringent defect layer of thickness d, respectively. Finally, Δn is the difference between the two refractive indices in the birefringent defect layer and $M(k, d, \Delta n)$ is the phase factor for single propagation of the light through a birefringent defect layer. It is assumed in deriving (7.7)–(7.8) that the external beam is only incident on the structure (Fig. 6.1) from above. In the presence of dielectric interfaces, there is light polarization conversion at the inner surfaces of the CLC layers in the DMS under reflection and transmission of light through a CLC layer, and the light field inside CLC layers cannot just be expressed as a superposition of two diffracting eigenmodes of the CLC (generally speaking, two non-diffracting eigenmodes are also present). The corresponding polarization vector inside the defect layer (after the light has crossed the interface between the CLC and defect layers), denoted by $\boldsymbol{\sigma}_e$, can be found (see [28, 29]), and the polarization vector $\boldsymbol{\sigma}_{ed}$ can be easily calculated if d and Δn are known. "The same can be said about finding the polarization of light exciting only the diffracting eigenmodes in a CLC layer when incident at the external CLC layer surface of the DMS." The corresponding polarization in the presence of dielectric interfaces is referred to as the diffracting polarization here. Polarization orthogonal to the diffracting polarization is referred to as a non-diffracting polarization. When incident on the DMS, light with a non-diffracting polarization excites only non-diffracting CLC eigenmodes in the CLC layers of the DMS. The polarization vectors $\boldsymbol{\sigma}_e$, $\boldsymbol{\sigma}_r$, and $\boldsymbol{\sigma}_{ed}$ can be presented in the form

$$\boldsymbol{\sigma}_{\mathbf{i}} = (\cos\alpha_{\mathrm{i}}\mathbf{e}_x + \mathrm{e}^{i\beta_{\mathrm{i}}}\sin\alpha_{\mathrm{i}}\mathbf{e}_y), \tag{7.9}$$

where $\mathbf{e}_x$ and $\mathbf{e}_y$ are the unit vectors along the x and y axes and α_{i}, β_{i} are the parameters determining the polarization. For example, $\alpha_{\mathrm{i}} = \pi/4$ and $\beta_{\mathrm{i}} = \pi/2(-\pi/2)$ correspond to right and left circular polarizations.

In the general case, for a DMS with a birefringent defect layer, the transmitted and reflected beams do not correspond to the diffracting circular polarization, and therefore there is reflection and transmission of light with the non-diffracting polarization, even for incident light with the diffracting polarization. Neglecting multiple scattering of light with the non-diffracting polarization, we obtain the reflection, $R(d, L)^-$, and transmission, $T(d, L)^-$, coefficients of light with the non-diffracting circular polarization (for incident light with the diffracting circular polarization):

$$\begin{aligned}|T(d,L)^-|^2 = [T_{\mathrm{e}}T_{\mathrm{d}}^-\{M(k,d,\Delta n)(\boldsymbol{\sigma}_{\mathbf{e}}\boldsymbol{\sigma}_{\mathbf{ed}}^{\perp}*) + (\boldsymbol{\sigma}_{\mathbf{r}}\boldsymbol{\sigma}_{\mathbf{ed}}*)(\boldsymbol{\sigma}_{\mathbf{e}}\boldsymbol{\sigma}_{\mathbf{ed}}*)\\ \times(\boldsymbol{\sigma}_{\mathbf{r}}\boldsymbol{\sigma}_{\mathbf{ed}}^{\perp}*)M^2(k,d,\Delta n)/[1 - M^2(k,d,\Delta n)(\boldsymbol{\sigma}_{\mathbf{r}}\boldsymbol{\sigma}_{\mathbf{ed}}*)^2(R_{\mathrm{d}}R_{\mathrm{u}}]\}|^2,\end{aligned} \tag{7.10}$$

$$\begin{aligned}|R(d,L)^-|^2 = |\{R_{\mathrm{e}}^- + R_{\mathrm{d}}T_{\mathrm{e}}T_{\mathrm{u}}^-M^2(k,d,\Delta n)\\ \times(\boldsymbol{\sigma}_{\mathbf{e}}\boldsymbol{\sigma}_{\mathbf{ed}}*)(\boldsymbol{\sigma}_{\mathbf{r}}\boldsymbol{\sigma}_{\mathbf{ed}}^{\perp}*)/[1 - M^2(k,d,\Delta n)(\boldsymbol{\sigma}_{\mathbf{r}}\boldsymbol{\sigma}_{\mathbf{ed}}*)^2R_{\mathrm{d}}R_{\mathrm{u}}]\}|^2,\end{aligned} \tag{7.11}$$

where R_{e}^- is the reflection coefficient of the CLC layer for light with the non-diffracting circular polarization, taking into account dielectric interfaces, for incident light with the diffracting polarization, and T^- is the transmission coefficient of the CLC layer for light with the non-diffracting circular polarization, taking into account dielectric interfaces, for incident light with the non-diffracting polarization. Further, $\boldsymbol{\sigma}_{\mathbf{ed}}^{\perp}$ is the polarization vector orthogonal to $\boldsymbol{\sigma}_{\mathbf{ed}}$. We note that the amplitude transmission coefficients T_{d}^- and T_{u}^- are approximately equal to $\exp[ikLn_-/n_0]$, where n_- is the refractive index of light with the non-diffracting circular polarization in the CLC layer.

The reflection and transmission coefficients can be calculated in the general case using (7.7), (7.8), (7.10), and (7.11), but these calculations are rather cumbersome. This is why we study the case of low birefringence in detail below and present expressions for $|T(d,L)|^2$ and $|R(d,L)|^2$ taking into account only the polarization transformation in the defect layer and neglecting transformations of the polarizations at the interfaces and small deviations in the diffracting and non-diffracting polarizations from the circular ones, since this allows simple analytical calculations.

With this simplification and under the assumption that the refractive indices of the DMS external media coincide with the average CLC refractive indices, the refractive indices of the defect layer can be expressed by the formulas

$$n_{\max} = n_0 + \Delta n/2, \quad n_{\min} = n_0 - \Delta n/2, \tag{7.12}$$

where n_0 coincides with the average CLC refractive index and Δn is small. The phase factor $M(k, d, \Delta n)$ is given by

$$M(k, d, \Delta n) = \exp[ikd] \cos(\Delta\varphi/2), \tag{7.13}$$

where the phase difference between the two beam components with different eigen-polarizations at the defect layer thickness is $\Delta\varphi = \Delta nkd/n_0$, $k = \omega n_0/c = \omega\varepsilon_0^{1/2}/c$.

Finally, in the case of low birefringence, inserting (7.13) into (7.7) and (7.8), we obtain explicit expressions for the reflection and transmission coefficients of light with a circular diffracting polarization for the incident beam with a circular diffracting polarization:

$$|T(d, L)|^2 = |[T_e T_d \exp[ikd] \cos(\Delta\varphi/2)]/[1 - \exp[i2kd] \cos^2(\Delta\varphi/2) R_d R_u]|^2, \tag{7.14}$$

$$|R(d, L)|^2 = |\{R_e + R_d T_e T_u \exp[i2kd] \cos^2(\Delta\varphi/2)/[1 - \exp[i2kd] \times \cos^2(\Delta\varphi/2) R_d R_u]\}|^2 \tag{7.15}$$

If $\Delta\varphi/2\pi$ is an integer, (7.14) and (7.15) are identical to the corresponding equations for the DMS with an isotropic defect layer (see Chap. 6 [12, 13]) and there is no conversion of the diffracting polarization into a non-diffracting one; but if $\Delta\varphi/2\pi$ is not an integer, there is conversion of the diffracting polarization into the non-diffracting one, so light leaks from the DMS, and in particular, the DM lifetime is less than for the case of the corresponding DMS with an isotropic defect layer. This dependence of the DM properties on the phase shift between the eigenwaves when they cross the defect layer opens up ways to control the DM properties. The simplest such possibility involves varying the thickness of the defect layer.

The polarization conversion has the effect of adding the non-diffracting components to the transmitted and reflected beams. For low birefringence, which corresponds to the condition $\Delta n/n_0 < \delta$, the amplitude transmission and reflection coefficients for light with the non-diffracting polarization (and incident light with the diffracting polarization) are given by

$$T(d, L)^- = [T_e \exp[ikLn_-/n_0] \exp[ikd] \sin(\Delta\varphi/2)]/[1 - \exp[i2kd] \times \cos^2(\Delta\varphi/2) R_d R_u], \tag{7.16}$$

$$R(d, L)^- = 1/2 R_u T_e \exp[ikLn_-/n_0] \exp[i2kd] \sin(\Delta\varphi)/[1 - \exp[i2kd] \times \cos^2(\Delta nkd/2n_0) R_d R_u], \tag{7.17}$$

where n_- is the refractive index of light with the non-diffracting circular polarization in the CLC layer.

The calculation results for the transmission coefficients $|T(d, L)|^2$ of light with the diffracting polarization for the case of low birefringence are presented in Fig. 7.6

for various values of the birefringent phase factor $\Delta\varphi$ related to single propagation of light in the birefringent defect layer. Figure 7.6 shows that at low values of the phase shift between eigenwaves when they cross the defect layer ($\Delta\varphi < \pi/2$), the shape of the transmission curve is very similar to the one for a DMS with an isotropic defect layer (for $\Delta\varphi$ equal to an integer multiple of 2π or zero, it coincides with the shape of the corresponding curve for the case of an isotropic defect layer). But as $\Delta\varphi$ approaches $\pi/2$ (see Fig. 7.6e–g), the increased transmission at the defect mode frequency, typical for an isotropic defect layer, gradually disappears and does not appear at all at $\Delta\varphi = \pi/2$ (Fig. 7.6g). This may be regarded, in particular, as a hint that the DM lifetime decreases with increasing shift between the eigenwaves when they cross the defect layer and that the DM does not exist at all at some value of the shift.

Taking into account the partial conversion of a circular non-diffracting incident polarization into a diffracting one, the picture of the transmission spectrum does not change radically. In Fig. 7.7, the transmission spectra for the total light intensity crossing the DMS (the sum of the intensities for both circular polarizations) calculated using (7.16) and (7.17) show a general decrease in transmission at the DM frequency ω_d as $\Delta\varphi$ increases, but it is much slower than for the diffracting polarization, and only for $\Delta\varphi$ close to $\pi/2$ does the transmission practically vanish (which demonstrates the polarization conversion in the birefringent layer).

It is well known [9] that the position of the defect mode frequency in the stop band is determined by the frequency of the transmission (reflection) coefficient maximum (minimum), and therefore the calculations of the transmission spectra carried out here (Figs. 7.6 and 7.7) determine the real component of the DM frequency. But because the DM is a quasi-stationary mode, the imaginary component of the DM frequency is not zero [12, 13]. A direct way to find the imaginary component of the DM frequency is to solve the dispersion equation. In the case of a birefringent defect layer, this equation can be found similarly to the case of an isotropic defect layer [12, 13], and if multiple scattering of light with the non-diffracting polarization is neglected, it can be written

$$\begin{aligned}\{M^2(k, d, \Delta n)\sin^2 qL - \exp(-\mathrm{i}\tau L)[(\tau q/\kappa^2)\cos qL \\ + \mathrm{i}((\tau/2\kappa)^2 + (q/\kappa)^2 - 1)\sin qL]^2/\delta^2\} = 0.\end{aligned} \quad (7.18)$$

In the general case, the solution of (7.18) has to be found numerically, and in the case of an isotropic defect layer, a detailed discussion of this can be found in [12, 13]. Some simplification of (7.18) occurs in the case of low birefringence, when the phase factor in (7.18) is given by (7.13).

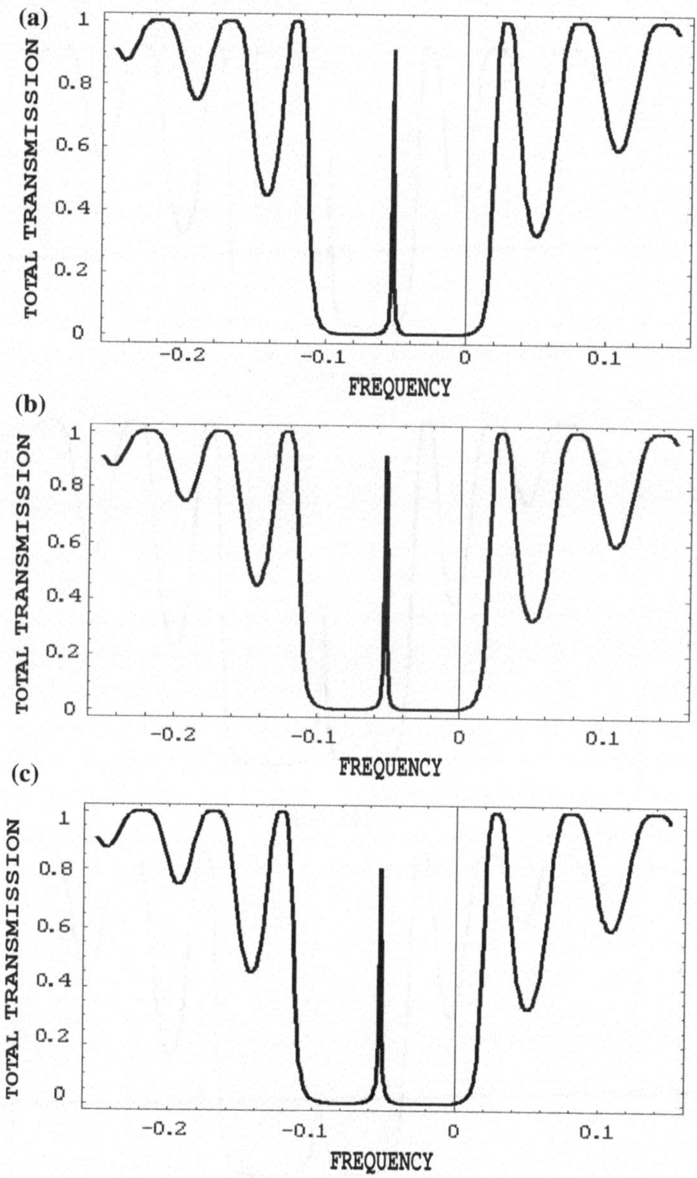

Fig. 7.7 Calculated total intensity transmission coefficient for a low birefringence defect layer versus frequency for diffracting incident polarization and for a birefringent phase shift at the defect layer equal to $\Delta\varphi = \pi/20$ (**a**), $\pi/16$ (**b**), $\pi/12$ (**c**), $\pi/8$ (**d**), $\pi/6$ (**e**), $\pi/4$ (**f**), $\pi/2$ (**g**), for a non-absorbing CLC with $d/p = 0.25$

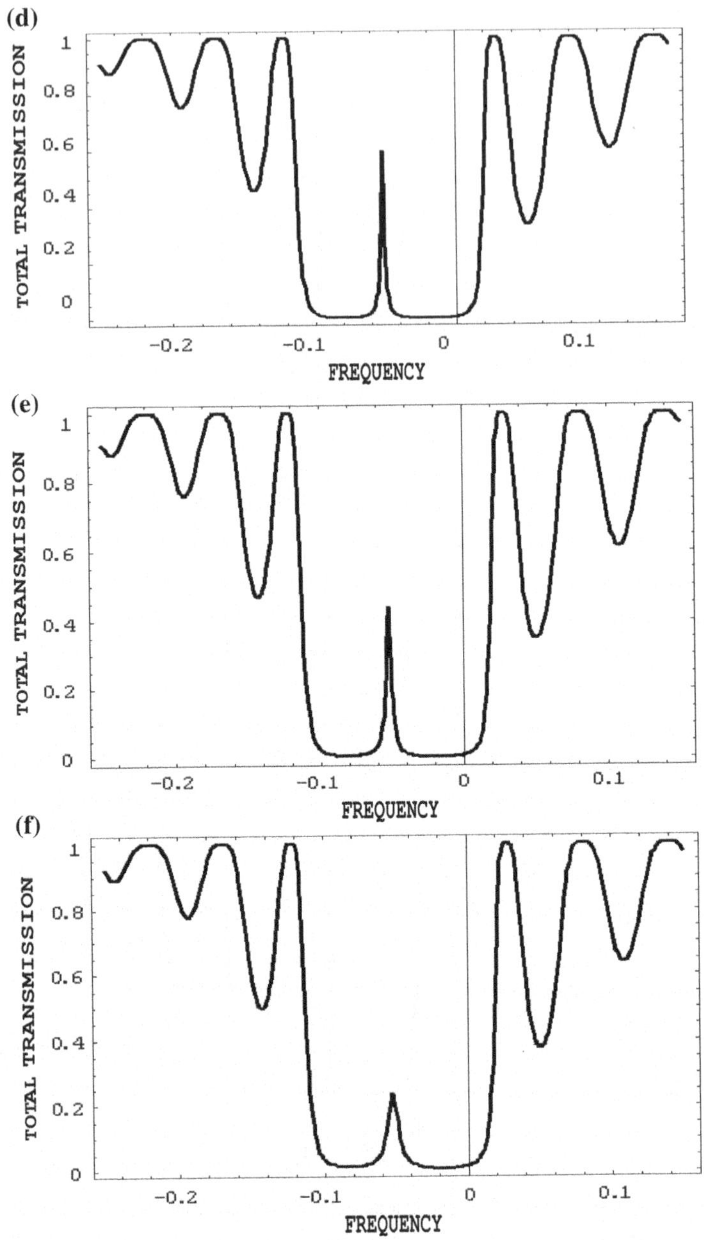

Fig. 7.7 (continued)

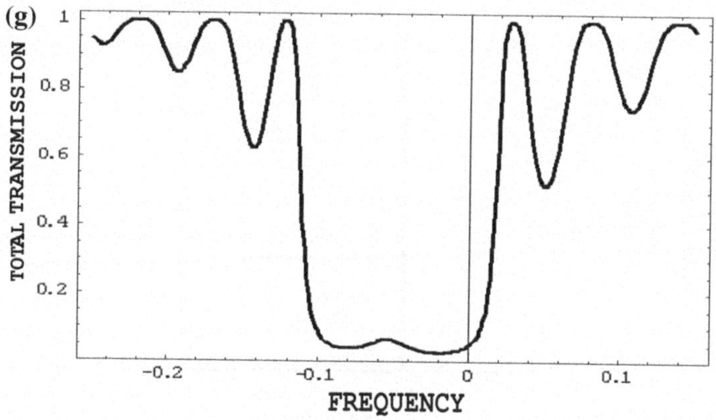

Fig. 7.7 (continued)

7.3.2 Amplifying and Absorbing CLC Layers

As experiment [3] and theory [12, 13] show, the unusual optical properties of DMSs at the DM frequency ω_D (abnormally strong absorption for an absorbing CLC and abnormally strong amplification for an amplifying CLC [12, 13, 28, 33]) can be effectively used to enhance DFB lasing. It is quite natural to study the way the birefringent defect layer influences the abnormally strong amplification and abnormally strong absorption effects. To do so, we assume, as in [12, 13], that the average dielectric constant of the CLC contains an imaginary term, i.e., $\varepsilon = \varepsilon_0(1 + 2i\gamma)$, where positive γ corresponds to an absorbing medium and negative γ to an amplifying medium. (We note that in real situations $|\gamma| \ll 1$.) As mentioned above, the value of γ can be found by solving the dispersion Equation (7.18). Another option (see [12, 13]) is to study the reflection and transmission coefficients (7.7)–(7.8), (7.14), and (7.15) as functions of γ close to $R(d, L)$ and $T(d, L)$ at the DM frequency.

For an amplifying CLC, the value of γ corresponding to a divergence in the DMS reflection and transmission coefficients determines the solution of the dispersion Equation (7.18) and also determines the threshold DFB lasing gain in the DMS (see [12, 13]). Therefore, the threshold value of γ can be found by calculating the DMS reflection and transmission coefficients at various values of γ and finding its value at the points where the DMS reflection and transmission coefficients diverge.

This procedure, performed here for a birefringent defect layer at various values of the birefringent phase factor $\Delta\varphi$ related to single propagation of the light in a birefringent defect layer, can be used to establish the dependence of the threshold lasing gain (γ) on the birefringent phase factor $\Delta\varphi$. Figure 7.8 presents values of the DMS transmission coefficient close to their divergence points, showing the increase in the threshold DFB lasing gain ($|\gamma|$) with increasing birefringent phase factor $\Delta\varphi$, and even disappearance of the divergence at the defect mode frequency when $\Delta\varphi = \pi/2$. This is in good agreement with the transmission spectra calculated in

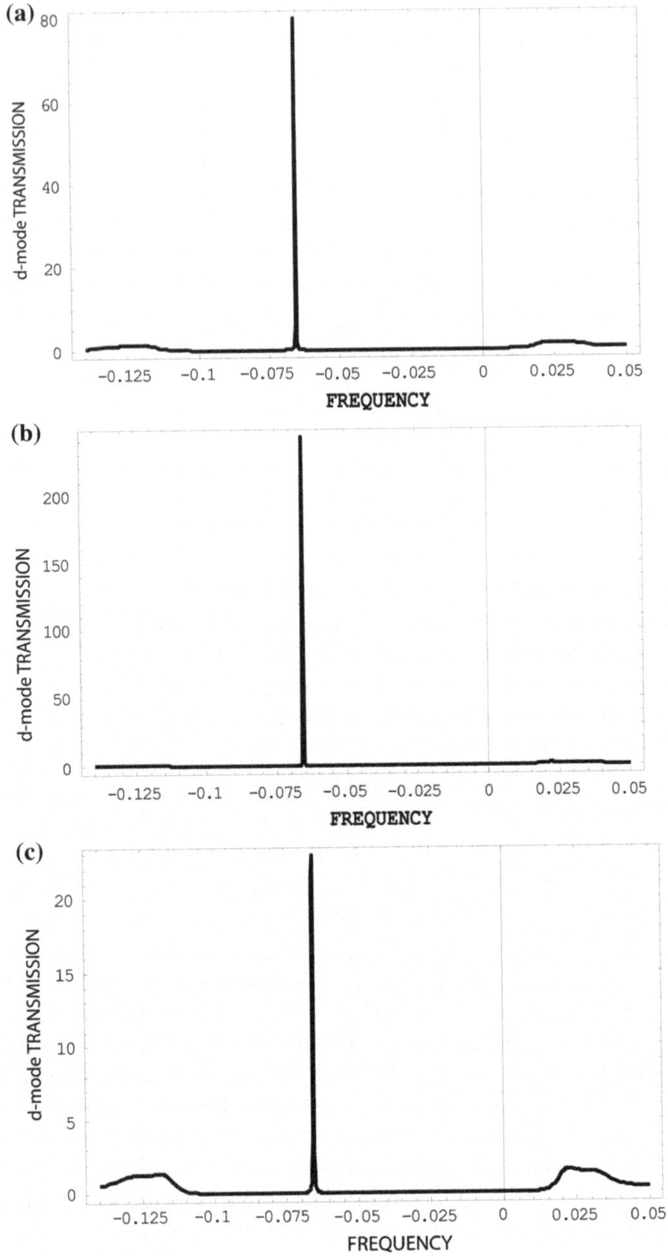

Fig. 7.8 Calculated transmission intensity coefficients of a low birefringence defect layer for an amplifying CLC layer versus frequency close to their divergence points as a function of γ for diffracting incident polarization with birefringent phase shift at the defect layer equal to $\Delta\varphi = \pi/20$, $\gamma = -0.00075$ (**a**), $\pi/16$, $\gamma = -0.00085$ (**b**), $\pi/12$, $\gamma = -0.00100$ (**c**), $\pi/8$, $\gamma = -0.00150$ (**d**), $\pi/6$, $\gamma = -0.002355$ (**e**), $\pi/4$, $\gamma = -0.003555$ (**f**), $\pi/2$, $\gamma = -0.004500$ (**g**), and $\Delta\varphi = 0$, $\gamma = -0.000675$ (**h**) corresponding to an isotropic defect layer; $d/p = 2.25$

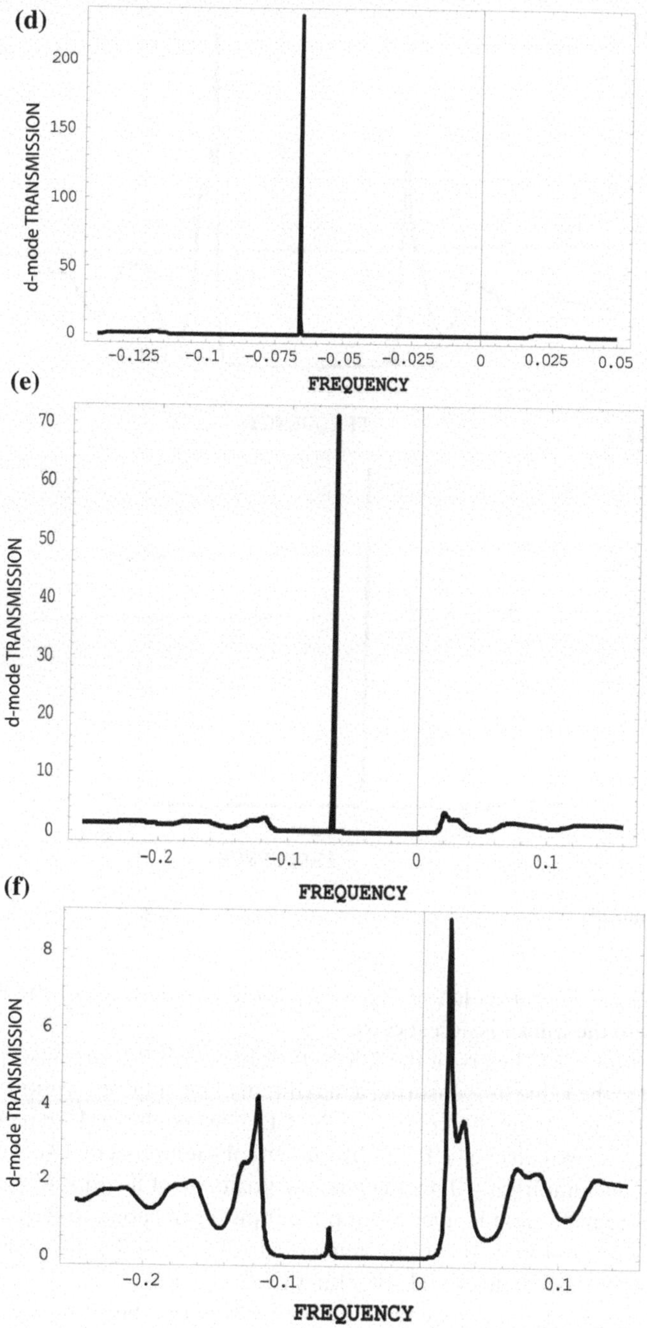

Fig. 7.8 (continued)

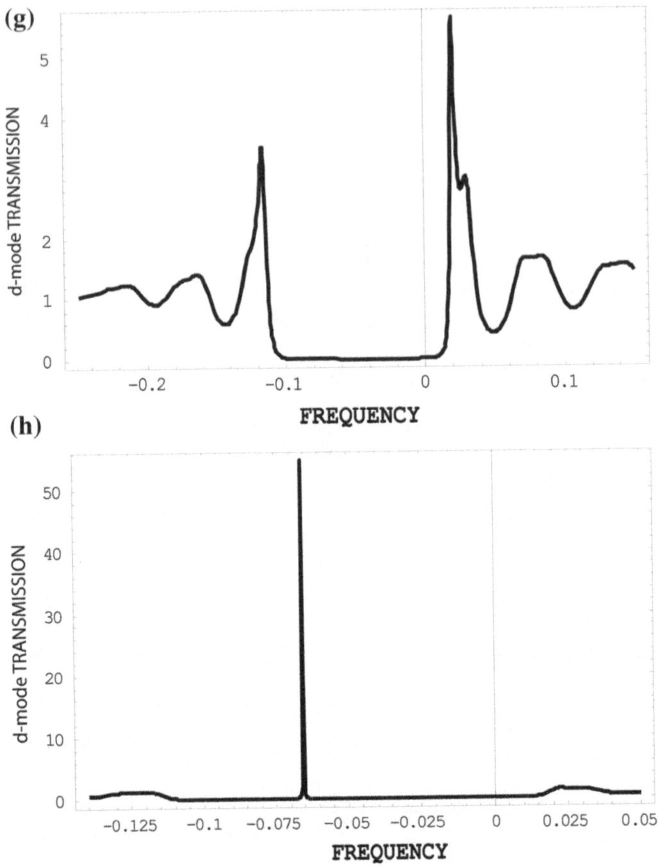

Fig. 7.8 (continued)

Figs. 7.6 and 7.7. In particular, at $\Delta\varphi = \pi/2$, there is no trace of the typical DM peculiarities in the transmission spectra.

For absorbing CLC layers in the DMS, the abnormally strong absorption effect reveals itself at the value of γ ensuring a maximum of the total absorption in the DMS (see [12, 13]). For a finite thickness L of the CLC layers, the DM frequency ω_D is a complex quantity, which can be found by numerical solution of (7.18). As in the case of absorbing and amplifying defect layers, the positions of the dips in reflection and the spikes in transmission inside the stop band just correspond to Re[ω_D], and this observation turns out to be useful for numerical solution of the dispersion equation for a birefringent defect layer and absorbing CLC layers.

We note that the results obtained here for DMs with a birefringent defect layer open up new options for varying the DM characteristics. An important result relating to DFB lasing at the DMS with a birefringent defect layer may be formulated as follows. The lasing threshold gain increases with an increase in the optical path difference of two eigenwaves at the defect layer thickness. A similar result relates to the effect of the anomalously strong absorption phenomenon, where the value of maximal absorption is dependent on the optical path difference at the defect layer thickness.

7.4 Defect Structure with Dielectric Jump

An isotropic defect layer whose dielectric constant differs from the average dielectric constant ε_0 of the CLC layers can also be effectively related to the case of an active defect layer. This is due to polarization conversion at the surfaces, which makes this case similar to the case of a birefringent defect layer. If the dielectric constant of the medium external to the DMS is different from the average dielectric constant ε_0 of the CLC layers, polarization conversion also occurs at the external DMS surfaces, but as we shall see below, the polarization conversion at the external DMS surfaces does not affect the DM properties as much as the polarization conversion at the defect layer surfaces. There are no major difficulties in obtaining the DM dispersion equation from the boundary conditions in the general case of dielectric jumps at all interfaces of the DMS. But the DM dispersion equation is rather complicated in the general case (it is connected with a system of 12 linear equations). Therefore, we first demonstrate the role of dielectric jumps for a localized mode for the simplest case of an edge mode (EM) which is related to a CLC layer with dielectric jumps at its surfaces.

7.4.1 Dielectric Jumps at Single CLC Layer

In accordance with the foregoing, we study the transmission and reflection of light by a CLC layer surrounded by media with dielectric constants differing from the average CLC dielectric constant ε_0 for light propagation along the helical axis (see the schematic of the boundary problem in Fig. 7.9). Following the approach in Chap. 3 [28, 29, 34], from the boundary conditions, we obtain [slightly modifying the notation in (1.20)] the system of equations for the amplitudes E_j^+ of eigenwaves in the layer

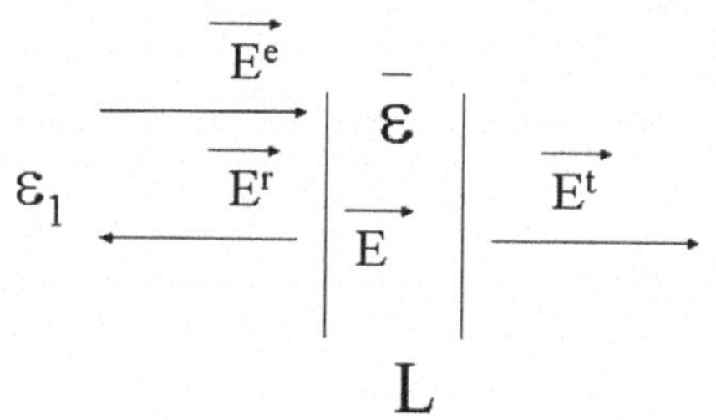

Fig. 7.9 Schematic of the CLC edge mode structure with dielectric jumps at the interfaces

excited by an external wave incident at the layer:

$$\sum_{j=1}^{4}(1+K_j^+/\kappa_{e,1})E_j^+ = 2E_e^+,\quad \sum_{j=1}^{4}e^{iK_j^+L}(1-K_j^+/\kappa_{e,2})E_j^+ = 0$$
$$\sum_{j=1}^{4}\xi_j(1+K_j^-/\kappa_{e,1})E_j^+ = 2E_e^-,\quad \sum_{j=1}^{4}\xi_j e^{iK_j^-L}(1-K_j^-/\kappa_{e,2})E_j^+ = 0 \quad (7.19)$$

where the incident, reflected, and transmitted waves and the wave inside the CLC have been written as follows:

$$\vec{E}^{e} = e^{i(\kappa_{e,1}z-\omega t)}(E_e^+\vec{n}_+ + E_e^-\vec{n}_-)$$
$$\vec{E}^{r} = e^{-i(\kappa_{e,1}z-\omega t)}(E_r^+\vec{n}_- + E_r^-\vec{n}_-)$$
$$\vec{E}^{t} = e^{i(\kappa_{e,2}z-\omega t)}(E_t^+\vec{n}_+ + E_t^-\vec{n}_-)$$
$$\vec{E} = e^{-i\omega t}\sum_{j=1}^{4}E_j^+(e^{iK_j^+z}\vec{n}_+ + \xi_j e^{iK_j^-z}\vec{n}_-)$$

Here, $\mathbf{n}_\pm$ are the left and right circular polarization vectors [see (7.9)] and we use the labeling of CLC eigenwaves proposed in Chap. 1. (The subscripts "1" and "4" correspond to the non-diffracting eigenwaves propagating in opposite directions and the subscripts "2" and "3" correspond to the diffracting eigenwaves.)

The wave vectors inside the CLC layer are $K_1^+ = \tau/2 + q_+$, $K_4^+ = \tau/2 - q_+$, $q_+ = \kappa\{1 + (\tau 2\kappa)^2 + [(\tau/\kappa)^2 + \delta^2]^{1/2}\}^{1/2}$, $K_2^+ = \tau/2 + q$, $K_3^+ = \tau/2 - q$ and q is determined by (5.11), $K_j^- = K_j^+ - \tau$, $\kappa_{e,1} = \frac{\omega\sqrt{\varepsilon_1}}{c}$, $\kappa = \omega\varepsilon_0^{1/2}/c$, $\kappa_{e,2} = \frac{\omega\sqrt{\varepsilon_2}}{c^-}$ and $\xi_i = \frac{\delta}{(K_i^+/\kappa - \tau/\kappa)^2 - 1}$, and the dielectric constants ε_1 and ε_2 are determined in Fig. 7.9.

The amplitudes of the reflected and transmitted waves are expressed in terms of $E_j^\pm$ by

$$E_r^+ = \frac{1}{2}\sum_{j=1}^{4}\xi_j\left(1-\frac{K_j^-}{\kappa_{e,1}}\right)E_j^+,\quad E_r^+ = \frac{1}{2}\sum_{j=1}^{4}e^{i(K_j^+-\kappa_{e,2})}\left(1+\frac{K_j^+}{\kappa_{e,2}}\right)E_j^+$$
$$E_r^- = \frac{1}{2}\sum_{j=1}^{4}\left(1-\frac{K_j^+}{\kappa_{e,1}}\right)E_j^+,\quad E_r^- = \frac{1}{2}\sum_{j=1}^{4}\xi j e^{i(K_j^- -\kappa_{e,2})}\left(1+\frac{K_j^-}{\kappa_{e,2}}\right)E_j^+ \quad (7.20)$$

It is convenient to introduce the parameters $r_1 = \varepsilon_0^{1/2}/\varepsilon_1^{1/2} = k/k_{e,1}$, $r_2 = \varepsilon_0^{1/2}/\varepsilon_2^{1/2} = k/k_{e,2}$ reducing the ratios $K_j^\pm/k_{e,i}$ in (7.19), (7.20) to $r_i K_j^\pm/k$. For the sake of generality, the case of different dielectric constants in the media surrounding the CLC layer is shown in Fig. 7.9, and accordingly, (7.19) and (7.20) relate to the case of different media at either side of the CLC layer.

Examples of calculations, performed using (7.19) and (7.20), which demonstrate the influence of dielectric jumps at the layer surfaces on the transmission and reflection coefficients are presented in [35]. Here, we shall not discuss the transmission and reflection of light by a layer, but concentrate on the influence of dielectric jumps at the layer surfaces on the EM properties. The EMs are determined by the homogeneous system corresponding to (7.19) and the EM dispersion equation for the EM frequency follows from the solvability condition for this homogeneous system [36]. It is known [36] that the real part of the EM frequency coincides approximately with the frequency positions of the minima of the reflection coefficient, and hence the solution of the above-mentioned homogeneous system at the frequency of such a minimum gives the amplitudes of all four eigenwaves in the layer composing the EM in the case where the dielectric constants of the media surrounding the CLC layer are different from ε_0. We recall that the EM in the absence of jumps in the dielectric constant at the interfaces is composed only of two diffracting eigenwaves [36]. Owing to the rather cumbersome form of the solution of the homogeneous system, we first use the consecutive approximation approach to solve the system. If the layer is thick enough, the known solution [36] in the absence of dielectric constant jumps can be used as the zeroth approximation. In this approximation, the homogeneous system under consideration reduces to a system of two equations for the amplitudes of two non-diffracting eigenwaves E_1^+ and E_4^+. The solution of the homogeneous system found using this method at the EM frequency shows that the amplitudes of the two non-diffracting eigenwaves in the solution for the EM decrease in inversely proportion to the layer thickness L. This result shows that, if the CLC layer is thick enough, the influence of the dielectric constant jumps at the layer surfaces is small, and in the limit of an infinitely thick CLC layer, the EM properties are the same as in the absence of the jumps in the dielectric constant. In Fig. 7.10a, the calculated variations in the EM lifetime versus the layer thickness L are presented for the case where dielectric jumps are absent, and for two values of the dielectric jump (Fig. 7.10b zooms in on a small part of the curve in Fig. 7.10a). The calculations of the EM lifetime versus the layer thickness L presented in Fig. 7.10 confirm the above statement that, as L increases, the EM lifetime (Imω) approaches, with decaying oscillations, the value corresponding to the absence of dielectric jumps.

7.4.2 Dielectric Jumps at Defect Layer

We return to the case of a DMS with an isotropic defect layer and a dielectric constant that differs from the average dielectric constant ε_0 of the CLC layers. In the general case of dielectric jumps at all interfaces in the DMS (see Fig. 6.1), we have to determine the 12 amplitudes of the eigenwaves propagating in the DMS (four amplitudes in each CLC layer and four amplitudes for waves propagating in the isotropic defect layer in both directions and with opposite circular polarizations). To simplify the problem, we assume that there are no dielectric jumps at the external DMS surfaces. As we have seen, the dielectric jumps at external DMS surfaces for

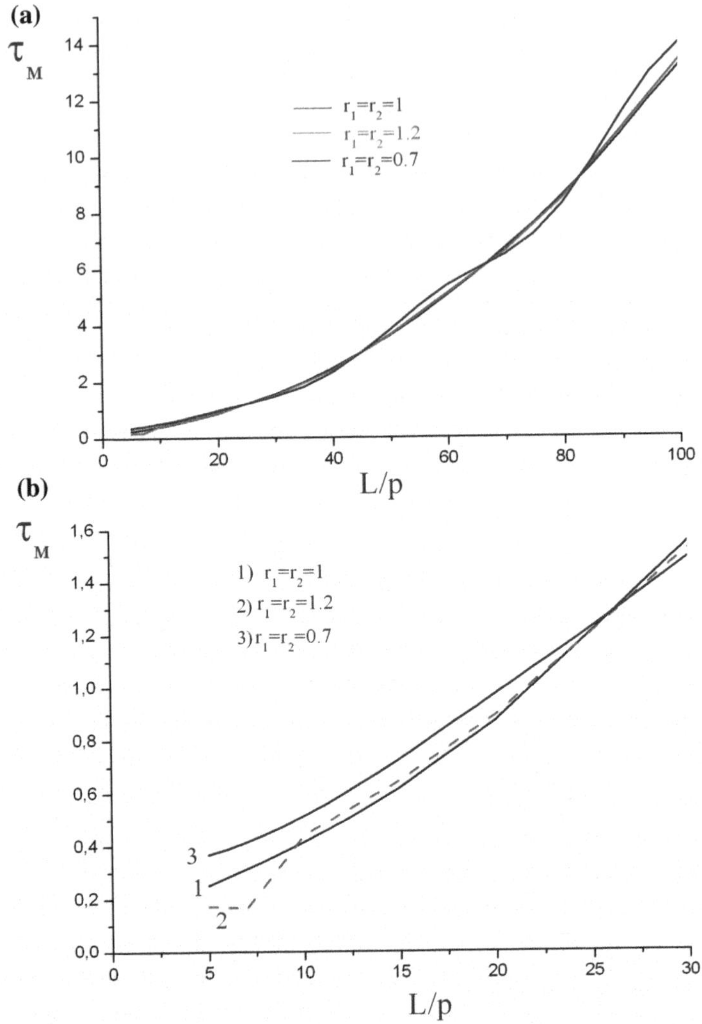

Fig. 7.10 **a** Calculated EM lifetime versus the CLC layer thickness normalized by the CLC layer flight time $\varepsilon_0^{1/2}$ L/c for several values of the dielectric jump at the CLC layer surface. **b** Zoom on part of (**a**)

thick CLC layers do not significantly affect polarization conversion. We therefore take the dielectric jumps into account only at the interfaces with the defect layer. Taking into account the form of the DM solution in the absence of dielectric jumps [12, 13], we have to determine only eight amplitudes of eigenwaves propagating in the DMS (two amplitudes in each CLC layer and four amplitudes for waves propagating in the isotropic defect layer).

If we accept the labeling of the eigenwaves used in Chap. 1 and specify them by superscripts "*u*" and "*d*" for the top and bottom CLC layers, respectively, in Fig. 6.1, then the corresponding system includes E_2^{u}, E_4^{u}, E_1^{d}, E_2^{d}, the amplitudes of the eigenwaves in the CLC, and $C_{\mathrm{R}}^{\pm}$, $C_{\mathrm{L}}^{\pm}$, the amplitudes of the right (left) polarized waves in the defect layer with two ($\pm$) possible propagation directions. We assume for definiteness that the diffracting circular polarization is the right-hand one. If we accept the ordering E_2^{u}, E_4^{u}, C_{R}^{+}, C_{R}^{-}, C_{L}^{+}, C_{L}^{-}, E_1^{d}, and E_2^{d} of the amplitudes mentioned above in the equations obtained from the boundary conditions, then the elements of the matrix a_{ik} of the corresponding system of equations are as follows:

$$a_{ik} = 0 \quad \text{for}\, i = 5, 6, 7, 8 \quad \text{and} \quad k = 1, 2; \quad a_{ik} = 0 \quad \text{for}\, i = 1, 2, 3, 4 \quad \text{and} \quad k = 7, 8;$$
$$a_{11} = \exp(\mathrm{i}K_2^+ L_-) - \exp(\mathrm{i}K_3^+ L_-), \quad a_{12} = \exp(\mathrm{i}K_4^+ L_-), \quad a_{13} = \exp(\mathrm{i}k_{\mathrm{d}} L_-),$$
$$a_{1k} = 0 \quad \text{for}\, k = 4, 5; \quad a_{16} = \exp(-\mathrm{i}k_{\mathrm{d}} L_-); \quad a_{21} = \zeta_2 \exp(\mathrm{i}K_2^- L_-) - \zeta_3 \exp(\mathrm{i}K_3^- L_-),$$
$$a_{22} = \zeta_4 \exp(-\mathrm{i}K_4^- L_-), \quad a_{2k} = 0 \quad \text{for}\, k = 3, 6; \quad a_{24} = \exp(-\mathrm{i}k_{\mathrm{d}} L_-),$$
$$a_{25} = \exp(\mathrm{i}k_{\mathrm{d}} L_-), \quad a_{25} = \exp(\mathrm{i}k_{\mathrm{d}} L_-); \quad a_{31} = -K_2^+ \exp(\mathrm{i}K_2^+ L_-) - K_3^+ \exp(\mathrm{i}K_3^+ L_-),$$
$$a_{32} = -K_4^+ \exp(\mathrm{i}K_4^+ L_-), \quad a_{33} = -k_{\mathrm{d}} \exp(\mathrm{i}k_{\mathrm{d}} L_-), \quad a_{3k} = 0 \quad \text{for}\, k = 4, 5;$$
$$a_{35} = k_{\mathrm{d}} \exp(-\mathrm{i}k_{\mathrm{d}} L_-); \quad a_{41} = K_2^- \zeta_2 \exp(\mathrm{i}K_2^- L_-) - \zeta_3 K_3^- \exp(\mathrm{i}K_3^- L_-),$$
$$a_{42} = -\zeta_4 K_4^- \exp(-\mathrm{i}K_4^- L_-), \quad a_{4k} = 0 \quad \text{for}\, k = 3, 6; \quad a_{44} = -k_{\mathrm{d}} \exp(-\mathrm{i}k_{\mathrm{d}} L_-),$$
$$a_{45} = k_{\mathrm{d}} \exp(\mathrm{i}k_{\mathrm{d}} L_-); \quad a_{53} = \exp(\mathrm{i}k_{\mathrm{d}} L_+), \quad a_{5k} = 0 \quad \text{for}\, k = 4, 5; \quad a_{56} = \exp(-\mathrm{i}k_{\mathrm{d}} L_+),$$
$$a_{57} = \exp(\mathrm{i}K_1^+ L_+), \quad a_{58} = \exp(\mathrm{i}K_2^+ L_+) - r_{32} \exp(\mathrm{i}K_3^+ L_+); \quad a_{6k} = 0 \quad \text{for}\, k = 3, 6;$$
$$a_{64} = \exp(-\mathrm{i}k_{\mathrm{d}} L_+), \quad a_{67} = \zeta_1 \exp(\mathrm{i}k_{\mathrm{d}} L_+), \quad a_{68} = \zeta_2 \exp(\mathrm{i}K_2^- L_+) - r_{32}\zeta_3 \exp(\mathrm{i}K_3^- L_+),$$
$$a_{74} = -k_{\mathrm{d}} \exp(\mathrm{i}k_{\mathrm{d}} L_+), \quad a_{7k} = 0 \quad \text{for}\, k = 4, 5; \quad a_{77} = K_1^+ \exp(\mathrm{i}K_1^+ L_+),$$
$$a_{78} = -K_2^+ \exp(\mathrm{i}K_2^+ L_+) + r_{32} K_3^+ \exp(\mathrm{i}K_3^+ L_+); \quad a_{8k} = 0 \quad \text{for}\, k = 3, 6;$$
$$a_{84} = -k_{\mathrm{d}} \exp(-\mathrm{i}k_{\mathrm{d}} L_+), \quad a_{85} = k_{\mathrm{d}} \exp(\mathrm{i}k_{\mathrm{d}} L_+), \quad a_{87} = \zeta_1 K_1^- \exp(-\mathrm{i}K_1^- L_+),$$
$$a_{88} = K_2^- \zeta_2 \exp(\mathrm{i}K_2^- L_+) - r_{32}\zeta_3 K_3^- \exp(\mathrm{i}K_3^- L_+), \quad \text{where}\ r_{32} = (\zeta_2/\zeta_3) \exp(4\mathrm{i}qL).$$

The dispersion equation for the DM frequency ω_{D}, and in particular, the DM lifetime ($\mathrm{Im}\omega_{\mathrm{D}}$), for a DMS with dielectric jumps only at the interfaces with the defect layer, is determined by the equation obtained by setting the determinant of the above matrix equal to zero. This equation must be solved numerically. However, a simple estimate of the DM lifetime can be obtained.

As is well known, the DM lifetime for a DMS with no dielectric jumps at the interfaces is determined by energy leakage through the external DMS surfaces [13], and the lifetime increases with increasing thickness of the CLC layer, becoming infinite for an infinite thickness. The changes in the DM lifetime for a DMS with dielectric jumps at the interfaces compared to the case without such jumps are mainly due to the conversion of the diffracting polarization into the non-diffracting one and the free escape of light with the non-diffracting polarization from the DMS. If the CLC layer is thick enough, this mechanism predominates over the leakage of light with the diffracting polarization through the external DMS surfaces. This is why, if the CLC layer in the DMS is thick enough, the DM lifetime is mainly determined by polarization conversion at the interfaces with the defect layer. Hence, to estimate the DM lifetime for a DMS with dielectric jumps at the interfaces, we can use the formula for the DM lifetime due to energy leakage through the external DMS surfaces for the case of no dielectric jumps at the interfaces (formula (7.22) in [13]), with the amplitude of the wave with converted polarization at the defect layer surface inserted instead of the diffracting wave amplitude leaking through the external DMS surface. The amplitude of the wave with converted polarization can be found approximately if, when solving the given homogeneous system, we assume that the field in the CLC layers is the same as for the DMS without dielectric jumps. This means that the amplitudes E_2^{u}, C_{R}^{+}, C_{R}^{-}, and E_2^{d} are the same as for the DM in a DMS without dielectric jumps and $E_1^{\mathrm{d}} = E_4^{\mathrm{u}} = 0$. However, we must still find C_{L}^{+} and C_{L}^{-}. This is straightforward using the expressions for the DM field in [13]. The next step is to express the nonzero E_1^{d} and E_4^{u}, which determine the field with non-diffracting circular polarization escaping from the DMS through the external surfaces, in terms of the known E_2^{u}, C_{R}^{+}, C_{R}^{-}, E_2^{d}, and the values found for C_{L}^{+}, C_{L}^{-}. A rather more crude estimate may be obtained without finding E_1^{d} and E_4^{u}, by calculating the direct polarization conversion at the interface with the defect layer for light with the diffracting polarization (for the DM field in the DMS without dielectric jumps). To estimate the polarization conversion, we can apply the formulas for polarization conversion at the interface of the CLC and an isotropic medium presented in Chap. 1 [28, 29, 34]. The reflection coefficient for light with the diffracting circular polarization into light with the non-diffracting circular polarization at a semi-infinite CLC layer, denoted R^{+-}, and the transmission coefficient for light with the non-diffracting circular polarization for incident light with the diffracting circular polarization, denoted T^{+-}, are given to zeroth order in δ by

$$R^{+-} = (1-r)^2/(1+r)^2, \quad T^{+-} = 4r(1-r)^2/(1+r)^4, \tag{7.21}$$

where $r = \varepsilon_d^{1/2}/\varepsilon_0^{1/2}$ and ε_{d} is the dielectric constant of the defect layer.

Because the circular polarization conversion at the interface of the CLC and an isotropic medium is proportional to the square of the small parameter δ, even in the absence of dielectric jumps [28, 29, 34], polarization conversion at the interfaces should be taken into account if the dielectric jump is sufficiently large ($|r-1| > \delta$). Therefore, the expressions (7.21) are accurate enough under these conditions to estimate the influence of the dielectric jumps on the DM lifetime in this case. The results of the corresponding analysis are as follows. The DM lifetime for a DMS with dielectric jumps at the interfaces increases as the thickness of the CLC layers increases to the value for which energy leakage through the external surfaces and leakage due to conversion of light with the diffracting polarization into light with the non-diffracting polarization become approximately equal. With a further increase in the thickness of the CLC layers, the DM lifetime is determined almost exclusively by the polarization conversion at the defect layer surfaces, and becomes practically independent of the CLC layer thickness L or, more correctly, becomes a very slowly increasing function of L. If, following [13], we represent the DM lifetime τ_{dr} for the

DMS with dielectric jumps at the interfaces as the ratio of the optical field energy in the DMS to the energy flow of light of converted polarization through the defect layer surfaces, then the relation between $\tau_{\rm dr}$ and the DM lifetime for a DMS without dielectric jumps at the interfaces $\tau_{\rm d}$ can be estimated as

$$\begin{aligned}\tau_{\rm dr} &= (\varepsilon_0^{1/2}/c)\int |E(\omega_{\rm D}, z, t)|^2 dz/[2r(1-r)^2|E_{\rm dr}|^2/(1+r)^4] \\ &= \tau_d|E^{\rm out}|^2/[2r(1-r)^2|E^{\rm u}_{\rm dr}|^2/(1+r)^4], \end{aligned} \quad (7.22)$$

where r is determined in (7.21) and all other quantities in (7.22) are related to the DM at the DMS without dielectric jumps: $E(\omega_{\rm D}, z, t)$ is the EM field in the CLC layer, $E_{\rm dr}$ is the DM field at the defect layer surface of light propagating toward the CLC layer as a function of the z coordinate along the layer normal and the time t, $E^{\rm out}$ is the EM field of the light propagating out of the CLC layer at the external CLC layer surface, $\omega_{\rm D}$ is the DM frequency, and the integration over z is carried out over the thickness L of CLC layer. Equation (7.22) shows that the DM lifetime $\tau_{\rm dr}$ for a DMS with thick CLC layers and dielectric jumps at the interfaces, in contrast to the lifetime $\tau_{\rm d}$ of a DM in the DMS without dielectric jumps at the interfaces, does not increase exponentially with L. The exponential increase in $\tau_{\rm d}$ is compensated in (7.22) by the exponential increase in $|E^{\rm u}_{\rm dr}|^2$ (see [13]). In order to restore the exponential increase in $\tau_{\rm dr}$ with L, the sharp jumps in the dielectric constant should be replaced by a smooth variation of the dielectric constant at the defect layer surfaces. We note that sharp jumps at the interfaces have a negative effect on the possibilities for lowering the lasing threshold, so smoothing of the dielectric jumps opens up options for lowering the lasing threshold compared with the case of DMSs with jump-like variations in the dielectric parameters.

In general, the localized optical modes in chiral liquid crystals studied theoretically in this chapter for a structure with jumps in the dielectric properties at their interfaces reveal a significant influence of the dielectric jumps on the EM properties, and especially on the DM properties, in particular, its lifetime. The effects studied here pave the way to optimizing the DM parameters by means of a proper choice of the dielectric properties of the defect layer.

7.5 Conclusion

As we have seen, isotropic defect layers with dielectric properties differing from those of the CLC layers in the DMS can effectively be regarded as active defect layers. The analytic description of the defect modes at active defect layers (amplifying (absorbing), birefringent, with dielectric jumps) allow one to obtain a clear physical picture of these modes which applies to defect modes in general (see [16]). For example, a lower lasing threshold and stronger absorption (under the conditions of the anomalously strong absorption effect) when the defect mode frequency lies at the middle of the stop band, compared to the situation when the defect mode frequency lies close to the stop-band edge, are features of any periodic medium. The results obtained suggest numerous ways to influence the DM properties by varying the dielectric characteristics of the defect layer. For a special choice of the parameters in the experiment, the resulting formulas can be applied directly to experiment. Some results can provide a qualitative explanation of the observed effects. This relates, for example, to the circular polarization sense of the wave emitted from the defect structure above the lasing threshold observed in experiment [3], which is opposite to the polarization sense responsible for the existence of the defect mode. An obvious explanation for the "lasing" at the opposite (non-diffracting) circular polarization is as follows. Due to the polarization conversion of the generated wave into a wave with the opposite circular polarization, the converted wave with non-diffracting polarization freely escapes from the structure. As mentioned above, this polarization conversion phenomenon, due to both birefringence and dielectric jumps, also makes a contribution to the frequency width of the defect mode. However, in the general case, a quantitative description of the measurements involves taking into account all possible "active properties" of the defect layer using the above formulas.

We note that the results obtained for the DM in the DMS consisting of CLC layers are qualitatively applicable to the corresponding localized electromagnetic modes in any periodic medium, and can be regarded as a useful guide in any study of localized modes with an active defect layer.

References

1. Y.-C. Yang, C.-S. Kee, J.-E. Kim et al., Phys. Rev. E **60**, 6852 (1999)
2. V.I. Kopp, A.Z. Genack, Phys. Rev. Lett. **89**, 033901 (2003)
3. J. Schmidtke, W. Stille, H. Finkelmann, Phys. Rev. Lett. **90**, 083902 (2003)
4. P.V. Shibaev, V.I. Kopp, A.Z. Genack, J. Phys. Chem. B **107**, 6961 (2003)
5. M.A. Matranga, M.P. DeSanto, G. Petriashvili, A. Chanishvili, G. Chilaya, R. Barbery, Ferroelectrics **395**, 1 (2010)
6. H. Yoshida, C.H. Lee, Y. Miura, A. Fujii, M. Ozaki, Thin Solid Films **516**, 2358 (2008)
7. Y. Takanishi, Y. Ohtsuka, G. Suzaki, S. Nashimura, H. Takezoe, Opt. Express **18**(12), 12909 (2010)
8. A.V. Shabanov, S.Y. Vetrov, A.Y. Karneev, JETP Lett. **80**(#3), 181 (2004)
9. V.I. Kopp, Z.-Q. Zhang, A.Z. Genack, Prog. Quant. Electron. **27**(6), 369 (2003)
10. M. Becchi, S. Ponti, J.A. Reyes, C. Oldano, Phys. Rev. B **70**, 033103 (2004)
11. J. Schmidtke, W. Stille, Eur. Phys. J. **90**, 353 (2003)
12. V.A. Belyakov, Mol. Cryst. Liq. Cryst. **494**, 127 (2008)
13. V.A. Belyakov, S.V. Semenov, JETP **112**, 694 (2011)
14. V.A. Belyakov, Mol. Cryst. Liq. Cryst. **559**, 39 (2012)
15. V.A. Belyakov, Mol. Cryst. Liq. Cryst. **559**, 50 (2012)
16. V.A. Belyakov, Localized optical modes in optics of chiral liquid crystals, in *New Developments in Liquid Crystals and Applications*, ed. by P.K. Choundry (New York, Nova Publishers, 2013), p. 199. (Chapter 7)
17. V.A. Belyakov, *Optics of Photonic Crystals* (MIPT Publishing House, 2013) (in Russian)
18. E. Yablonovitch, T.J. Gmitter, R.D. Meade et al., Phys. Rev. Lett. **67**, 3380 (1991)

19. V.G. Arkhipkin, V.A. Gunyakov, S.A. Myslovets et al., Zhurn. Eksperim. Teor. Fiziki **139**, 666 (2011)
20. I.J. Hodkinson, Q.H. Wu, K.E. Thorn, A. Lakhtakia, W. McCall, Opt. Commun. **184**, 57 (2000)
21. F. Wang, A. Lakhtakia, Opt. Express **13**(n19), 7319 (2005)
22. M.H. Song, N.Y. Ha, K. Amemiya, B. Park, Y. Takanishi, K. Ishikaw, J.W. Wu, S. Nishimura, T. Toyooka, H. Takezoe, Adv. Mater. **18**, 193 (2006)
23. H. Yoshida, C.H. Lee, A. Fuji, M. Ozaki, Appl. Phys. Lett. **89**, 231913 (2006)
24. H. Yoshida, R. Ozaki, K. Yoshino, M. Ozaki, Thin Solid Films **509**, 197 (2006)
25. A.H. Gevorgyan, M.Z. Haratyunyan, Phys. Rev. E **76**, 031701 (2007)
26. M.H. Song, B. Park, Y. Takanishi et al., Jap. J. Appl. Phys. **44**, 8165 (2005)
27. V.A. Belyakov, S.V. Semenov, JETP **118**(5), 798–813 (2014)
28. V.A. Belyakov, V.E. Dmitrienko, Optics of chiral liquid crystals, in *Soviet Scientific Reviews/Section A, Physics Reviews*, vol. 13. ed. by I.M. Khalatnikov (Harwood Academic Publisher, 1989), pp. 1–203 (p. 54)
29. V.A. Belyakov, *Diffraction Optics of Complex Structured Periodic Media* (Springer, New York, 1992). (Chapter 4)
30. H. de Vries, Acta Crystallogr. **4**, 219 (1951)
31. E.I. Kats, Sov. Phys. JETP **32**, 1004 (1971)
32. P.G. Gennes, J. Prost, *The Physics of Liquid Crystals* (Clarendon Press, Oxford, 1993)
33. V.A. Belyakov, A.A. Gevorgian, O.S. Eritsian, N.V. Shipov, Zh. Tekhn. Fiz. **57**, 1418 (1987). [Sov. Phys. Techn. Phys. **32**(n7), 843–845 (1987), English translation; Sov. Phys. Crystalography. **33**(n3), 337 (1988)]
34. V.A. Belyakov, A.S. Sonin, *Optika kholestericheskih zhidkih kristallov* (Nauka, Moscow, 1982)
35. V.A. Belyakov, S.V. Semenov, in *Program of 24th International Liquid Crystal Conference* (Mainz, 2012), p. 92, PIII-148
36. V.A. Belyakov, S.V. Semenov, JETP **109**, 687 (2009)

Chapter 8
Localized Modes in Optics of Photonic Liquid Crystals with Local Anisotropy of Absorption

8.1 Introduction

Recently, there has been much activity in the field of localized optical modes, in particular, edge modes (EM) and defect modes (DM) in chiral liquid crystals (CLCs), mainly due to the possibilities for reaching a low lasing threshold for mirrorless distributed feedback (DFB) lasing [1–4] in CLCs. EMs and DMs existing as localized electromagnetic eigenstates with their frequencies close to the forbidden band gap and inside the forbidden band gap, respectively, were first investigated in periodic dielectric structures [5]. The corresponding EMs and DMs in CLCs, and more generally in spiral media, are very similar to the EMs and DMs in one-dimensional scalar periodic structures. They reveal anomalous reflection and transmission [1, 2] and allow distributed feedback (DFB) lasing at a low lasing threshold [3].

Almost all theoretical studies of the EMs and DMs in chiral and scalar periodic media were performed by means of a numerical analysis with the exceptions of [6, 7], where the known exact analytical expression for eigenwaves propagating along the helix axis [8–10] was used for a general study of DMs. The approach used in [6, 7] looks profitable because it can provide an easy understanding of DM and EM physics, and this is why it deserves further implementation in the study of these modes.

In the cited studies, optical absorption in CLCs was treated as isotropic. However, the case of isotropic absorption does not cover all options occurring in CLCs. For example, it is quite common for dye molecules to align with clearly represented absorption lines in liquid crystals. If the director distribution in a liquid crystal sample is not homogeneous (which is the case in CLCs), a local absorption anisotropy may manifest itself in the sample in some circumstances [11, 12]. The corresponding effects depend on the value of the liquid crystal order parameter and disappear if the order parameter is zero, i.e., at the liquid crystal to the liquid phase transition. The corresponding effects in the CLC transmission and reflection spectra, in particular

V. Belyakov, *Diffraction Optics of Complex-Structured Periodic Media*, Springer Series in Optical Sciences 203,
https://doi.org/10.1007/978-3-319-43482-7_8

the Borrmann effect, have been studied both experimentally [13, 14] and theoretically [11–13]. In the present chapter, we study theoretically the influence of local absorption anisotropy on the characteristics of localized modes. We present analytical solutions for the EMs and DMs (associated with a local absorption anisotropy in CLCs) and consider some limiting cases simplifying the problem [15].

8.2 Boundary Problem

To investigate EMs in a CLC, we have to consider a boundary problem, i.e., the transmission and reflection of light incident on a CLC layer along the spiral axis [10–12]. We assume that the CLC is represented by a planar layer with a spiral axis perpendicular to the layer surfaces (Fig. 5.1). We also assume that the average CLC dielectric constant coincides with the dielectric constant of the ambient medium. This assumption practically prevents the conversion of one circular polarization into another at the layer surfaces [11, 12] and allows one to take into account only two eigenwaves with diffracting circular polarization.

Although the procedure for obtaining the reflection, R, and transmission, T, coefficients for a CLC layer with a local absorption anisotropy is similar to the one for a non-absorbing CLC, some complications arise due to the fact that the elements of the CLC dielectric tensor are now complex quantities.

We begin by discussing the dielectric tensor of a substance with locally anisotropic absorption. The principal values of the corresponding dielectric tensor are complex and have different imaginary parts, depending on the case of CLCs on the liquid crystal order parameter S [10]. In the general case, all three imaginary parts are different. For a CLC, the imaginary parts are different for $S = 1$ and are equal for $S = 0$. Returning to the CLC, we assume for simplification of the problem that only one principal value of the dielectric tensor is complex at $S = 1$. This corresponds to the assumption that the absorption in CLCs is due only to dye molecules, and that at $S = 1$ a complete ordering of the dye molecules occurs. We assume also that the axis corresponding to the real principal value of the dielectric tensor is directed along the spiral axis and that the other two axes rotate around the spiral axis. These rotating axes determine the local direction of the absorption anisotropy, which depends on the coordinate along the spiral axis. Now, we have to insert the complex principal values of the dielectric tensor in the expressions for the dielectric anisotropy [11, 12, 16]. As a result, the dielectric anisotropy $\Delta\varepsilon$ (see below) becomes a complex quantity. Luckily, the expressions for the reflection and transmission coefficients for light propagating along the cholesteric axis in a CLC layer [11, 12, 16] are exact and are applicable to the case of anisotropic absorption under consideration here.

In the following calculations and the formulas for limiting values of the problem parameters, it is assumed that the most favorable situation occurs for the Bormann effect. "In particular, it is assumed that the CLC absorption is due only to the dye molecules 'dissolved' in the CLC, and that the corresponding molecular dye absorption oscillators are directed along the long molecular axes. Moreover, it is also assumed that their orientational ordering may be complete, i.e., the order parameter S can reach 1." Under these assumptions, the imaginary parts of the principal values of the dielectric tensor for S differing from unity are expressed in the following way via the imaginary part of the single complex principal value ε_1, or ε_2 for $S = 1$, in the cases where the absorption oscillator axis is parallel or perpendicular to the long molecular axis, respectively:

$$\mathrm{Im}[\varepsilon_{\|}] = \mathrm{Im}[\varepsilon_1](1 + 2S)/3, \quad \mathrm{Im}[\varepsilon_{\perp}] = \mathrm{Im}[\varepsilon_1](1 - S)/3 \tag{8.1a}$$

and the local dielectric anisotropy is

$$\Delta\varepsilon = \varepsilon_{\|} - \varepsilon_{\perp} = \mathrm{Re}[\varepsilon_{\|} - \varepsilon_{\perp}] + \mathrm{i}\mathrm{Im}[\varepsilon_1]S \tag{8.2a}$$

for the oscillator axis parallel to the long molecular axis;

$$\mathrm{Im}[\varepsilon_{\perp}] = \mathrm{Im}[\varepsilon_2](1 + 2S)/3, \quad \mathrm{Im}[\varepsilon_{\|}] = \mathrm{Im}[\varepsilon_2](1 - S)/3 \tag{8.1b}$$

and the local dielectric anisotropy is

$$\Delta\varepsilon = \varepsilon_{\|} - \varepsilon_{\perp} = \mathrm{Re}[\varepsilon_{\|} - \varepsilon_{\perp}] - \mathrm{i}\mathrm{Im}[\varepsilon_2]S. \tag{8.2b}$$

for the oscillator axis perpendicular to the long molecular axis.

In the following, we shall present the results related to the first option, for the case where the absorption oscillator axis lies parallel to the long molecular axis. If for the first option, the Borrmann effect occurs close to the high-frequency stop-band edge, for the second option, there is a similar occurrence of the Borrmann effect close to the low-frequency stop-band edge [11, 12].

Because the Borrmann effect in CLC was studied in [10–12, 16, 17], we state here only the final expressions for the amplitude transmission, T, and reflection, R, coefficients (6.5) for light incident on a CLC layer of thickness L, where

$$q = \kappa\left\{1 + (\tau/2\kappa)^2 - \left[(\tau/\kappa)^2 + \delta^2\right]^{1/2}\right\}^{1/2}$$

and, $\varepsilon_0 = (1 + \mathrm{i}\gamma)\mathrm{Re}[(\varepsilon_{\|} + \varepsilon_{\perp})]/2$, $\delta = \mathrm{Re}[(\varepsilon_{\|} - \varepsilon_{\perp})]/\mathrm{Re}[\varepsilon_0] + \mathrm{i}\varepsilon_1 S/\mathrm{Re}[\varepsilon_0]$ and $\gamma = \varepsilon_1(1 - S)/3\mathrm{Re}[\varepsilon_0]$ is the parameter describing the locally isotropic part of the absorption. Here δ is the dielectric anisotropy, while $\varepsilon_{\|}$ and $\varepsilon_{\perp}$ are the local principal values of the CLC dielectric tensor [10–12], $\kappa = \omega\varepsilon_0^{1/2}/c$, with c the speed of light, and $\tau = 4\pi/p$ with p the cholesteric pitch.

Note that, in the case of absorption due solely to the dye molecules "dissolved" in the CLC, only one principal value of the dielectric tensor is complex at $S = 1$ if the absorption oscillators of the dye molecules are directed along a single axis in the molecule and the degree of orientational order of the dye molecules is the same as for the CLC. In principle, the orientational order parameter for dye molecules may be different from the CLC order parameter S.

As would be expected, the CLC local absorption anisotropy reveals itself in the reflection and transmission spectra due to the Borrmann effect (well known in X-ray diffraction) [11–13], manifested in the suppression of absorption at one of the stop-band edges. Contrary to the case of isotropic absorption where the reflection and transmission spectra are symmetric relative to the stop-band center, in the case of locally anisotropic absorption in the CLC, the spectra are non-symmetric relative to the stop-band center. For a CLC with locally anisotropic absorption [11–13], absorption suppression also occurs at one of the stop-band edges, as in the X-ray case, and is due to the orthogonality of the whole optical electric field in the sample to the absorption oscillators of the dye molecules at the frequency of one stop-band edge. Figures 8.1 and 8.2 show the Borrmann effect in reflection and transmission, respectively, in a CLC layer with locally anisotropic absorption [for calculations performed using (6.5)] for the order parameter $S = 1$. However, here and below, the notations R and T are different from the ones in (6.5), referring to the reflection and transmission intensity coefficients.

The Borrmann effect is most clearly manifested in the total absorption spectra for a sample, given by 1-R-T (see Fig. 8.3).

Figures 8.1, 8.2, and 8.3 show increasing reflection and transmission at one stop-band edge compared to the other stop-band edge and a strong suppression of absorption at this stop-band edge. To illustrate the dependence of the Borrmann effect on the value of the order parameter S, Figs. 8.4 and 8.5 present the calculation results for the reflection and total absorption 1-R-T, respectively, for $S = 1, 0.5, 0.3, 0$. These figures show that the suppression of the absorption effect decreases with decreasing CLC order parameter S. Here it should be mentioned that, in the above calculations,

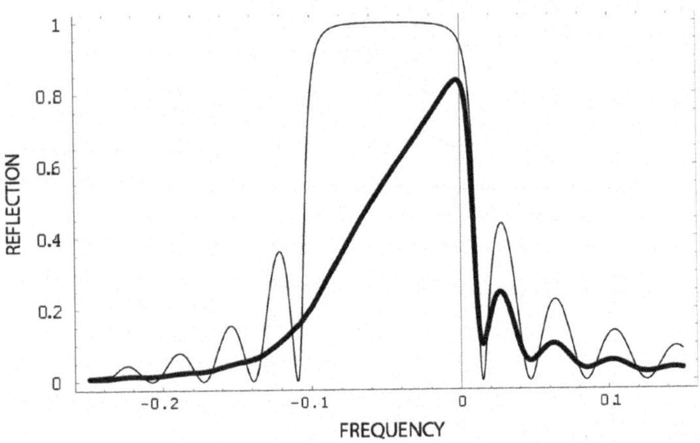

Fig. 8.1 Reflection versus dimensionless frequency for locally anisotropic absorption in a CLC layer when $\delta = 0.05 + 0.03i$, $l = 300$, with $l = L\tau = 2\pi N$, where N is the director half-turn number at the CLC layer thickness L. The thin curve corresponds to complete absence of absorption. Here and in all further figures in this chapter, the frequency is defined in the same way as in the preceding chapters (see Fig. 5.2)

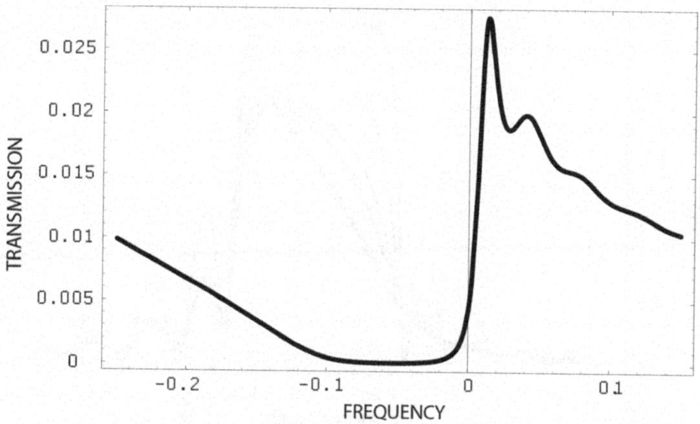

Fig. 8.2 Transmission versus frequency for locally anisotropic absorption in CLC layer (see Fig. 5.1) at $\delta = 0.05 + 0.015i, l = 300$

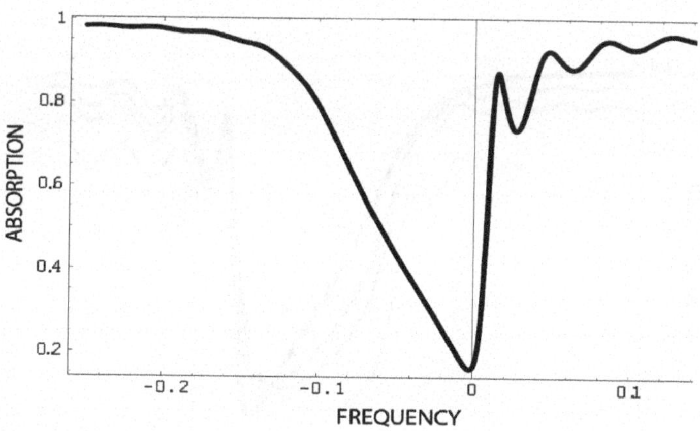

Fig. 8.3 Total absorption 1-R-T versus frequency for locally anisotropic absorption in CLC layer (see Fig. 5.1) at $\delta = 0.05 + 0.015i, l = 300$

it was assumed that the pitch value and δ are the same for all values of S, which means that the variations in the order parameter of the dye molecules can be different from the corresponding variations in the order parameter of the CLC molecules. In real situations, variations in the order parameter are usually connected with temperature variations (and as is well known, the pitch is a temperature dependent CLC parameter). Figure 8.6 shows that the total absorption at the EM frequency decreases with increasing thickness of the CLC layer (L in Fig. 8.6 is ten times larger than in Fig. 8.5). Figure 8.7 shows that the transmission T at the EM frequency decreases when the CLC order parameter S increases, and a more pronounced decrease occurs at the stop-band edge opposite to the edge where the reflection is increased due

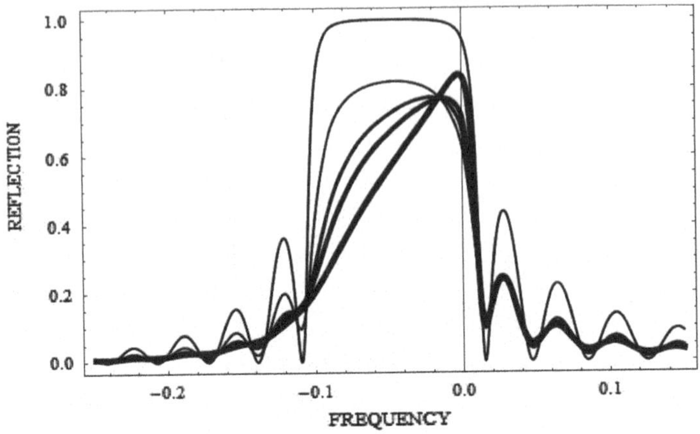

Fig. 8.4 Reflection for a CLC layer with local absorption anisotropy versus frequency for $S = 0$, 0.3, 0.5, and 1 (the thickness of a curve is increasing with increasing S) when $\mathrm{Im}[\varepsilon_1]/\mathrm{Re}[\varepsilon_0] = 0.03$, $\delta = 0.05$, $l = 300$ (the thinnest curve corresponds to complete absence of absorption)

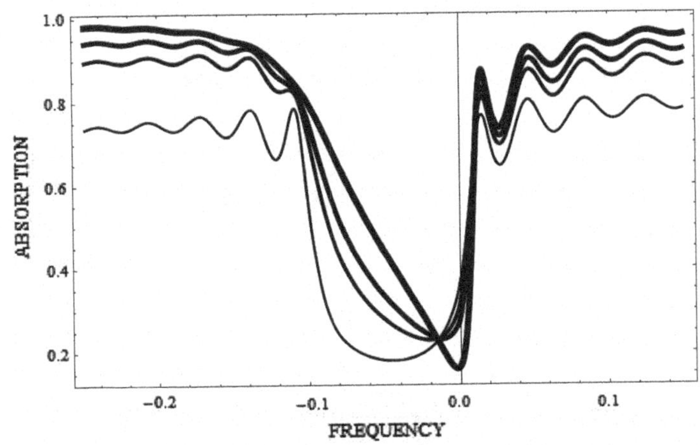

Fig. 8.5 Absorption for a CLC layer with local absorption anisotropy versus frequency for $S = 0$, 0.3, 0.5, and 1 (the thickness of a curve is increasing with increasing S) when $\mathrm{Im}[\varepsilon_1]/\mathrm{Re}[\varepsilon_0] = 0.03$, $\delta = 0.05$, $l = 300$

to the Borrmann effect. It is worth noting that the calculation results presented in Figs. 8.4, 8.5, 8.6, and 8.7 take into account the fact that all three principal values of the dielectric tensor are complex, and it is only when the order parameter $S = 1$ that a single principal value of the local dielectric tensor corresponding to the long molecular axes direction is complex.

The absorption spectra shown in Figs. 8.5 and 8.6 are directly related to the luminescence spectra if the dye absorption line overlaps the stop band. Because the absorbed optical photons of the incident beam are re-emitted as luminescence

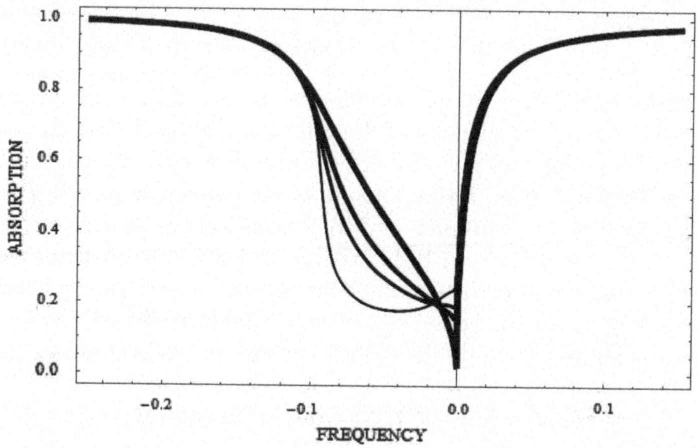

Fig. 8.6 Absorption for a CLC layer with local anisotropy of absorption versus the frequency for $S = 0, 0.3, 0.5, 1$ (the thickness of a curve is increasing with increasing S) at $\mathrm{In}[\varepsilon_1]\mathrm{Re}[\varepsilon_0] = 0.03$, $\delta = 0.05$, $l = 3000$

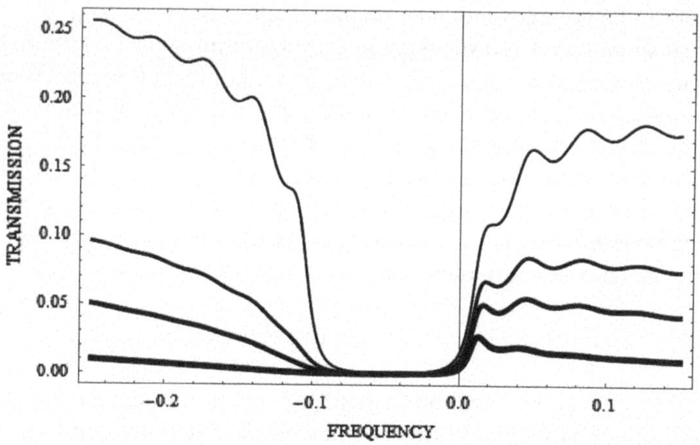

Fig. 8.7 CLC layer transmission versus frequency for locally anisotropic absorption in CLC layer for $S = 0, 0.3, 0.5, 1$ (the thickness of a curve is increasing with increasing S) at $\mathrm{Im}[\varepsilon_1]/\mathrm{Re}[\varepsilon_0] = 0.03$, $\delta = 0.05$, $l = 300$

photons, the absorption maximum naturally results in a luminescence intensity maximum, at a frequency that is shifted in the direction of lower frequencies. In this way, the luminescence spectra are dependent on the local absorption anisotropy, with a different luminescence intensity for the pumping wave frequency coincides with the opposite stop-band edge frequencies.

8.3 Edge Mode

In a non-absorbing CLC, $\gamma = 0$ in the general expression for the dielectric constant $\varepsilon = \varepsilon_0(1 + i\gamma)$. Calculations of the reflection and transmission coefficients as functions of the frequency give the well-known results [8–12] which differ from the curves presented in Figs. 8.1 and 8.2, in particular, when $\gamma = 0, T + R = 1$ for all frequencies. We now discuss the way a locally anisotropic absorption in the CLC influences the properties of EMs existing at discrete frequencies outside the stop band [15, 16], and in particular, the EM lifetime. To do this, we have to solve the corresponding dispersion equation. Following the reference [15], we obtain the dispersion equation in a form similar to the case for zero absorption, viz.,

$$\mathrm{tg}qL = i\left(q\tau/\kappa^2\right)/\left[(\tau/2\kappa)^2 + (q/\kappa)^2 - 1\right] \tag{8.3}$$

but the parameters entering this equation are as given in Sect. 8.2, i.e., they depend on some additional quantities compared to the case of zero absorption.

Solutions to (8.3) for EM frequencies ω_{EM} situated outside the stop-band edges are discrete and in the general case can only be found numerically. The EM frequencies ω_{EM} turn out to be complex and may be presented in the form $\omega_{FM} = \omega(1 + i\Delta)$, where in real situations Δ is a small parameter determining the EM lifetime. From a general analysis of (8.3), we conclude that the EM lifetimes at frequencies outside the opposite stop-band edges are different in the case of locally anisotropic absorption (there is no symmetry in their values relative to the stop-band center).

Fortunately, an analytic solution can be found for thick CLC layers and a sufficiently small Δ ensuring the condition $L\mathrm{Im}(q) \ll 1$. In this case, ω and Δ are determined by the conditions $qL = n\pi$, and the EM lifetimes for the case of isotropic absorption may be expressed in the form

$$\tau_{EM} = 1/\Delta\omega_{EM} = 2/\omega_{EM}\left[\delta(n\pi)^2/(\pi\delta L/p)^3 + \gamma\right], \tag{8.4}$$

where the integer n is the edge mode number [16] ($n = 1$ corresponds to the EM frequency (minimum of the reflection coefficient) closest to the stop-band edge) and γ is the parameter determining the isotropic fraction of the absorption in the CLC. So, in the case of isotropic absorption, the localized mode lifetime is bounded by $2/\omega_{EM}\gamma$.

In the case of locally anisotropic absorption in the CLC, the EM lifetime depends on the EM frequency position relative to the stop-band center and may exceed the value given by (8.4) due to the Borrmann effect. For example, if the CLC layer thickness L is sufficiently large and the condition $L\ \mathrm{Im}(q) \ll 1$ holds, the ratio of lifetimes at the opposite stop-band edge frequencies may be estimated by the expression

$$\tau_B/\tau_{AB} = \left[\omega_{EM}(\varepsilon_0\gamma + 2\varepsilon_0\mathrm{Im}\,\delta) + 1/\tau_m\right]/(\omega_{EM}\varepsilon_0\gamma + 1/\tau_m), \tag{8.5}$$

where τ_B, τ_{AB}, τ_m are the lifetimes at the stop-band edge where the Borrmann effect occurs, at the opposite edge, and at the edges in the case of a non-absorbing CLC (see (8.4) for $\gamma = 0$), respectively, and γ determines the isotropic component of the absorption. If γ tends to zero, which happens when the order parameter tends to 1 (if we neglect all sources of absorption except the dye), τ_B coincides with τ_m as given by (8.4) for $\gamma = 0$, which corresponds to the suppression of absorption for the EM at the stop-band edge frequency in the case of a CLC with local absorption anisotropy. Note that, to be quite accurate, a complete suppression of absorption for the EM is achievable only in the limit of an infinitely thick CLC layer. At the opposite stop-band edge frequency, the absorption is enhanced and the EM lifetime, being proportional to $1/2\varepsilon_0 \operatorname{Im} \delta$, is shorter than τ_m.

The results of numerical solutions to (8.3) for the EM frequencies ω_{EM} determining the EM lifetime at the low-frequency stop-band edge for "isotropic non-absorbing and locally anisotropic absorbing CLCs" are presented in Figs. 8.8 and 8.9. Figure 8.8 shows suppression of absorption (curve 4 for the absorption oscillators directed along axis 2) and absorption enhancement (curve 3 for the absorption oscillators directed along axis 1) resulting in a shorter lifetime for this case than for isotropic absorption (curve 2). Figure 8.9 shows the dependence of the lifetime ratio of the first EM at the high- and low-frequency stop-band edges on the value of the order parameter S. It shows that the growth of the EM lifetime at $S = 1$ due to the Borrmann effect at one stop-band edge compared to the opposite one disappears at $S = 0$.

Concerning the luminescence spectra if the dye absorption line overlaps with the stop band, the intensity of luminescence is enhanced for the pump wave frequency at

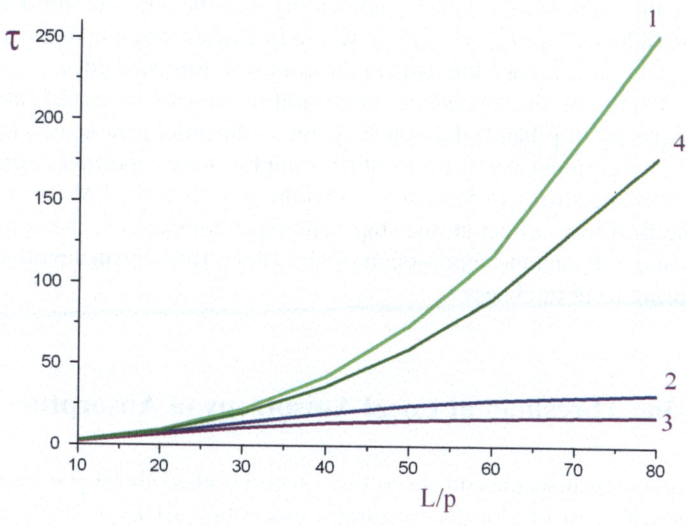

Fig. 8.8 Calculated lifetime of the first EM (in units $\varepsilon_0^{1/2} p/c$) versus the CLC layer thickness: (1) zero absorption; (2) isotropic absorption; (3) locally anisotropic absorption $\operatorname{Im}[\varepsilon_1] = 0.01$, $\operatorname{Im}[\varepsilon_2] = 0$; (4) locally anisotropic absorption $\operatorname{Im}[\varepsilon_1] = 0$, $\operatorname{Im}[\varepsilon_2] = 0.01$ [see (8.1) for $S = 1$], $\delta = 0.05$

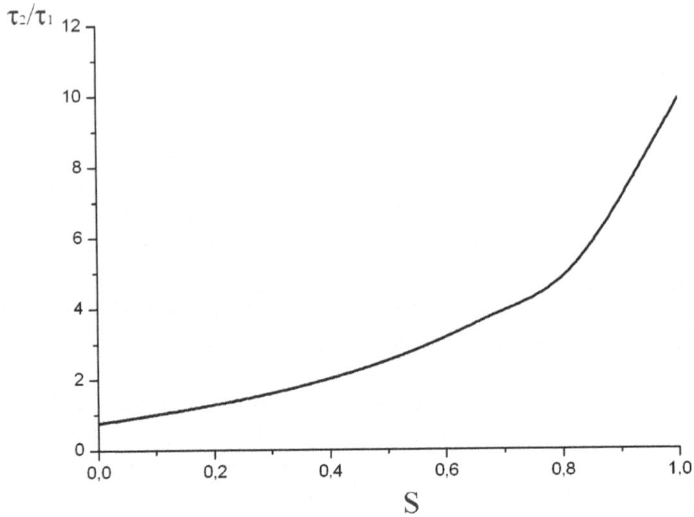

Fig. 8.9 Calculated lifetimes ratio of the first EM at the high- and low-frequency stop-band edges versus the value of the order parameter *S* for locally anisotropic absorption $\text{In}[\varepsilon_1] = 0$, $\text{Im}[\varepsilon_2] = 0.02$ [see (8.1)], $\delta = 0.05$, $L/p = 80$

the stop-band edge corresponding to the shorter EM lifetime and suppressed for the pump wave frequency at the stop-band edge corresponding to the longer EM lifetime. This gives an estimate of the pump wave intensity ratio ensuring equal luminescence intensity for the pump wave frequency coinciding with the opposite stop-band edges according to (8.5), i.e., $I_B/I_{AB} = \tau_B/\tau_{AB}$, where I_B/I_{AB} are the pump wave intensities ensuring equal luminescence intensity at the opposite stop-band edges.

Figure 8.10 presents the dependence of the lifetime ratio of the first EM at the high- and low-frequency stop-band edges on the value of the order parameter *S* for several values of the layer thickness (valid for thick samples if the condition $L\ \text{Im}(q) \ll 1$ holds), as calculated using (8.5). It shows that the growth of the EM lifetime at $S = 1$ due to the Borrmann effect at one stop-band edge compared to the opposite one disappears at $S = 0$, and the suppression of absorption (the Borrmann effect) grows with increasing layer thickness *L*.

8.4 Lasing Threshold at Local Anisotropy of Absorption

The reflection, transmission, and absorption spectra studied above give hints that the DFB lasing in the case of a locally anisotropic absorption in CLC is also significantly influenced by the anisotropy. To study the lasing threshold, we have to solve the dispersion equation at the EM frequency relative to the negative imaginary part of dielectric tensor.

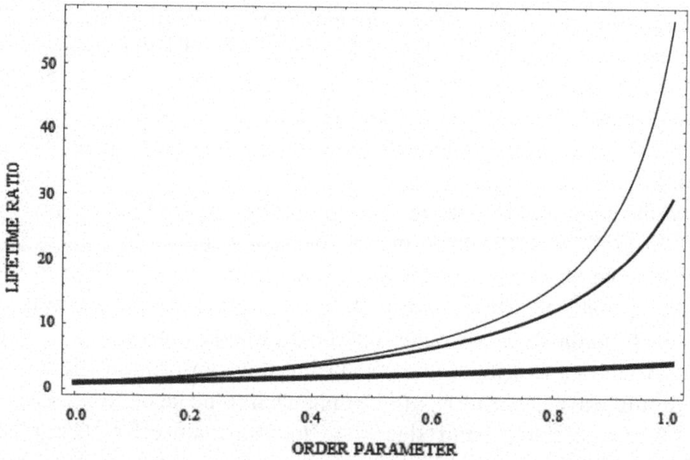

Fig. 8.10 Calculated lifetime ratio of the first EM at the high- and low-frequency stop-band edges versus the value of the order parameter S for locally anisotropic absorption $\mathrm{Im}[\varepsilon_1] = 0.015$, $\mathrm{Im}[\varepsilon_2] = 0$ [see (8.1)], $\delta = 0.05$, $L/(p/\delta) = 1; 2; 2.5$ (curves from bottom to top, respectively)

The imaginary addition to the dielectric tensor in the case of assumed absorption isotropy may be taken into account by introducing into the dielectric tensor a factor of the form $(1 - ig)$ where g is a small positive quantity. In the case of non-absorbing CLC, the threshold problem was studied in [16], and in the limit of a thick CLC layer the threshold is given by the following expression:

$$g = \delta(n\pi)^2/(\delta\, L\, \tau/4)^3. \tag{8.6}$$

In the case of an isotropically absorbing CLC in the same limit, the threshold is given by

$$g = \delta(n\pi)^2/(\delta\, L\, \tau/4)^3 + \gamma, \tag{8.7}$$

where γ is a small positive parameter determining the isotropic absorption in the CLC (see Chaps. 5 and 6).

So, naturally, the threshold gain becomes higher if there is absorption in the CLC. In the case of locally anisotropic absorption, the values of the threshold gain depend on the order parameter S and are different for the EM frequencies at the opposite stop-band edges. At one edge (where the Borrmann effect occurs), it may approach the value given by (8.6), while at the opposite stop-band edge the threshold gain may be significantly higher. If the absorption is due only to the dye molecules, the threshold gain becomes dependent on the order parameter S and its minimal value is given by

$$g = \delta(n\pi)^2/(\delta\, L\, \tau/4)^3 + \varepsilon_1(1 - S)/3\mathrm{Re}[\varepsilon_0], \tag{8.8}$$

where ε_1 is the imaginary part of the principal value of the dielectric tensor at $S = 1$ determined by absorption by the dye molecules, i.e., under the assumption that the absorption is due only to the dye molecules. It is interesting to note that the lasing thresholds corresponding to (8.7) and (8.8) relate to a vanishingly small population inversion of the lasing transition levels (vanishingly low amplification at the lasing frequency).

Note that the anomalously strong absorption effect [11, 18] at the pumping wave frequency may influence the lowering of the lasing threshold gain in the case of locally anisotropic absorption in the CLC even more strongly than in the case of isotropic absorption [19, 20]. In reality, if the absorption oscillators (at the pumping wave frequency) in the dye molecules and the oscillators corresponding to the lasing frequency have the same orientations, the lasing wave absorption will be suppressed and the pumping wave absorption will be enhanced if the lasing frequency coincides with the low-frequency stop-band edge where the Borrmann effect takes place and the pumping wave frequency coincides with the high-frequency stop-band edge where the absorption is enhanced.

8.5 Defect Mode

The defect mode studied recently by many authors [21–30] is a localized solution of the boundary problem for the structure (DMS) shown in Fig. 6.1. The solution to the boundary problem is carried out in a similar way to what was done for the CLC layer above. Below we give the final results for the case of locally anisotropic absorption in CLC layers (all the simplifications assumed above for the CLC layer are also implemented for the DMS).

There is an option to obtain formulas determining the optical properties of the structure depicted in Fig. 6.1 via the solutions found for a single CLC layer [15]. If we use the expressions for the amplitude transmission, $T(L)$, and reflection, $R(L)$, coefficients for a single cholesteric layer (6.5), the transmission and reflection intensity coefficients for the whole structure may be written

$$T(d, L) = \left|\left[T_e T_d \exp(ikd)\right]\left[\left[1 - \exp(2ikd) R_d R_u\right]\right|^2, \tag{8.9}$$

$$R(d, L) = \left|\left\{R_e + R_u T_e T_u \exp(2ikd)/\left[1 - \exp(2ikd) R_d R_u\right]\right\}\right|^2, \tag{8.10}$$

where d is the defect layer thickness, $R_e(T_e)$, $R_u(T_u)$, and $R_d(T_d)$ (Fig. 6.1) are the amplitude reflection (transmission) coefficients of the CLC layers (Fig. 5.1) for light incidences on the outer (top) layer surface, the inner top CLC layer surface from the inserted defect layer, and the inner bottom CLC layer surface from the inserted defect layer, respectively. It is assumed in deriving (8.9) and (8.10) that the external beam is only incident on the structure (Fig. 6.1) from above.

Because we are focusing on CLCs with locally anisotropic absorption, the parameters applied in the expressions for the DMS reflection (8.10) and transmission (8.9) coefficients have to be taken from (8.1) and (8.2). The cases of DMs with non-absorbing and isotropically absorbing CLCs were studied in Chap. 6 [21–23].

Similarly to the case for EMs, the DM frequency ω_D is determined by the zeros of the determinant of the system (dispersion equation) corresponding to the boundary problem solution for the structure depicted in Fig. 6.1 [21–23]. The corresponding dispersion equation is

$$\{\exp(2ikd)\sin^2 qL - \exp(-i\tau L)$$
$$[(\tau q/\kappa^2)\cos qL + i((\tau T/2\kappa)^2 + (q/\kappa)^2 - 1)\sin L]^2/\delta^2]\} = 0. \quad (8.11)$$

We recall that, for a non-absorbing CLC, at a finite length L, (8.11) does not reach zero for a real value of ω. However, it can reach zero when ω is complex.

As we would expect, the local absorption anisotropy reveals itself through the DMS reflection and transmission spectra due to the Borrmann effect [11–13]. Figure 8.11 illustrates the Borrmann effect in reflection for a DMS with local absorption anisotropy. This figure, which presents the calculation results, shows that the stop-band edges are no longer equivalent with regard to the scattering spectra (for non-absorbing CLCs or CLCs with isotropic absorption, the reflection is symmetric relative to the stop-band center).

The value of the reflection coefficient close to the frequencies of one stop-band edge is much larger than the value close to the frequencies of the other stop-band edge. Figure 8.12 illustrates the Borrmann effect in transmission for a DMS with local absorption anisotropy. Figure 8.12 shows that the stop-band edges are also inequivalent with regard to the transmission spectra.

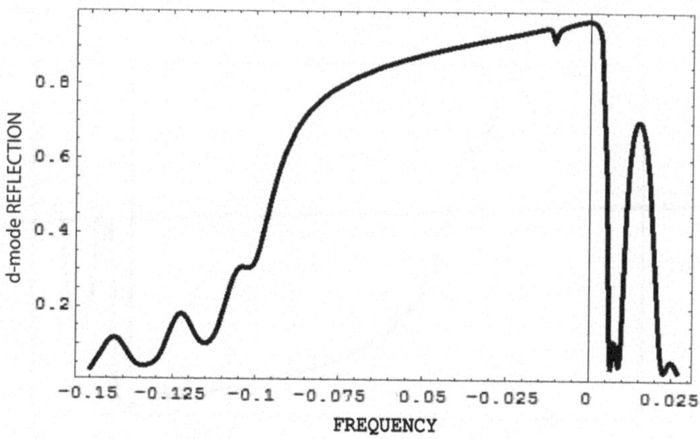

Fig. 8.11 Borrmann effect in reflection for the DMS (see Fig. 6.1) for a CLC with locally anisotropic absorption when $\delta = 0.05 + 0.003i$, $N = 75$, $d/p = 0.1$

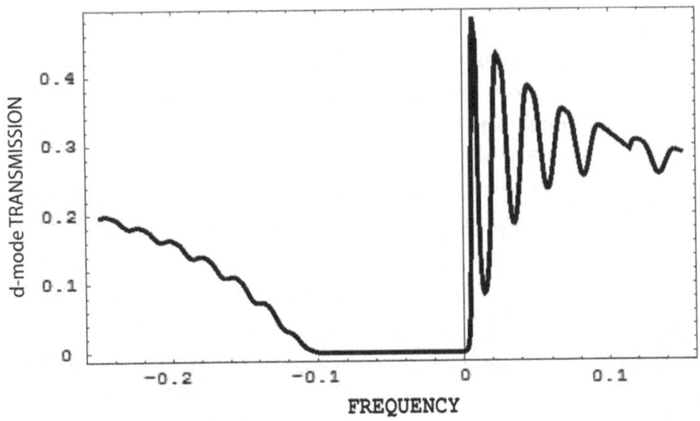

Fig. 8.12 Borrmann effect in transmission for the DMS for a CLC with locally anisotropic absorption when $\delta = 0.05 + 0.003\mathrm{i}$, $N = 75$, $d/p = 0.1$

The value of the transmission coefficient close to the same frequency as for the reflection stop-band edge showing an enhancement of reflection is much larger than close to the frequency of the opposite stop-band edge.

Naturally, the total absorption (1-*R*-*T*) has to be different at the stop-band edges.

Figure 8.13 shows that, due to the Borrmann effect in the transmission and reflection spectra for a CLC with local absorption anisotropy, a suppression of the total absorption occurs at frequencies close to one of the stop-band edges.

The strength of the discussed absorption suppression effect depends on the value of the order parameter *S* and the thickness of the CLC layers. The effect disappears when the order parameter is zero and reaches a maximum when it is equal to one (complete absorption suppression occurs for infinitely thick CLC layers).

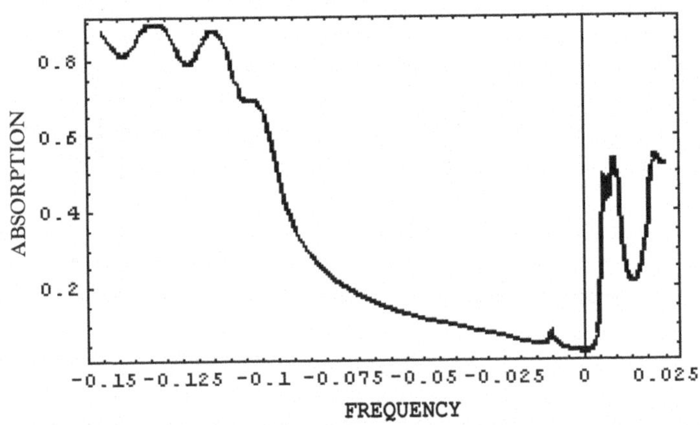

Fig. 8.13 Total absorption in the DMS for a CLC with locally anisotropic absorption (see Fig. 6.1) when $\delta = 0.05 + 0.003\mathrm{i}$, $N = 75$, $d/p = 0.1$

Similarly to the case of a CLC layer, the manifestation of the Borrmann effect is strongly dependent on the CLC order parameter S. The calculation results for the DMS reflection, transmission, and absorption spectra for different values of the order parameter ($S = 0, 0.3, 0.5, 1$), under the assumption that the absorption is entirely due to the dye molecules and that the degree of ordering of the dye molecules is relevant to S, are presented at Figs. 8.14, 8.15 and 8.16, respectively.

"The reflection and transmission spectra for the DMS show that the DM lifetime in the case of a local absorption anisotropy in the CLC depends on the position of the localized mode frequency relative to the stop-band edge frequency, growing as

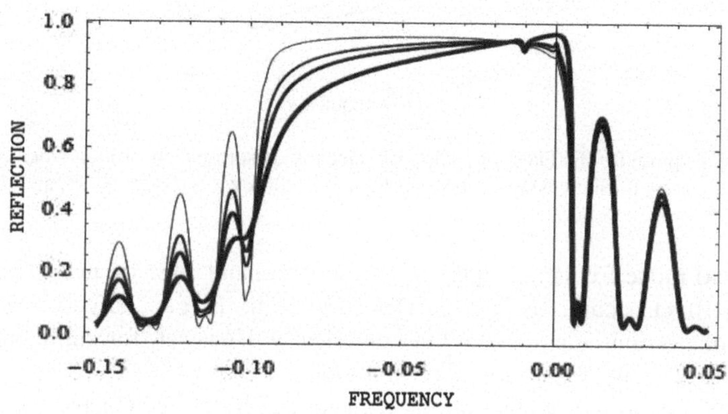

Fig. 8.14 Reflection for the DMS (see Fig. 6.1) of a CLC with locally anisotropic absorption when $S = 0, 0.3, 0.5, 1$ (thicker curves for increasing S), $\mathrm{Im}[\varepsilon_1]/\mathrm{Re}[\varepsilon_0] = 0.003$, $\delta = 0.05$, $N = 75$, and $d/p = 0.1$

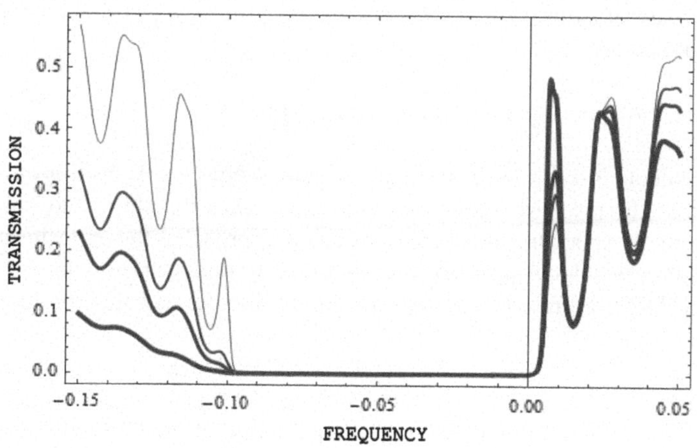

Fig. 8.15 Transmission for the DMS of a CLC with locally anisotropic absorption when $S = 0, 0.3, 0.5, 1$ (thicker curves for increasing S), $\mathrm{Im}[\varepsilon_1]/\mathrm{Re}[\varepsilon_0] = 0.003$, $\delta = 0.05$, $N = 75$, and $d/p = 0.1$

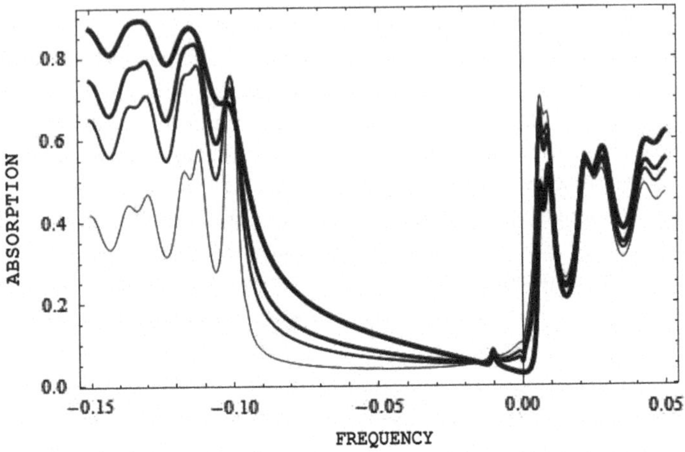

Fig. 8.16 Absorption for the DMS of a CLC with locally anisotropic absorption when $S = 0, 0.3, 0.5, 1$ (thicker curves for increasing S), $\mathrm{Im}[\varepsilon_1]/\mathrm{Re}[\varepsilon_0] = 0.003$, $\delta = 0.05$, $N = 75$, and $d/p = 0.1$

the localized mode frequency approaches the stop-band edge frequency where the Borrmann effect occurs. To find the DM lifetime in the general case, we have to solve the dispersion equations (8.11) numerically. However, for sufficiently thick CLC layers, an analytic solution may be found.

In the case of isotropic absorption characterized by γ (see Chaps. 5 and 6), the DM lifetime τ_{DM} decreases compared to the case of zero absorption ($\gamma = 0$), and for sufficiently thick CLC layers in the DMS, an analytic expression for the DM lifetime may be found (see [21]). Because the corresponding expression is rather complicated for a general position of the DM frequency ω_{DM} inside the stop band, we give τ_{DM} here for the case of isotropic CLC absorption when the ω_{DM} position coincides with the stop-band center:

$$\tau_{DM} = 2/\omega_{DM}\big[(4/3\pi)(p/\delta L)\exp[-2\pi\delta L/p] + \gamma\big]. \tag{8.12}$$

As the calculations show in the case of anisotropic local absorption, the DM lifetime grows as the DM frequency ω_{DM} approaches the stop-band edge where the Borrmann effect occurs and decreases as ω_{DM} approaches the opposite stop-band edge. For this case of local anisotropic absorption, we find the following S-dependent bound on the DM lifetime when the ω_{DM} position coincides with the stop-band center:

$$\tau_{DM} < 2/[\omega_{DM}\varepsilon_1(1 - S)/3\mathrm{Re}[\varepsilon_0]], \tag{8.13}$$

where ε_1 is the imaginary part of the principal value of the CLC dielectric tensor at $S = 1$ determined by the dye molecule absorption (under the assumption that absorption in the CLC is entirely due to the dye molecules).

Regarding the DFB lasing threshold for a CLC with locally anisotropic absorption, this may be found in a similar way to what was done for the EM, from a numerical solution of the dispersion (8.11). Nevertheless, some qualitative conclusions about the increase in the threshold compared to the case of non-absorbing CLCs may be formulated from the analytic solution found in the case of thick CLC layers. In particular, the DFB lasing threshold at the DM frequency ω_{DM} increases differently for different positions of ω_{DM} in the stop band and is strongly dependent on the value of the order parameter S. For $S = 1$, the DFB lasing threshold may approach its value for the case of non-absorbing CLCs [22] if the ω_{DM} position is close to the stop-band edge where the Borrmann effect occurs.

8.6 Conclusion

In this chapter, we began by studying the influence of the Borrmann effect (studied previously for CLC transmission and reflection spectra [11–14]) on the properties of the localized modes in CLCs with local absorption anisotropy. The analytical description of the EMs and DMs (neglecting polarization mixing) brings out typical EM and DM features related to the existence of a local absorption anisotropy in the CLC, e.g., the dependence of the localized mode lifetime and lasing threshold on the position of localized mode frequency relative to the stop-band edges and the sample thickness (the thickness dependence has recently been studied experimentally in [31]).

The results obtained here for EMs and DMs (see also [32]) clarify the physics of the absorption suppression due to the Borrmann effect and the dependence of the localized mode lifetimes on their frequency positions relative to the stop-band edges. The predictions formulated above (e.g., regarding the different lasing thresholds at the lasing frequencies at opposite stop-band edges) are open for experimental verification and may be used to optimize experiments on DFB lasing in CLCs. One specific possibility in the corresponding experiments in CLCs would be to study lasing at the opposite stop-band edges without actually changing the lasing frequency, changing instead the CLC pitch by temperature variations or by application of magnetic or electric fields applied to the CLC.

The influence of the Borrmann effect on the CLC layer absorption spectra studied above predicts new options for experimental investigation of this effect using luminescence techniques (or conversely, studying luminescence under conditions when the Borrmann effect is realized).

Note that the obtained results are qualitatively applicable to the corresponding localized electromagnetic modes in any periodic media, and may be regarded as a useful guide in studies of localized modes under conditions where the Borrmann effect occurs. For example, the results for CLCs with local anisotropy of absorption may be useful for optimizing DFB lasing in general. In fact, the corresponding theoretical predictions show which one of the two stop-band frequencies will be preferable in order to obtain the lowest lasing threshold.

References

1. Y.-C. Yang, C.-S. Kee, J.-E. Kim et al., Phys. Rev. E **60**, 6852 (1999)
2. V.I. Kopp, A.Z. Genack, Phys. Rev. Lett. **89**, 033901 (2003)
3. J. Schmidtke, W. Stille, H. Finkelmann, Phys. Rev. Lett. **90**, 083902 (2003)
4. P.V. Shibaev, V.I. Kopp, A.Z. Genack, J. Phys. Chem. B **107**, 6961 (2003)
5. E. Yablonovitch, T.J. Gmitter, R.D. Meade et al., Phys. Rev. Lett. **67**, 3380–3380 (1991)
6. M. Becchi, S. Ponti, J.A. Reyes, C. Oldano, Phys. Rev. B **70**, 033103 (2004)
7. J. Schmidtke, W. Stille, Eur. Phys. J. **90**, 353 (2003)
8. H.deVries, Acta Crystallogr. **4**, 219 (1951)
9. E.I. Kats, JETP **32**, 1004 (1971)
10. P.G deGennes, J. Prost, *The Physics of Liquid Crystals* (Clarendon Press, Oxford, UK, 1993)
11. V.A. Belyakov, V.E. Dmitrienko, Optics of chiral liquid crystals, in *Soviet Scientific Reviews/Section A: Physics Reviews*, ed. by I.M. Khalatnikov (Harwood Academic Publisher, London, UK, 1989)
12. V.A. Belyakov, *Diffraction Optics of Complex Structured Periodic Media* (Springer, New York, 1992), p. 80 (Chap. 4)
13. S. Chandrasekhar, *Liquid Crystals* (Cambridge Unversity Press, Cambridge, 1977)
14. S.N. Aronishidze, V.E. Dmitrienko, D.G. Khoshtariya, G.S. Chilaya, Pis'ma ZhETF **32**, 19 (1980)
15. V.A. Belyakov, S.V. Semenov, JETP **122**, 932 (2016)
16. V.A. Belyakov, S.V. Semenov, JETP **109**, 687 (2009)
17. V.I. Kopp, Z.-Q. Zhang, A.Z. Genack, Prog. Quantum Electron. **27**, 369
18. V.A. Belyakov, A.A. Gevorgian, O.S. Eritsian, N.V. Shipov, Zhurn. Tekhn. Fiz. **57**, 1418 (1987) [Sov. Phys. Tech. Phys. **32**, 843 (1987), English translation; Sov. Phys. Crystallogr. **33**, 337 (1988)]
19. V.A. Belyakov, Mol. Cryst. Liquid Cryst. **453**, 43–51 (2006); Ferroelectrics **344**, 163 (2006)
20. Y. Matsuhisa, Y. Huang, Y. Zhou, et al., Appl. Phys. Lett. **90**, 091114 (2007)
21. V.A. Belyakov, Mol. Cryst. Liquid Cryst. **494**, 127 (2008); Ferroelectrics **364**, 33 (2008)
22. V.A. Belyakov, S.V. Semenov, JETP **112**, 694 (2011)
23. V.A. Belyakov, in *New Developments in Liquid Crystals and Applications*, ed. by P.K. Choundry (Nova Publishers, New York, 2013), p. 199 (Chap. 7)
24. V.G. Arkhipkin, V.A. Gunyakov, S.A. Myslovets, et al. JETP **112**, 577 (2011)
25. M. Faryad, A. Lakhtakia, M.W. McCall, Adv. Opt. Photonics **6,** 225
26. F. Wang, A. Lakhtakia, Opt. Exp. **13**, 7319 (2005)
27. A.H. Gevorgyan, M.Z. Harutyunyan, Phys. Rev. E **76**, 031701 (2007)
28. V.A. Belyakov, Mol. Cryst. Liquid Cryst. **559**, 50 (2012)
29. V.A. Belyakov, Mol. Cryst. Liquid Cryst. **559**, 39 (2012)
30. V.A. Belyakov, S.V. Semenov, JETP **118**, 798 (2014)
31. G. Sanz-Enguita, J. Ortega, C.L. Folcia et al., J.Appl.Phys. **119**, 073102 (2016)
32. V.A. Belyakov, Mol. Cryst. Liquid Cryst. **612**, 81 (2015)

Chapter 9
Localized Modes and Enhancement of Some Optical Phenomena in Photonic Crystals

9.1 Introduction

As shown in Chaps. 5–8, the localized modes influence essentially the linear optics of spiral photonic crystals and photonic crystals in general, which occur at frequencies in and near the stop band. In addition, recent studies have shown that many optical phenomena that have not been considered or only partially considered in this book also reveal some unusual properties at the frequencies in the stop-band range. For example, an enhancement of the nonlinear second harmonic generation (SHG) in chiral liquid crystals (which are spiral photonic crystals) at the edge of the stop band [1–6] and SHG at the frequency of the DM [7–9] have been studied. The corresponding enhancement phenomena have also been studied in conventional photonic crystals [10–14].

In addition to the phenomena mentioned above, there is a general case of nonlinear frequency conversion, sum frequency generation, Cherenkov radiation, etc. Some of the phenomena discussed have been observed in spiral media, so the theory developed in the preceding chapters is directly applicable, and some have been observed in conventional photonic crystals where the theory can be applied as a qualitative guide. In general, the situation with the theoretical description of localized mode effects in conventional photonic crystals appears to be as follows. Almost all publications reporting experimental results on the localized modes are accompanied by numerical calculations of the measured quantities showing good agreement with the measured results. However, in many cases, the theoretical interpretation of the observed results does not provide a clear physical image of the observed phenomenon. Indeed, the most common explanation is of the type "the effect is enhanced near to the stop-band edge."

In the following, we interpret the experimental results on localized modes in terms of the parameters introduced in Chaps. 5–8 for spiral photonic crystals, and using the generality of the results as regards spiral photonic crystals, we also interpret the experimental results for conventional photonic crystals in the same

V. Belyakov, *Diffraction Optics of Complex-Structured Periodic Media*, Springer Series in Optical Sciences 203,
https://doi.org/10.1007/978-3-319-43482-7_9

terms. This approach can be regarded as a proof of the usefulness of the analytic results obtained under the given simplifying assumptions.

9.2 Nonlinear Second Harmonic Generation in Photonic LC

An enhancement of nonlinear second harmonic generation (SHG) in photonic liquid crystals has been reported under some circumstances in a number of papers [1–3]. A typical result is the observation of a sharp maximum in the temperature dependence of the SHG intensity for a definite temperature value (see Fig. 9.1). A similar enhancement can be observed in the dependence of the SHG intensity on some external influence, for example, an electric field. This happens when the SHG or pumping wave frequency coincides with the stop-band edge. If the changes in the dielectric constant at boundaries within the structure under consideration are negligible, effects due to localized modes are only revealed for light with the diffracting polarization. For example, in the case of a collinear geometry, this polarization is right or left circular polarization depending on the chirality of the spiral medium.

The theory explaining the observed SHG intensity enhancement [15] and predicting discrete enhanced SHG frequencies says nothing about the localized modes. However, if the expressions for enhanced SHG frequencies [15] are compared with the expressions for EM frequencies [16], it so happens that the frequencies coincide. We may therefore claim that SHG enhancement occurs just at the EM frequencies. This conclusion is quite predictable from very general arguments. In particular, (1.8) and (4.10) for the propagation of light and for nonlinear frequency conversion, respectively, have the same homogeneous parts. This is why the frequencies of specific features of the solution in periodic media are the same for both phenomena. The same can be seen from the corresponding experimental curves: the observed enhanced SHG frequency coincides with a minimum in the linear reflection spectrum. However, as shown in Chap. 5 the EM frequencies coincide exactly with the minima in the reflection spectrum.

If changes in the dielectric constant at the boundaries of the structure under consideration are non-negligible, nonlinear frequency conversion effects due to the localized modes can be further enhanced because a standing wave is formed at the pumping frequency [4, 5].

A similar SHG enhancement can be obtained without changes in the dielectric constant at boundaries, for a certain complex pumping wave configuration, namely in the form of two counter-propagating plane pumping waves [3] (see Fig. 9.2).

And finally, an even stronger SHG enhancement can be observed if the fundamental and SHG waves simultaneously experience diffraction of different orders in the structure under consideration. The experimental observation of this diffraction SHG enhancement [6] has been shown to confirm the theoretical predictions [17–19] (Fig. 9.3).

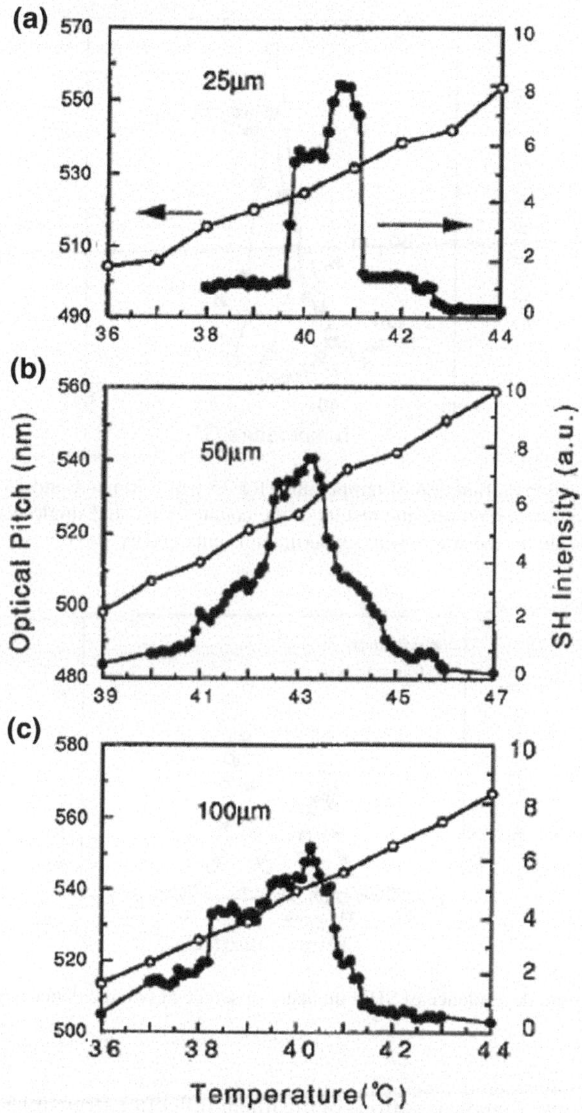

Fig. 9.1 **a–c** Temperature dependence of the SH-light intensity and optical pitch in three cells of different thickness; **a** 25 μm; **b** 50 μm; and **c** 100 μm. The increase of the SH-light intensity is observed when the optical pitch is about 532 nm, the SH-light wavelength [2]

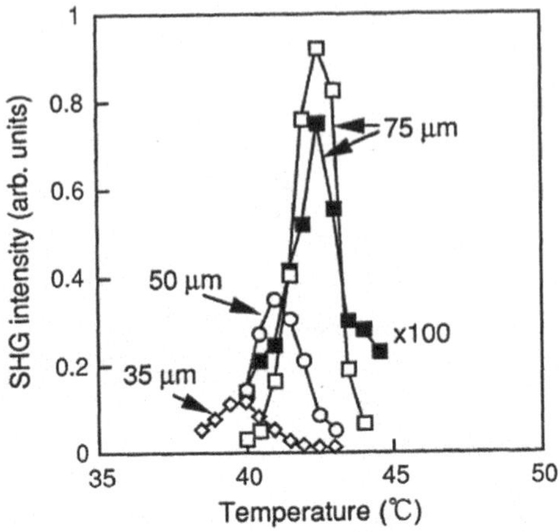

Fig. 9.2 SH intensity as a function of temperature for 35-μm-, 50-μm-, and 75-μm-thick cells. Open and closed symbols represent the results of the counter-wave and single-wave experiments, respectively. The scale for the single-wave experiment is enlarged by 100 times (!) [3]

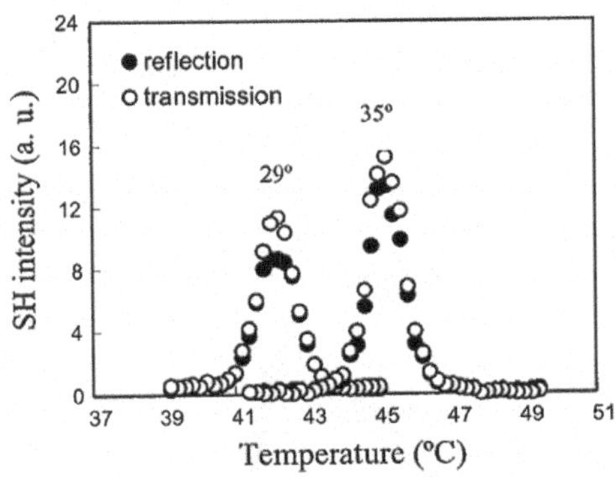

Fig. 9.3 Temperature dependence of SHG intensity observed at two incidence angles, 29° and 35° [6]

Figure 9.4 shows that the positions of the linear reflection (transmission) minimum and SHG enhancement are connected in a similar manner as for the EM, i.e., the pumping wave reflection minimum coincides with the SHG intensity maximum (see also [17]).

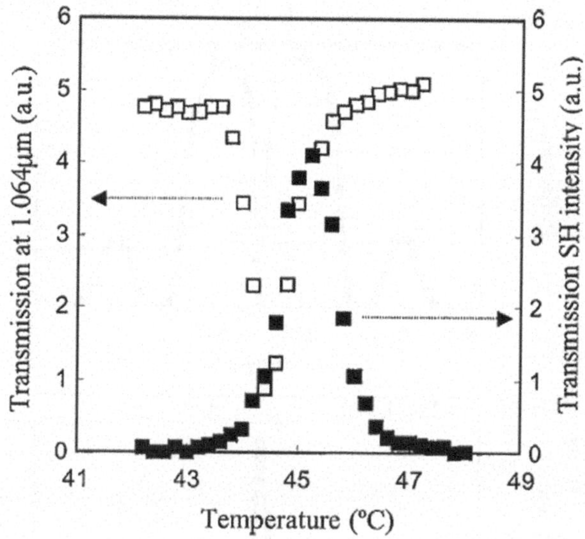

Fig. 9.4 Relation of the full-pitch band (transmittance loss) to SHG enhancement when the fundamental wave is obliquely incident at 35° [6]

9.3 Nonlinear Frequency Conversion at Defect Mode Structure

An enhancement of nonlinear second harmonic generation (SHG) in defect photonic liquid crystal structures has been predicted in a number of theoretical papers [7–9]. The enhancement occurs at SHG frequencies coinciding with the DM frequencies. The situation for nonlinear frequency conversion enhancement at the DM frequencies is similar to the case of nonlinear conversion at the EM frequencies studied in the last section.

In the non-depleted pumping wave approximation, the field $\mathbf{E}_s$ at the SHG frequency is described by similar equations to (6.1) for the DM boundary problem (with the amplitudes of the external fields replaced by values proportional to the corresponding components of the nonlinear polarization $\mathbf{P}(r,t)_{NL}$). The equations obtained as mentioned for SHG in the defect structure shown in Fig. 6.1 were solved numerically in [8]. The main results of these calculations concerning the spectral dependencies of the SHG intensity and linear reflection are presented in Fig. 9.5.

Similarly to the case of SHG at the EM frequency, the SHG enhancement in defect photonic liquid crystal structures occurs when the SHG frequency coincides with the DM frequency. As in the case of SHG at the EM frequency, the location of the SHG frequency inside the stop band corresponding to the SHG enhancement coincides with the minimum of the reflection coefficient of the defect structure. The authors of the paper [9] directly relate the SHG enhancement to the defect-localized mode.

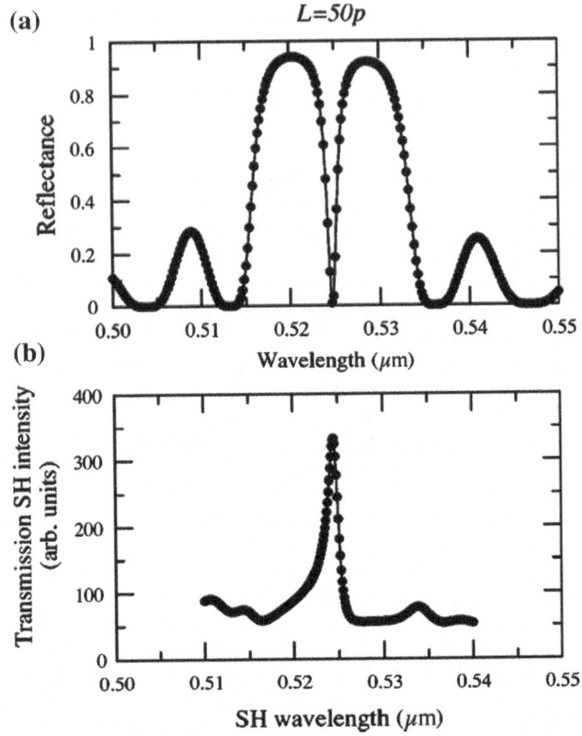

Fig. 9.5 **a** Linear reflectance and **b** transmission SHG spectra for RCP incident wave. $L = 50p$ [8]

9.4 Cherenkov Radiation at Localized Mode Frequency

Some peculiarities of Cherenkov radiation in spiral media related to both its angular distribution and its intensity distribution and polarization properties have been theoretically predicted in [20, 21]. We shall reproduce here the main results of [20, 21] and show that the key features of Cherenkov radiation in spiral media are connected to the edge-localized modes studied in Chap. 5.

In Chap. 3, we considered a variety of spiral media (in particular, CLCs) which sometimes reveal absolutely unique optical properties [22–24]. It is natural that these unusual properties manifest themselves in Cherenkov radiation, i.e., in the radiation from charged particles in cholesteric liquid crystals. As mentioned in Chap. 3, the properties of cholesteric crystals have a one-dimensional spatial periodicity with period in the range of optical wavelength. The periodicity of the dielectric properties is due to the spatial variations in the orientation of the principal axis of the dielectric permittivity tensor. We have already seen that the spatial periodicity of the dielectric tensor results in the diffraction of light at certain specific frequencies and, as a consequence, produces some very interesting peculiarities in the optical properties at these frequencies (wavelengths). For the same reason, Cherenkov radiation at these frequencies has certain peculiarities and unusual polarization properties.

In addition to the well-known Cherenkov radiation cone, whose axis lies along the particle velocity, there is a diffraction cone in chiral liquid crystals whose axis differs from the former [20, 21]. The spectrum of radiation emitted into the diffraction cone is essentially different from that of the Cherenkov cone, since it contains only frequencies which experience diffraction in the liquid crystal. Furthermore, the polarization properties of this radiation are rather unusual. In general, Cherenkov radiation from chiral liquid crystals is elliptically polarized, the ellipticity depending on the angle between the helical axis and the direction of emission. The differential parameters of the radiation (relative to frequency and angle) [20, 21] also differ significantly from those of Cherenkov radiation in homogeneous media.

Emission from a Sample with Finite Dimensions. Below we consider the coherent radiation induced by charged particles in cholesteric crystals. Let us consider the Cherenkov radiation of a relativistic charged particle with charge e which moves with constant velocity through a cholesteric crystal, the particle velocity v being greater than the phase velocity of light c_p in this medium.

The specific properties of Cherenkov radiation in cholesteric crystals are connected to the form of the dielectric permittivity tensor (1.6). Generally speaking, in addition to the wave with wave vector k_1 whose propagation direction lies on the Cherenkov cone, there are more waves in a cholesteric crystal. These waves are related to the former wave by the Bragg condition $k_2 = k_1 + \tau$. Since the dielectric anisotropy δ is small, we may to a first approximation take into account only one such wave. Hence, we look for the radiation field (its temporal and spatial Fourier components) as a superposition of two plane waves $E(k_1, k_2) = \left(E_1 e^{ik1r} + E_2 e^{ik2r}\right) e^{i\omega t}$ where k_1 lies in the Cherenkov cone and $k_2 = k_1 + \tau$. The equations for E_1 and E_2 are

$$
\begin{aligned}
&(\varepsilon_0 - k_1^2/q^2)\mathbf{E}_1 + \varepsilon_{-1}\mathbf{E}_2 = (ie/2\pi^2\omega)\left[\mathbf{v} - k_1(k_1 v)/q^2\right]\delta(\omega - k_1 v) \\
&\varepsilon_1\mathbf{E}_1 + (\varepsilon_0 - k_2^2/q^2)\mathbf{E}_2 = 0
\end{aligned}
\tag{9.1}
$$

where ε_i are the harmonics of the Fourier expansion of the dielectric tensor (2.15).

The radiation field for a finite sample is a superposition of the solutions of (9.1) and the solutions of a homogeneous system obtained from (9.1) by canceling the right-hand sides. The coefficients in this superposition are to be determined from the boundary conditions.

For an arbitrary direction of the particle velocity with respect to the cholesteric axis, the solutions of (9.1) and the boundary problem cannot be represented explicitly in analytical form. Below we analyze a case allowing simplification of the general relationships.

We consider the case when a particle moves at an angle to the surface of a sample, "with thickness L," whose cholesteric axis is perpendicular to the surface and where there is a direction on the Cherenkov cone which coincides with the cholesteric axis (Fig. 9.6). We find for this direction that $\omega_B = \tau c_P/2$ and the vector k_2 are also directed

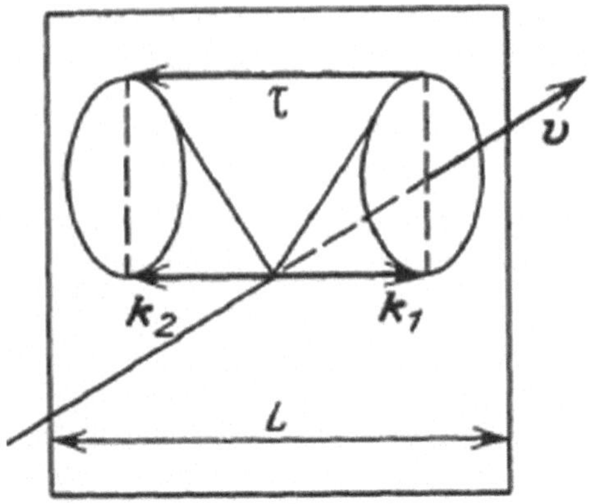

Fig. 9.6 Geometry of Cherenkov radiation which provides backward diffraction scattering

along the cholesteric axis, but opposite to k_1. Near these values of ω_B, k_1, and k_2, the system (9.1) splits into two independent systems (each with two equations), one of them describing the wave with diffracting circular polarization and the other a wave with the opposite circular polarization which does not experience diffraction in the CLC [20, 21]. The solution of (9.1) for the diffracting polarization may be written explicitly in terms of the boundary conditions [21]. The amplitudes E_i oscillate and depend strongly on the frequency, with their minima occurring at $\nu = 0$ and their maxima at $|\nu| = \delta/2$. The intensity of radiation leaving the crystal in the directions of the Cherenkov and diffraction cones, I_1 and I_2, is given by the following expressions (see also Fig. 9.7):

$$I_1 = I_e\left\{\nu^2 - \left[(\delta/2)^2/2\right][1 + (\sin 2kLq)/(2kLq)]\right\}/\left[\nu^2 - (\delta/2)^2 \cos^2 kLq\right]$$
$$I_2 = I_e(1/2)\left\{\left[(\delta/2)^2\right][1 - (\sin 2kLq)/(2kLq)]\right\}/\left[\nu^2 - (\delta/2)^2 \cos^2 kLq\right] \quad (9.2)$$

where I_e is the spectral density of radiation for the corresponding homogeneous sample,

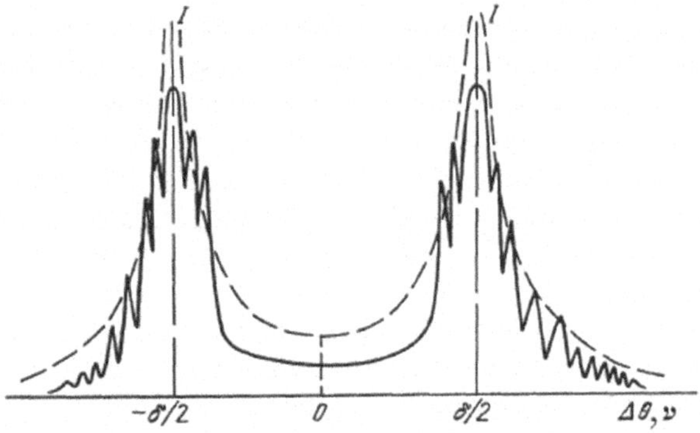

Fig. 9.7 Angular (frequency) dependence of the intensities of waves emitted into the diffraction cone in an unlimited cholesteric crystal (dashed line) and in a sample with finite dimensions (solid line) [18]

$$k = \omega_B/c_p,\ q = [\nu^2 - (\delta/2)^2]^{1/2},\ \text{and}\ \nu = 2(\omega - \omega_B)/\omega_B,\ \omega_B = c\tau/2\varepsilon_0^{1/2}$$

It follows from the above formulas that the intensities of radiation at the Cherenkov and diffraction cones vary rapidly as functions of frequency near ω_B and experience sharp beats for $|\nu| \sim \delta/2$, because of the diffractive scattering of light with the diffracting polarization. At $|\nu| \gg \delta$, the radiation intensity tends to I_e in the Cherenkov cone and zero in the diffraction cone. The radiation at the diffraction cone is circularly polarized and corresponds to the diffracting polarization; whereas, the radiation in the Cherenkov cone is determined by interference between the Cherenkov radiation fields due to the two eigenpolarizations and the changes with frequency, tending to the linear polarization at $|\nu| \gg \delta$. Note that the maximum intensity of radiation (the quantity which is differential in frequency and azimuthal angle) is reached, practically speaking, at the boundary of the selective reflection band at $\nu^2 - (\delta/2)^2 = (\pi/kL)^2$ for both the Cherenkov and the diffraction cone, this maximal intensity being equal to $I_1 = I_2 = (1/2)I_e(\delta kL/2\pi)^2$. Hence, it is $(\delta kL/2\pi)^2$ times greater than the corresponding value for a homogeneous medium. Naturally, we must bear in mind that this result is only valid for a sample whose thickness does not lead to significant absorption. Thus, the periodicity of cholesteric crystals changes the Cherenkov radiation qualitatively near certain selected frequencies and directions. The angular and frequency ranges in which the above-mentioned changes take place are of the order of δ. Hence, we can say that, not only the integral parameters of radiation, but also the differential parameters are accessible to experimental observation.

It follows from the above discussion that the maxima in the intensity of the beats at the diffraction Cherenkov cone just correspond to the EM frequencies. This is why the frequency widths of the corresponding maxima are different, being determined by the lifetime of the corresponding EM. This means it would be possible to apply a time-delay technique to determine the EM lifetimes directly. If Cherenkov radiation is detected from an individual charged particle, the delay in the time of observation of Cherenkov radiation photons from different maxima at the diffraction Cherenkov cone has to be different because it is determined by the corresponding EM lifetime. Hence, an experimental observation of the Cherenkov radiation in the diffraction Cherenkov cone of a CLC can be used to measure the EM lifetime in photonic liquid crystals.

9.5 Effects of Localized Modes in Conventional Photonic Crystals

Here, we discuss some enhanced phenomena [10–12] observed experimentally in conventional photonic crystals and propose an interpretation in terms of localized modes. In particular, we describe enhancement phenomena in conventional photonic crystals for SHG [10], third harmonic generation (THG) [12], and sum frequency

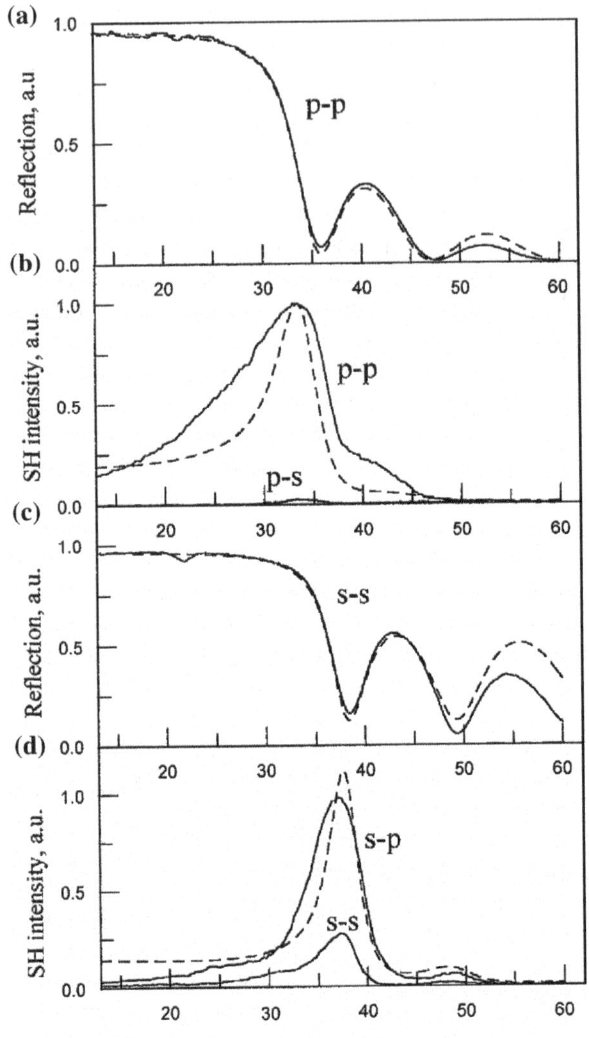

Fig. 9.8 Angular tuning curves of (**a**), (**c**) linear Bragg reflection and (**b**), (**d**) the reflected SH signal for different sets of polarization. The dashed curves are calculated from the theory, and the solid curves are the experimental data [10]

generation [11], selecting examples from many experimental studies where enhancement effects at the stop-band edge frequency are attributed to the decrease in the light group velocity, the increase in the density of states, and the growth of the wave field at this frequency [13, 14].

The enhancement of SHG generation in a $ZnS–SrF_2$ periodic structure near the photonic band edge [10] was observed when femtosecond laser pumping pulses were applied to the sample. The SHG was measured as a function of the pumping wave incidence angle. The SHG intensity enhancement (see Fig. 9.8) corresponded exactly to the pumping wave incidence angle coinciding with the first minimum in the linear reflection curve for the pumping frequency. As we know from Chap. 5, this frequency corresponds to the first EM frequency.

Enhancement of THG generation in a periodic silicone structure [12] was observed when the pumping wave incidence angle coincided with the first minimum in the linear reflection curve for the fundamental frequency. The authors attribute the observed enhancement (see Fig. 9.9) to an increase in the pumping wave field in the structure. This explanation is in full agreement with an interpretation attributing the observed effect to a localized mode. In reality (see Fig. 5.4), the pumping wave for this frequency is concentrated inside the periodic structure and the anomalously strong absorption effect occurs for the pumping wave, resulting in an enhancement of THG generation.

Enhancement of sum frequency generation in a ZnS–SrF_2 periodic structure [11] was observed when both pumping wave frequencies were close to the opposite stop-band edge frequencies. In the non-depleted pumping wave approximation, the sum frequency field $\mathbf{E}_s$ is given by

$$\text{rotrot}\,\mathbf{E}(r,t)_s + \left(1/c^2\right)\partial^2\mathbf{D}(r,t)_s/\partial t^2 = -\left(4\pi/c^2\right)\partial^2\mathbf{P}(r,t)_{NL}/\partial t^2, \qquad (9.3)$$

where $\mathbf{D}(r,t)_s = \varepsilon(z)\mathbf{E}(r,t)_s$ and the nonlinear polarization $\mathbf{P}(r,t)_{NL}$ are proportional to the product of the amplitudes of the pumping waves with different frequencies. Hence, sum frequency generation is enhanced if the pumping wave frequency coincides with the EM frequency and its amplitude increases correspondingly. Naturally, a maximum increase occurs in the sum frequency field $\mathbf{E}_s$ "if both pumping wave frequencies coincide with the EM frequencies at the opposite stop-band edges." Figure 9.10 shows that, in the experiment, the maximal enhancement of sum frequency generation corresponds exactly to 25° when, according to the calculations, both pumping wave frequencies coincide with the EM frequencies.

The explanations of the enhancement effects proposed by the authors of the publications [10–12] do not contradict the explanation in terms of localized modes. In particular, all the phenomena discussed here, i.e., the decrease in the light group velocity, increase in the density of states, and growth of the wave field, are naturally revealed by the localized modes.

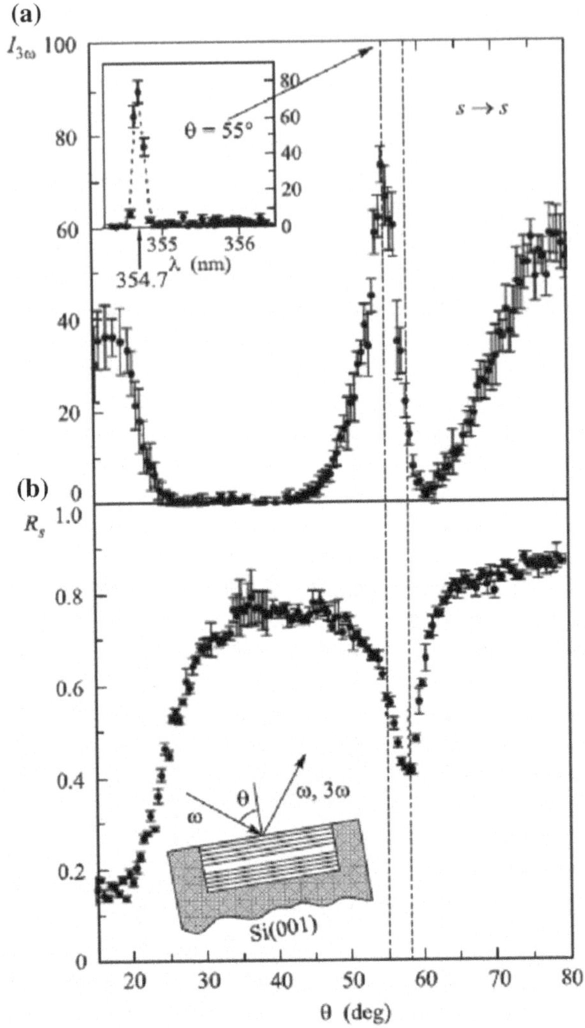

Fig. 9.9 **a** Intensity of s-polarized TH $I_{3\omega}$ versus θ for s-polarized pumping wave ($\lambda_\omega = 1064$ nm) measured in with $\lambda_{mc} = 1300$ nm. The dashed lines visualize the angular shift between TH resonance and the resonance mode. Inset: Spectrum of optical signal close to $\lambda_{3\omega}$ measured for the angle of incidence corresponding to the TH resonance angle ($\theta = 55°$). **b** Reflection coefficient R_s for the s-polarized pumping wave at the micro resonator versus the angle of incidence [12]

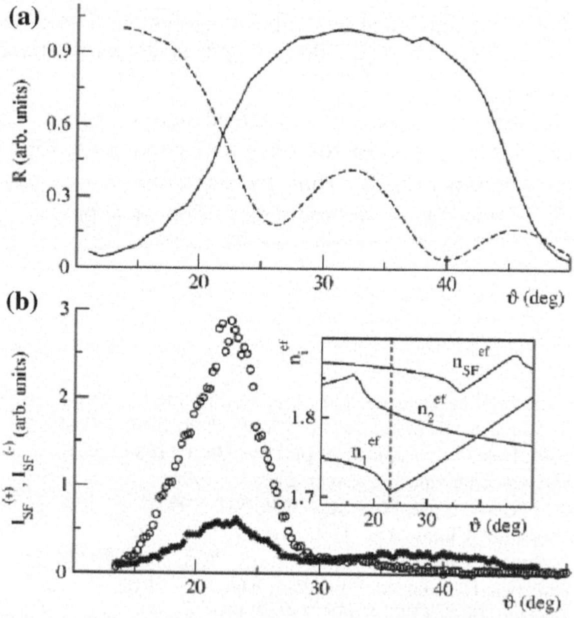

Fig. 9.10 **a** Experimental measurements of the linear reflection of light with wavelengths $\lambda_1 =$ 736 nm (solid line) and $\lambda_2 = 5813$ nm (dashed line) versus the angle of incidence on the structure. **b** Experimental measurements of the sum frequency transmitted intensity (filled symbols) and reflected sum frequency intensity (empty symbols) signals. In the inset, the calculated effective refractive indexes are given [11]

9.6 Conclusion

The above analytical investigations of the influence of optical localized modes on certain optical phenomena do not cover all options in the field, and there is therefore room for further investigations here. In particular, there are other kinds of localized mode than the two considered here. The best known example is a well-known analog of the acoustic whispering-gallery mode existing for electromagnetic radiation [25]. As shown in [25], the observed electromagnetic whispering-gallery modes in liquid crystal droplets exhibit sharp discrete resonances which look promising in applications, like the localized optical modes.

In the above, we studied localized optical modes and the prospects for using localized optical modes to enhance certain processes [26, 27]. Therefore, the shift in the developed optics approach toward the short wavelength range (X-rays) seems very attractive, and a broadening of the wavelength range has already been proposed in [28]. Recent developments of X-ray sources (synchrotron radiation sources, free-electron lasers, etc.) show that experimental studies of X-ray localized modes, similar to the optical ones, are quite achievable. Indeed, the closest goal may be the realization of a super-high X-ray monochromator based on X-ray localized modes [29, 30].

Another problem not discussed above is the development of time-delay techniques for the study of localized modes, something which seems particularly urgent for X-ray localized modes.

And finally, it is interesting to note that localized modes similar to the X-ray modes or optical EMs and DMs, also exist for corpuscular radiation, for example, in the case of low energy neutrons [31, 32]. Thus, there is a reason to expect investigations of localized modes to become an important part of physical optics.

References

1. K. Kajikawa, T. Isozaki, H. Takezoe et al., Jpn. J. Appl. Phys. Lett. **31**, L679 (1992)
2. T. Furukawa, T. Yamsada, K. Ishikawa et al., Appl. Phys. B **60**, 485 (1995)
3. J. Yoo, S. Choi, H. Hoshi et al., Jpn. J. Appl. Phys. **36**, L1168 (1997)
4. M. Copic, I. Drevensek-Olenik, Liq. Cryst. **21**, 233 (1996)
5. I. Drevensek-Olenik, M. Copic, Phys. Rev. **E56**, 581 (1997)
6. K.C. Shin, H. Hoshi, D. Chung et al., Opt. Lett. **27**, 128 (2002)
7. H. Hoshi, K.C. Shin, D. Chung et al., Jpn. J. Appl. Phys. **40**, L970 (2001)
8. H. Hoshi, K. Ishikawa, H. Takezoe, Phys. Rev. **E68**, 020701 (2003)
9. H. Hoshi, H. Takezoe, Phys. Lett. A **358**, 242 (2006)
10. A.V. Balakin, V.A. Bushuev, N.I. Koroteev et al., Opt. Lett. **24**, 793 (1999)
11. A.V. Balakin, V.A. Bushuev, B.I. Mantsyzov et al., Phys. Rev. **E63**, 046609 (2001)
12. T. Dolgov, A.I. Maidykovski, M.G. Martem'yanov et al., Pis'ma ZhETF. **75**, 17 (2002)
13. J.P. Dowling, M. Scalora, M.J. Bloemer et al., J. Appl. Phys. **75**, 1896 (1994)
14. M. Scalora, M.J. Bloemer, A.S. Manka, J.P. Dowling et al., Phys. Rev. A **56**, 3166 (1997)
15. V.A. Belyakov, N.V. Shipov, ZhETF **82**, 1159 (1982); Sov. Phys. –JETP. **55**, 674 (1982)
16. V.A. Belyakov, S.V. Semenov, JETP **109**, 687–698 (2009)
17. V.A. Belyakov, JETP Lett. **70**(12), 811 (1999)
18. V.A. Belyakov, M. Copic, Mol. Cryst. Liq. Cryst. **328**, 151 (1999)
19. V.A. Belyakov, M. Copic, Ferroelectrics **245**, 197 (2000)
20. V.A. Belyakov, V.E. Dmitrienko, V.P. Orlov, Uspekhi Fiz. Nauk **127**, 221 (1979); Sov. Phys. - Usp. **22**, 63 (1979)
21. N.V. Shipov, V.A. Belyakov: ZhETF **75**, 1589 (1978); SOy. Phys. - JETP **48**,802 (1978)
22. V.A. Belyakov, A.S. Sonin, *Optics of Cholesteric Liquid Crystals* (Nauka, Moscow, 1982). (in Russian)
23. V.A. Belyakov, V.E. Dmitrienko, Uspekhi Fiz. Nauk, **146**, 369 (1985); Sov. Phys. - Usp. **28**, 535 (1985)
24. V.A. Belyakov, Localized optical modes in optics of chiral liquid crystals (Chap. 7), in *New Developments in Liquid Crystals and Applications*, ed. by P.K. Choundry (Nova Publishers, New York, 2013), p. 199
25. M. Humar, M. Ravnik, S. Pajk, I. Musevich, Nat. Photonics **3**, 595 (2009)
26. V.A. Belyakov, J. Laser Opt. Photonics **4**, 153 (2017). https://doi.org/10.4172/2469-410X.1000153
27. V.A. Belyakov, Mol. Cryst. Liq. Cryst. **612**, 81–97 (2015)
28. V.A. Belyakov, in *Proceedings of the XVI International Symposium on Nanophysics and Microelectronics, Nizhnii Novgorod,* vol. 2, p. 542 (2012)
29. V.A. Belyakov, in *Proceedings of the XX International Symposium on Nanophysics and Microelectronics, Nizhnii Novgorod*, vol. 1, p. 347 (2016)

30. V.A. Belyakov, in *Proceedings of the XXI International Symposium on Nanophysics and Microelectronics, Nizhnii Novgorod*, vol. 1, p. 355 (2017)
31. V.A. Belyakov, N. Kaputkina, in *Proceedings of the XIV International Symposium on Nanophysics and Microelectronics, Nizhnii Novgorod*, vol. 2, p. 327 (2010)
32. V.A. Belyakov, JETP **124**, 873 (2017)

Index

V. Belyakov, *Diffraction Optics of Complex-Structured Periodic Media*, Springer Series in Optical Sciences 203,
https://doi.org/10.1007/978-3-319-43482-7

The manufacturer's authorised representative in the EU is Springer Nature Customer Service Centre GmbH, Europaplatz 3, 69115 Heidelberg, Germany. If you have any concerns regarding our products, please contact ProductSafety@springernature.com

Printed and bound by CPI Group (UK) Ltd, Croydon, CR0 4YY
22/09/2026
02220990-0007